Advances in the Modelling of Thermodynamic Systems

Elhoucine Essefi
University of Gabes, Tunisia

Ibtissem Jendoubi
Faculty of Sciences of Bizerte, Tunisia

A volume in the Advances in Chemical and Materials Engineering (ACME) Book Series

IGI Global
PUBLISHER of TIMELY KNOWLEDGE

Published in the United States of America by
IGI Global
Engineering Science Reference (an imprint of IGI Global)
701 E. Chocolate Avenue
Hershey PA, USA 17033
Tel: 717-533-8845
Fax: 717-533-8661
E-mail: cust@igi-global.com
Web site: http://www.igi-global.com

Library of Congress Cataloging-in-Publication Data

Names: Essefi, Elhoucine, 1976- editor. | Jendoubi, Ibtissem, 1983- editor.

Title: Advances in the modelling of thermodynamic systems / Elhoucine Essefi, and Ibtissem Jendoubi, editors.
Description: Hershey, PA : Engineering Science Reference, an imprint of IGI Global, [2022] | Includes bibliographical references and index. | Summary: "This book offers an interdisciplinary vision of thermodynamics, from recent advances in modeling of thermodynamic systems as well as the state of the art of many disciplines including human made industrial processes (materials studies, renewable energy, heat and mass transfer, and heat pump and air conditioning) and natural processes taking place globally"-- Provided by publisher.
Identifiers: LCCN 2021024332 (print) | LCCN 2021024333 (ebook) | ISBN 9781799888017 (h/c) | ISBN 9781799888024 (s/c) | ISBN 9781799888031 (ebook)
Subjects: LCSH: Thermodynamics--Mathematics. | Heating--Mathematical models. | Cooling--Mathematical models.
Classification: LCC TJ265 .A265 2022 (print) | LCC TJ265 (ebook) | DDC 621.402/1--dc23
LC record available at https://lccn.loc.gov/2021024332
LC ebook record available at https://lccn.loc.gov/2021024333

This book is published in the IGI Global book series Advances in Chemical and Materials Engineering (ACME) (ISSN: 2327-5448; eISSN: 2327-5456)

British Cataloguing in Publication Data
A Cataloguing in Publication record for this book is available from the British Library.

For electronic access to this publication, please contact: eresources@igi-global.com.

Advances in Chemical and Materials Engineering (ACME) Book Series

J. Paulo Davim
University of Aveiro, Portugal

ISSN:2327-5448
EISSN:2327-5456

MISSION

The cross disciplinary approach of chemical and materials engineering is rapidly growing as it applies to the study of educational, scientific and industrial research activities by solving complex chemical problems using computational techniques and statistical methods.

The **Advances in Chemical and Materials Engineering (ACME) Book Series** provides research on the recent advances throughout computational and statistical methods of analysis and modeling. This series brings together collaboration between chemists, engineers, statisticians, and computer scientists and offers a wealth of knowledge and useful tools to academics, practitioners, and professionals through high quality publications.

COVERAGE

- Biomaterials
- Numerical Techniques
- Wear of Materials
- Metallic Alloys
- Ductility and Crack-Resistance
- Materials to Renewable Energies
- Polymers Engineering
- Computational methods
- Industrial Chemistry
- Artificial Intelligence Methods

IGI Global is currently accepting manuscripts for publication within this series. To submit a proposal for a volume in this series, please contact our Acquisition Editors at Acquisitions@igi-global.com or visit: http://www.igi-global.com/publish/.

The Advances in Chemical and Materials Engineering (ACME) Book Series (ISSN 2327-5448) is published by IGI Global, 701 E. Chocolate Avenue, Hershey, PA 17033-1240, USA, www.igi-global.com. This series is composed of titles available for purchase individually; each title is edited to be contextually exclusive from any other title within the series. For pricing and ordering information please visit http://www.igi-global.com/book-series/advances-chemical-materials-engineering/73687. Postmaster: Send all address changes to above address.

Titles in this Series

For a list of additional titles in this series, please visit: http://www.igi-global.com/book-series

The Classes of Higher Dimensional Polytopes in Chemical, Physical, and Biological Systems
Gennadiy Vladimirovich Zhizhin (Russian Academy of Natural Sciences, Russia)
Engineering Science Reference • © 2022 • 315pp • H/C (ISBN: 9781799883746) • US $195.00

Handbook of Research on Tribology in Coatings and Surface Treatment
Amirhossein Pakseresht (FunGlass – Centre for Functional and Surface Functionalized Glass, Alexander Dubček University of Trenčín, Slovakia) and Omid Sharifahmadian (FunGlass – Centre for Functional and Surface Functionalized Glass, Alexander Dubček University of Trenčín, Slovakia)
Engineering Science Reference • © 2022 • 470pp • H/C (ISBN: 9781799896838) • US $285.00

Advanced Manufacturing Techniques for Engineering and Engineered Materials
R. Thanigaivelan (Muthayammal Engineering College, India) N. Rajan (Vinayaga Mission Kripananda Variyar Engineering College, India) and T.G. Argul (St. Mary's Engineering College, India)
Engineering Science Reference • © 2022 • 351pp • H/C (ISBN: 9781799895749) • US $245.00

Handbook of Research on Green Synthesis and Applications of Nanomaterials
Rajni Garg (Rayat Bahra University, India) Rishav Garg (Galgotias College of Engineering and Technology, India) and Nnabuk Okon Eddy (Department of Pure and Industrial Chemistry, University of Nigeria, Nsukka, Nigeria)
Engineering Science Reference • © 2022 • 569pp • H/C (ISBN: 9781799889366) • US $295.00

Handbook of Research on Nanoemulsion Applications in Agriculture, Food, Health, and Biomedical Sciences
Karthikeyan Ramalingam (B.S. Abdur Rahman Crescent Institute of Science and Technology, India)
Engineering Science Reference • © 2022 • 671pp • H/C (ISBN: 9781799883784) • US $295.00

IGI Global
PUBLISHER of TIMELY KNOWLEDGE

701 East Chocolate Avenue, Hershey, PA 17033, USA
Tel: 717-533-8845 x100 • Fax: 717-533-8661
E-Mail: cust@igi-global.com • www.igi-global.com

Table of Contents

Detailed Table of Contents

identify the loads operating on these systems' supporting structures. In this chapter, computational fluid dynamics (CFD) analysis is largely used to simulate, analyze, and grasp the impacts of wind forces on solar panels utilizing high-speed computer capabilities. Furthermore, unlike previous studies in this field, this study focuses on the sequential arrangement of ground-mounted solar panels and the effects of the sheltering effect. The steady state SST k-omega turbulence model is used in the CFD analysis. 3D studies are carried out, and it is discovered that for the solar panel configurations investigated in this chapter, they give accurate findings with robust modelling and solutions in terms of computing time.

The thermodynamic theoretical modeling of brines from the Chott Djerid, Southeastern Tunisia and Sebkha Bazer, Northeast Algeria between the initial solution and evaporation of 100% shows different geochemical evolutions and variable precipitated mineral species. This is due to different initial geochemical compositions, which are basically related to different geological and climatic contexts. A rigorous thermodynamic model has been presented for determining the crystallization sequence during the different stages of the evaporation process. A geochemical software program (PHREEQC) was used. Based on the analyses, PHREEQC appears the precipitation of a sequence of minerals. These results are confirmed by the mineralogical analyses (XRD results). The gradual evaporation process of brine showed that thenardite, gypsum, halite, and huntite are the feasible salts that could be extracted. Halite is the most abundant mineral along with the evaporation experiment of the two sebkhas.

Sorption isotherms of prickly pear seeds were determined by static gravimetric method at temperatures 45°, 60°, and 70°C, over a relative moisture range of 5-95%. Sorption isotherms are important to define dehydration limits of the product, estimate moisture content alterations under environment conditions, and to acquire moisture content values for safe storage. Four mathematical models were applied to analyze the experimental data. Equilibrium moisture contents of prickly pear seeds decreased with temperature increment at a constant value of relative humidity. The GAB model showed the best fitting to the experimental data. Isosteric heat and differential entropy, determined by applying the Clausius-Clapeyron and Gibbs-Helmholtz equations respectively, decreased strongly as the moisture content increased and could be well adjusted by an empirical exponential relationship. Enthalpy-entropy compensation theory is valid for the sorption of prickly pear seeds, in which the water sorption mechanism in seeds can be considered to be enthalpy controlled.

In this chapter, in order to find the optimum coefficient of performance (COP) and exergy performance (COPex) of desiccant cooling system operating on ventilation and recirculation cycles under hot and dry climate, a simple theoretical model has been developed based on the first and second laws of thermodynamics. Then, the model was implemented in the MATLAB software. The obtained theoretical results were compared with those of the literature and showed a good agreement. Moreover, results showed that COP of ventilation and recirculation cycles are 1.89 and 1.13 respectively, greater than the corresponding COPex, which is equal to 0.7 and 0.38 respectively. In addition, the maximum destruction exergy percentages are provided by the desiccant wheel and the heat source, which are respectively 57% and 24.67% in ventilation cycle and 33.08% and 38.83% in recirculation cycle. Finally, the sensitivity of exergy destruction desiccant wheel and heat source with reference (dead-state) were explored.

This chapter is mainly focused on the minimization of the total entropy generation in a thermodynamic system, which concerns the heating of water when it passes through a saturated porous media. The heating process is assumed by an array of heating tubes immerged on the porous media and perpendicular to the water flow direction. This irreversibility calculation and minimization is carried out in dimensional form in order to have a real idea about the entropy production in the system. Also, the authors focus on several points regarding the rigor of entropy generation calculation, when they pass from dimensional form to dimensionless form.

This chapter provides a specific study of the performance of thermal systems, principally heat exchangers, which are applied in several industrial applications such as chemical industry, energetic industry, industrial lasers, and so on. These thermodynamics systems were critical in transferring heat from a higher to a lower temperature fluid. They have been used for several years and are available currently for various designs. Thermodynamic properties influence the heat transfer and the performance of heat exchangers. Therefore, it is important during the design of heat exchangers to select primary the accurate operating conditions in terms of thermodynamics to provide a minimum amount of entropy generation in the system. In this study, the concept of entropy is used to analyze heat transfer processes from the thermodynamic viewpoint through the second law of thermodynamics. To assess heat exchanger performance, investigations are given for entropy generation, entropy generation number, and efficiency. These studies offer a new way to obtain well-designed heat exchangers.

The present work reports a numerical simulation of entropy generation and heat transfer in a lid-driven porous cavity filled with a nanofluid using Darcy-Forchheimer model. Given the large number of dimensionless parameters related to this problem, some of them are kept constant and therefore the other governing dimensionless number such as number, the Hartman number, and the nanoparticles volume fraction, 0£Ha£50, 2%£φ£8%, respectively. The effects of the nanoparticles volume fraction and Hartman number on the different irreversibilities are studied. Results show that the entropy generation is strongly affected by the increase of Hartmann number and the volume fraction. Results reveal that the irreversibility in the nanofluid decrease with the nanoparticle volume fraction for different Hartmann numbers.

The objective of this work was to microscopically and macroscopically interpret entropy within the framework of quantum mechanics: quantum computer, Coulomb crystal, chaos, and cosmology. Indeed, in quantum physics, the concept of information is the very basis of the minimal interpretation of the concept of state vector as a contextual prediction tool. The Coulomb crystal is the basic element for the development of a quantum computer. For example, the Coulomb crystal represents the basic element of high precision clocks, provides a favorable environment for the detailed study of chemical reactions, and constitutes an original technology for the development of a quantum computer. In addition, the combination of chaos with the recent definition of entropy allows us to understand very small systems at the atomic and quantum microscopic level, as well as very large systems at the macroscopic level of galaxies and black holes.

This work aimed to study the coefficient of performance (COP) of an active magnetic regenerative refrigeration (AMRR) system by new analytical approaches of magnetic work Wm(B,x,y) and magnetocaloric effect MCE (T,B). Those approaches were applied to a permanent magnet magnetic refrigerator. The studied refrigeration system consisted of four regenerators, each of which was formed by parallel plates of gadolinium, a circulation pump, a rotating magnet, and two heat exchangers. The heat transfer fluids used were water and gallium. A resolution of the continuity equation, the amount of movement equation, and the heat equation were carried out in order to study the temperature profile in both the regenerator and the fluid. Furthermore, the authors deduced the temperatures at the inlet and the outlet of the heat exchangers in order to establish a thermal balance.

Chapter 11

Saoussen El Aguel, Institut Supérieur des Sciences Appliquées et Technologies de Gabès, Tunisia
Zina Meddeb, Institut Supérieur des Sciences Appliquées et Technologies de Gabès, Tunisia
Mohamed Razak Jeday, National Engineering School of Gabes, Tunisia

Several factors influence the functioning of the heat exchangers in phosphoric acid concentration units and significantly affect their energy performance such as the cleanliness of the products, the operating conditions, the deposit of fouling on the walls. During the phosphoric acid concentration operation, fouling leads to a significant drop in the overall heat transfer coefficient, which is highly dependent on the thermal efficiency of the heat exchanger. This chapter presents the statistical study of the experimental database of two concentration units in order to study the variability of the system and identify outliers using principal component analysis. According to Hotelling's (T2) test, the authors identified 148 outliers for the two heat exchangers. Two reduced models of the thermal efficiencies were obtained by projection to latent structures (PLS) method. The application of the PLS regression method resulted in reliable correlation coefficients R2 equal to 0.9 for both configurations of heat exchangers.

Chapter 12

Zina Meddeb, Institut Supérieur des Sciences Appliquées et Technologies de Gabès, Tunisia
Saoussen El Aguel, National Engineering School of Gabes, Tunisia
Mohamed Razak Jeday, National Engineering School of Gabes, Tunisia

The purpose of this work is to establish mathematical models for monitoring the fouling of heat exchangers and programming shutdowns for the cleaning of this equipment. To achieve this objective, the authors have adopted an approach comprising essentially the identification of the operating parameters involved in the fouling of the heat exchangers used in the concentration of phosphoric acid, collection of the technical characteristics of the heat exchangers studied, collection and sorting of the operational data of the exchangers studied over a period of two and a half years, establishment of a database on the cycles of operation/shutdown, for the cleaning of the heat exchangers studied, and development of preliminary models for monitoring the fouling and to help the operator decide when to shut down for cleaning the heat exchangers.

In this study, a coupling model between the water tank with immersed condenser coil and the vapor compression system was developed in ANSYS fluent for simulating the heat transfer between the refrigerant in condenser and water in the storage tank. Further study was performed to analyze the effect of condenser coil location on the heating process. The results indicated that, when the condenser coil is placed in a lower part of the water tank, a higher water velocity can be observed. From the testing results, when the condenser coil is placed in the lower part of the water tank, the convective heat transfer is better than the other positions.

Lithium has worldwide been proven of great energetic interest. One of the origins of lithium is salt lakes brine. Tunisia is marked by the presence of many saline systems containing lithium ranging from 20 mg/L to 50 mg/L. Nonetheless, extracting lithium from natural brine is really finding a needle in a haystack. This difficulty is basically due to the nuclear, electronic, and thermodynamic features of lithium as well as the other ions present in the brine. In this chapter, the authors study the technical and thermodynamic methods leading to the enrichment of lithium in brine, which in turn leads to its easy extraction. For this reason, techniques of extraction and their limitation were reviewed. In addition, the evaporation process of the brine is thermodynamic methods leading to an enrichment of with lithium in the brine due to the extraction of water molecules. Also, the precipitation of minerals including ions representing a noise of the lithium, especially magnesium, paves the way for easy extraction within lithium window.

Preface

APPLICATION OF THERMODYNAMICS IN DIFFERENT FIELD OF STUDIES

Based on a bibliometric analysis, it may be noticed that thermodynamics is a common field of studies of different specialties including physics, chemistry, mechanics, geology, and cosmology. As a matter of fact, thermodynamics is useful for human made industrial processes related to materials studies, renewable energy, heat and mass transfer, heat pump and air conditioning. Furthermore, thermodynamics proves vital for understanding natural macro-processes related to geology, areology (Martian geology) and cosmology as well as micro-processes related to quantum thermodynamics. This book will collect recent advances in modeling of thermodynamic systems as well as the state of the art of many disciplines including human made industrial processes (materials studies, renewable energy, heat and mass transfer, and heat pump and air conditioning) and natural processes taking place on Earth and the universe on the one hand and the atom on the other. This book reveals an interdisciplinary vision of thermodynamics from the very big to the very small, from the natural processes of geology to human made phenomena in mechanics. For this reason, this book covers a huge audience interested in natural processes of geology and cosmology and industrial applications in solar energy and cooling.

ORGANIZATION AN OVERVIEW OF CONTRIBUTIONS

The first chapter studies the thermodynamic parameters of the external geodynamics of Mars. As matter of fact, the thermodynamics on Mars surface is basically controlled by the solar forcing and the internal geodynamics of the planet. On the one hand, the Sun provides with the necessary energy to maintain the temperature governing the geological processes of the Martian external geodynamics. On the other hand, the internal dynamics of the planet, which is basically related to the composition and structure, influence the composition and the thermodynamic parameters (Pressure and Temperature) of the atmosphere controlling the external geodynamics through the dynamics and the physical state (solid, liquid, and gas) of the principle agent: *water.*

In Chapter 2, a computational fluid dynamics (CFD) analysis is largely used to simulate, analyze, and grasp the impacts of wind forces on solar panels utilizing high-speed computer capabilities. This study focuses on the sequential arrangement of ground-mounted solar panels and the effects of the sheltering effect.

Chapter 3 deals with the thermodynamic theoretical modeling of brines from the Chott Djerid, southeastern Tunisia and sebkha Bazer, northeast Algeria between the initial solution and evaporation of 100% shows different geochemical evolutions and variable precipitated mineral species. Based on our analyses PHREEQC appears as the precipitation of a sequence of minerals. These results are confirmed by the mineralogical analyses.

Chapter 4 shows that sorption isotherms of prickly pear seeds were determined by static gravimetric method at temperatures 45, 60, and 70 °C, over relative moisture range of 5-95%. In this work, four mathematical models were applied to analyze the experimental data. Isosteric heat and differential entropy, determined by applying the Clausius-Clapeyron and Gibbs-Helmholtz equations respectively, decreased strongly as the moisture content increased and could be well adjusted by an empirical exponential relationship.

Chapter 5 aimed to find the optimum coefficient of performance (COP) and exergy performance (COPex) of desiccant cooling system operating on ventilation and recirculation cycles under hot and dry climate. A simple theoretical model was developed based on the first and second laws of thermodynamics. Then, the model was implemented in the MATLAB software. The obtained theoretical results were compared with those of the literature and showed a good agreement.

Chapter 6 is mainly focused on the minimization of the total entropy generation in a thermodynamic system, which concerns the heating of water when it passes through a saturated porous media. The irreversibility calculation and minimization are carried out in dimensional form, in order to have a real idea about the entropy production in the system.

Chapter 7 provides a specific study of the performance of thermal systems principally heat exchangers, which are applied in several industrial applications such as chemical industry, energetic industry, industrial lasers and so on. These thermodynamics systems were critical in transferring heat from a higher to a lower temperature fluid. In this study, the concept of entropy is used to analyze heat transfer processes from the thermodynamic viewpoint throw the second law of thermodynamics.

Chapter 8 reports a numerical simulation of entropy generation and heat transfer in a lid-driven porous cavity filled with a nanofluid using Darcy-Forchheimer model. The effects of the nanoparticles volume fraction and Hartman number on the different irreversibilities are studied. Results reveal that the irreversibility in the nanofluid decreases with the nanoparticle volume fraction, for different Hartmann numbers.

Chapter 9 microscopically and macroscopically interprets entropy within the framework of quantum mechanics: Quantum Computer, Coulomb Crystal, Chaos and Cosmology. The Coulomb crystal is the basic element for the development of a quantum computer. In addition, the combination of chaos with the recent definition of entropy allows us to understand very small systems at the atomic and quantum microscopic level, as well as very large systems at the macroscopic level of galaxies and black holes.

Chapter 10 aimed to study the coefficient of performance (COP) of an active magnetic regenerative refrigeration (AMRR) system by new analytical approaches of magnetic work Wm(B,x,y) and magneto caloric effect MCE (T,B). The resolution of the continuity equation, the amount of movement equation and the heat equation were carried out in order to study the temperature profile in both the regenerator and the fluid.

Chapter 11 presents the statistical study of the experimental database of two concentration units in order to study the variability of the system and identify outliers using Principal Component Analysis. According to Hotelling's (T2) test, we identified 148 outliers for the two heat exchangers. Two reduced models of the thermal efficiencies were obtained by Projection to Latent Structures (PLS) method. The application of the (PLS) regression method resulted in reliable correlation coefficients R2 equal to 0.9 for both configurations of heat exchangers.

Chapter 12 establishes mathematical models for monitoring the fouling of heat exchangers and programming shutdowns for the cleaning of this equipment. It aims the establishment of a database on the cycles of operation/shutdown, for the cleaning of the heat exchangers studied, development of preliminary models for monitoring the fouling and to help the operator decide when to shut down for cleaning the heat exchangers.

Chapter 13 presents a coupling model between the water tank with immersed condenser coil and the vapor compression system was developed in ANSYS Fluent for simulating the heat transfer between the refrigerant in condenser and water in the storage tank. Further study was performed to analyze the effect of condenser coil location on the heating process. From the testing results, when the condenser coil is placed in the lower part of the water tank, the convective heat transfer is better than the others' positions.

Chapter 14 studies the technical and thermodynamic methods leading to the enrichment of lithium in brine, which in terns leads to its easy extraction. For this reasons, techniques of extraction and their limitation were reviewed. In addition, the evaporation process of the brine is thermodynamic methods leading to an enrichment of with lithium in the brine due to the extraction of water molecules. Also, the precipitation of minerals including ions representing a noise of the lithium, especially magnesium paves the way for and easy extraction within lithium window.

This work brings together contributions from researchers, scholars, and practitioners from global communities, enriching and giving visibility to their application of thermodynamic principles and equation.

Elhoucine Essefi
University of Gabes, Tunisia

Ibtissem Jendoubi
Faculty of Sciences of Bizerte, Tunisia

Introduction

Thermodynamic systems may be natural as well as human made. Understanding the functioning of these complex systems necessitates the deal on advanced mathematical modeling. In addition, computer sciences have provided with some useful software helping in the modeling to explain the current functioning and predicting the future behavior. In this vein, IGI Global Publisher presents this product titled *Advances in the Modelling of Thermodynamic Systems* to satisfy its customers by a book dealing with recent advances in thermodynamics as a whole. As it is meant, the book has grouped researches from different field of studies: geologists, physicists, chemists, and so on. This multidisciplinary approach attracted a heterogeneous audience. The edited volume is a collection of chapters with up-to-date information, original research, reviews, and discussions by several authors from different parts of the world and different field of studies, presenting a global picture of what is presently being done in the field of thermodynamics.

Chapter 1
Thermodynamics of the External Geodynamics of Mars Water Phases and Weathering Processes

Elhoucine Essefi
University of Gabes, Tunisia

ABSTRACT

This chapter studies the thermodynamic parameters of the external geodynamics of Mars. As matter of fact, the thermodynamics on Mars surface are basically controlled by the solar forcing and the internal geodynamics of the planet. In relation to the physical and chemical characteristics of water on Mars, we can build the Martian chronology. First, the Phyllosian is the phyllosilicates thermodynamics era. Second, the Theiikian is the age of sulfate thermodynamics. Third, the Siderikian, according to siderikos (ferric in Greek), is the era of anhydrous ferric oxides thermodynamics. The Martian chronology may be also built on catering. So, the meteoritic bombardment is linked to increasing disorder in the solar systems. That is to say, it is quite linked to the thermodynamics of the solar system. As direct repercussions of variable thermodynamics during the Mars history, the authors investigate the sedimentology and stratigraphy in different localities on Mars: Arabia Terra, Meridiani Planum, Terby Crater, and Gale Crater.

DOI: 10.4018/978-1-7998-8801-7.ch001

INTRODUCTION

The thermodynamics on Mars surface is basically controlled by the solar forcing (Appelbaum & Flood, 1990; Khatib et al., 2020) and the internal geodynamics of the planet (Phillips et al., 2001). On the one hand, the Sun provides with the necessary energy to maintain the temperature governing the geological processes of the Martian external geodynamics. This control allows the chemistry (Selco, 1995) and the physical state water to unsure the weathering of Mars surface (Chevrier et al., 2007; Lu & Kieffer, 2009). On the other hand, the internal dynamics of the planet, which is basically related to the composition and structure Mars, influences the composition and the thermodynamic parameters (Pressure and Temperature) of the atmosphere (green house effect) controlling the external geodynamics through the dynamics and the physical state (solid, liquid, and gas) of the principle agent: water. Giving a special care to the water history and its effect on the external geodynamics, this chapter is meant to investigate the variation of the thermodynamic parameters controlling the weathering, transport and sedimentation during the geological history of Mars.

ASTRONOMIC CONTROL OF MARS THERMODYNAMICS

Formed 4.6 billion years ago, the Solar System consists of the Sun and the planets of its planetary system (Fig. 1), their moons, and other non-stellar objects. Traditionally, the solar system consists of eight planets. Nevertheless, the discovering of new extraterrestrial bodies provides continuous updating of this list of planets (Tab. 1). After Venus, Mars is the second closest to Earth. It is slightly more than the half of Earth size. Belonging to the solar system, Mars is the outermost of the four terrestrial planets (Mercury, Venus, Earth, and Mars) (Fig. 1).

The main mass of the system is concentrated in Sun, and the remaining part is contained in Jupiter. The four smaller inner planets, Mercury, Venus, Earth and Mars, are primarily rocky. Some characteristics of the eight planets are summarized in Table 1. These features control the thermodynamic parameters on planets, including Mars.

Table 2. shows the characteristics of the planets that are controversially considered as part of the solar system. Some of their characteristics are still unknown. For instance, Ceres, Pluto and Eris are celestial bodies sometimes given the rank of planets. In spite of their dimension, these so-called planets influence the atmosphere of the others by weak attractions.

The setting of Mars within these astronomic conditions of the solar systems gives the planet specific thermodynamic features. The dimension of Mars, its distance and orbiting trajectory around Sun control the radiation flux and the atmosphere structure and composition that controls greenhouse Mischna et al., 2013; Jakosky et al., 2015).

Figure 1. Schematic presentation of the solar system (https://photojournal.jpl.nasa.gov/)

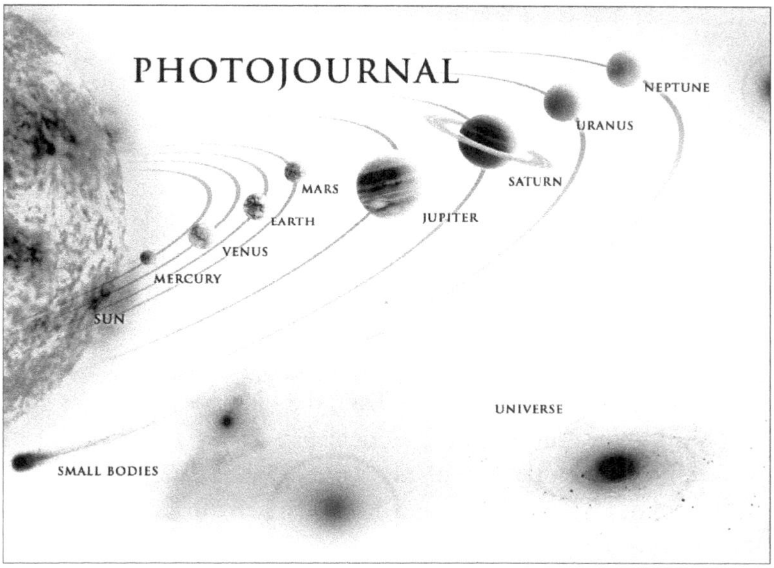

Table 1. Characteristics of traditional components of the solar system http://www.solstation.com/stars/sol-sum.htm

	Orbital Distance (AU)	Mass (earths)	Diameter (earths)	Rotational Period (days)	Orbital Period (years)	Mean Density (earths)	Surface Gravity (earths)	Moons
Sun	0.0	330,000	109.2	25.4	...	1.42	28	...
Mercury	0.4	0.06	0.38	59	0.24	0.98	0.38	0
Venus	0.7	0.81	0.95	243	0.62	0.95	0.90	0
Earth	1.0	1.00	1.00	1.00	1.0	1.00	1.00	1
Mars	1.5	0.11	0.53	1.03	1.9	0.71	0.38	2
Jupiter	5.2	317.8	11.2	0.42	11.9	0.24	2.34	63
Saturn	9.5	95.2	9.4	0.44	29.4	0.12	1.16	60
Uranus	19.2	14.5	4.0	0.72	83.7	0.23	1.15	27
Neptune	30.1	17.2	3.9	0.67	163.7	0.30	1.19	13

Table 2. Characteristics of the new discovered planets the solar system http://www.solstation.com/stars/sol-sum.htm

	Orbital Distance (AU)	Mass (earths)	Diameter (earths)	Rotational Period (days)	Orbital Period (years)	Mean Density (earths)	Surface Gravity (earths)	Moons
(Ceres)	2.8	0.00015	0.07	0.38	4.6	0.38	0.03	0
(Pluto)	39.4	0.002	0.18	6.40	248.0	0.37	0.04	3
(Eris)	67.7	0.002?	0.18	~8	557	0.42	?	1

MARTIAN GEOLOGY AND THERMODYNAMICS

As a standalone discipline, the martian geology is the study of the surface (e.g., Boynton et al., 2002), the crust (e.g., Connerney et al., 2001), and the internal content (e.g., Zuber et al., 2000) of Mars. This study takes into account the thermodynamic parameters controlling the geological processes (Gainey et al., 2022; Ahrens et al., 2021). It deals on the study of the composition (e.g., Harry & McSween, 2012), the structure (e.g., Johnston & Toksöz, 1977), the history (Bibring et al., 2006; & references therein), and the physico-chemical processes (Bishop et al., 2002; & references therein) shaping the red planet. Divided into different sub-disciplines, martian geology argues full similarity with the field of terrestrial geology. In planetary sciences, the term geology is used in its broadest sense to mean the study of the solid parts of planets and moons. As a synonym for Mars' geology, the term *areology* (Katterfel'd et al., 1968) (from the Greek word *Arēs*, i.e. Mars) sometimes appears in the popular media and works of science fiction but the term is rarely, if ever, used by professional geologists and planetary scientists. The term Mars geology fully incorporates all branches of terrestrial geology such as geophysics (e.g., De Vera et al., 2012), geochemistry (e.g., Niles et al., 2013; & references therein), mineralogy (e.g., Battler et al., 2012; & references therein), geodesy (e.g., Hirt et al., 2012; & references therein), hydrology (e.g., Robertson & Bish, 2012; & references therein), hydrogeology (e.g., Jouannic et al., 2012; & references therein), tectonics (e.g., Vaz et al., 2012; & references therein), geodynamics (Wakabayashi & Shervais, 2012; & references therein), climatology (Sefton-Nash et al., 2013; & references therein) cartography (de Pablo Hernández & Carrillo, 2012), sedimentology (Essefi, 2013), and stratigraphy (Limaye et al., 2012; Essefi et al., 2014). Thermodynamics is a master peace in all forementioned fields of studies.

PURPOSES OF MARS EXPLORATION

The challenges facing the Humanity made into question the huge budget spent to explore extraterrestrial bodies. Many reasons of spatial activity are still to be understood. However, explorations strategy may help inferring reasons of martian exploration. In the broadest sense, the declared scientific goals of Mars exploration are to determine: (1) evidences of present or past life on Mars; (2) the origin of the solar system; (3) Mars formation and evolving to its present state; (4) the history and current state of the atmosphere and climate and how are they changing; (5) the atmospheric dynamics compared to weather on Earth; (6) the interaction of solar wind with Mars. To reach these purposes scientists should take into account all thermodynamic parameters in order to avoid all misunderstanding of data explored from remote sensing and missions (McMahon & Cosmidis, 2022).

Life on Mars

A long-range objective of Mars exploration is to determine if life ever started there (e.g., Gross, 2012; Fernández-Remolar, 2012; Nicholson et al., 2013). Due to thermodynamic reasons, liquid water is unstable under present climatic conditions (e.g., El Maarry et al., 2012) everywhere within a few hundred meters of the surface. So, evidences of present-day life on the red planet are extremely small. However, abundant evidence for the presence of liquid water at the surface in the past suggests that climatic conditions could have been more hospitable for life. There are three main lines to proceed in studying the evidence of life. The first line is to better understand the climatic history (e.g., Kaltenegger, 2013) of the planet and to search for evidence of existing subsurface water ice reservoirs (e.g., Nicholson et al., 2013). The second line is looking for direct signatures of life that may have existed. The third line of thinking deals on terrestrial analogues to deeply understand the geological, geophysical, geochemical conditions favorable for life proliferation (Stivaletta et al., 2009; Barbieri et al., 2011; Barbieri & Stivaletta, 2012; Barbieri, 2013). To conclude, any form of life on Mars necessitates specific geochemical and thermodynamic conditions for the proliferations of organisms (Carrier et al., 2020).

Origin of the Solar System

The thermodynamics of the solar system has been a matter of interest to find and extension for universe formation and evolution: cosmology (Nicolaou et al., 2020; Mayhew, 2020). Mars is a key planet (Williamson et al., 2013) to be studied to understand the mode of formation of the Solar System. Events during accretion established the starting thermodynamic conditions for the subsequent evolution of the planets (Righter et al., 2020). The observations made on Mars can be compared to those collected on Earth, and later, on Mercury and Venus (e.g., Clark et al., 2013). These results would enable us to conclude some essential facts about the origin and formation of the inner Solar System planets and provide us with more evidence about the history of the entire Solar System (Klima, 2013).

Formation and Thermodynamic Evolution of Mars

The evolution of the interior and the thermodynamic parameters of Mars should have been very different from that of Earth. The Earth's core has remained molten, and the mantle has stirred surface materials to depths of at least 700 km and probably to the base of the mantle at 2900 km. Isotopic patterns of oxygen showing different reservoirs of oxygen in SNC meteorites (Agee et al., 2013) suggest that the Mars mantle is stably layered. However, it is unlikely that the Mars mantle is laterally homogeneous in view of the extreme localization of volcanic activity in Tharsis (Yin, 2012) and Elysium (Pasckert, 2012).

Climate Change

One of key aspects of Mars evolution is the possibility of major climate changes. The widespread erosional surfaces by water and ice can make a case. However, a better understanding of climatic changes needs a better insight into thermodynamic processes involved in the climatic cycle such as the chaotic variability of the tilt (e.g., Brasser & Walsh, 2011). In addition, of great importance is the understanding of some external factors responsible for climate change such as the direct interaction of the Solar wind with martian atmosphere (Solomon & Qian, 2012) and the evacuation of the upper atmosphere due to various pick-up and acceleration processes (Moores et al., 2012). Detection of thick carbonate deposits would support the supposition that the Mars atmosphere was thicker in the past, having hence a stronger green house effect (Niles et al., 2013). Samples of past atmospheres trapped in surface materials would provide direct evidence of past climate (Cassata et al., 2012). More indirect evidence is provided by the isotopes of volatile species (Ehlmann et al., 2013) such as nitrogen and the noble gases, and how these isotopes have changed with time. In addition to long term climate changes, Mars may experience shorter term, more modest changes as a result of obliquity changes. The best record of such changes is probably preserved in the polar layered terrains (Portyankina et al., 2012). Subsequently, sampling through sections of the polar layered terrains is of considerable importance. Traditionally, scientists make a clear distinction between recent and old Mars climates.

Since the 1970s, climatic change on Mars has been a subject of great interest to planetary scientists. Nevertheless, most of the attention has been oriented toward understanding the setting of thermodynamic conditions favourable for a flowing of liquid water on Early Mars surface (i.e., circa 3.5 Ga). Unfortunately, huge as they can be, the considerable body of work performed on this subject reached no clear consensus on the nature of the early martian climatic system. Consequently, some scholars (Haberle et al., 2013) argued the tractability of recent (i.e., 20 Ma) Mars climate based on several arguments. (1) The geologic record and evidence for climate change are better preserved. (2) Compared to Earth, the probable cause of climate change (spin axis/orbital variations) is more pronounced on Mars. For instance, within the 20 Ma, the obliquity is believed to have varied from a low of 15° to a high of 45° with a regular oscillation time scale of 105 years while the amplitude of the corresponding variations for Earth is typically less than 2. (3) The general circulation models (GCMs) for Mars have reached a level of sophistication that justifies their application to the study of spin axis/orbitally forced climate change. Thus, evidence for geologically recent climate change on Mars is strong (Haberle et al., 2013). The main driver for recent climate change is spin axis/orbital variations. The lack of a stabilizing Moon and the proximity to Jupiter lead to large variations in Mars' spin axis/orbit parameters. The most recent published calculations of these parameters were recently updated (Fig. 2) (Haberle et al., 2013). Hundred times thinner than the atmosphere of Earth (e.g., Cassata et al., 2012; & references therein), the Mars atmosphere is primarily composed of carbon dioxide (e.g., Stanley et al., 2012; & references therein). On Mars, temperatures are generally cold; the mean annual surface temperature is approximately -50C° at the equator and close to -130 degrees C° at the poles. Due to the thin atmosphere **(e.g.,** Stanley et al., 2012; & references therein), the range of diurnal temperature is large, greater than -100 degrees C° at the equator. Summer temperatures rise above 0 degrees C° at midday despite the low diurnal mean. Analogically to earth, Mars experiences distinct seasonal weather patterns due to the angle between the rotational axis and the ecliptic. A particularly spectacular seasonal event is the annual dust storm (Dust Devil) (Klose & Shao, 2013; & references therein). During summer, in the southern hemisphere, large dust storms develop and obscure the totality of its surface. Another regular seasonal event is the formation of clouds of carbon dioxideice particles in the Polar Regions during the fall as gas starts to condense out of the atmosphere onto the growing cap (Gudipati & Cooper, 2013; & references therein). Primitive Mars climate started out much like the earth's early climate, but it evolved differently. Once warm enough to support flowing water, Mars is now so cold that carbon dioxide freezes at the poles every winter.

Figure 2. (a) Obliquity (degrees), (b) eccentricity, and (c) insolation (W m2) at the North Pole for the past 20 My and the next 10 My (Laskar et al., 2004)

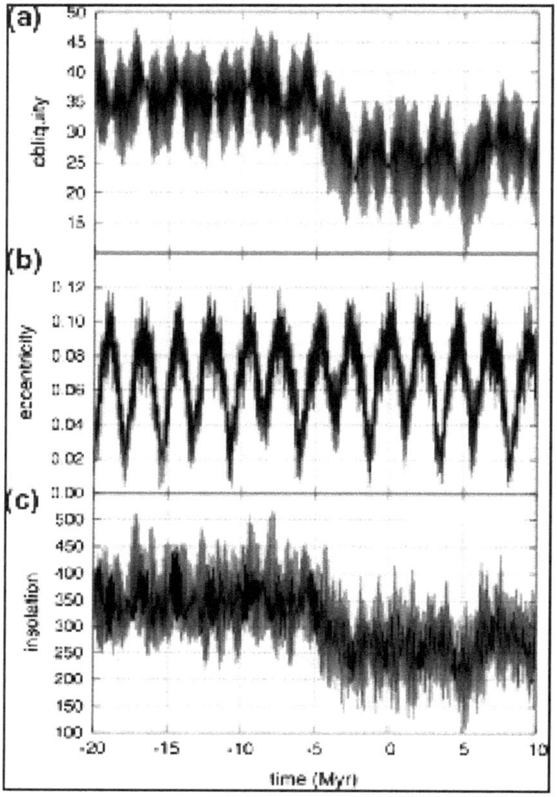

GENERAL OVERVIEW

Earth and Mars

Due to their different internal dynamics (Šrámek et al., 2012) and effects of their atmospheres (Goody, 2012) and oceans (Oehler & Allen, 2012), the geologic histories of Mars and Earth are quite different. These different conditions provide different thermodynamic features on surfaces of both planets. Whereas plate tectonics dominates the geology of Earth, Mars displays little, if any, evidence of plates' activity (Foley et al., 2012). Subsequently, the martian crust appears very stable. Long linear mountain chains and subduction zones noticed on Earth are quasi absent; compressional and extentional features of any kind are rare (Essefi et al., 2013a).

Mars Geomorphology and Geography

The martian surface has some characteristics of Earth surface, some of the Moon, and some unique features. The planet is very asymmetric in appearance. Most of the southern hemisphere is densely cratered and superficially resembles the lunar highlands. In contrast, the northern hemisphere is relatively sparsely cratered and has many large volcanoes that have no lunar counterparts (Fig. 3).

Figure 3. Cartography of Mars (Hargitai & Gide, 2009)

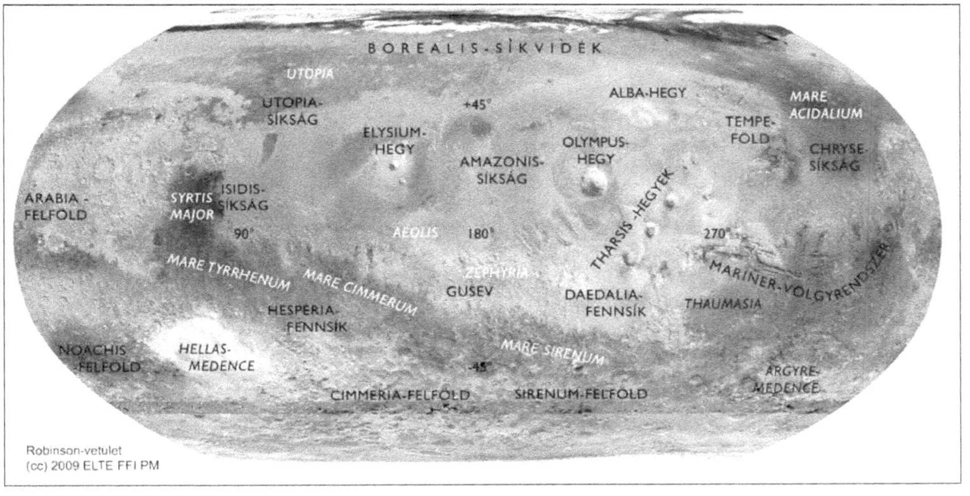

The most obvious aspect of the martian geography is the North-South dissymmetry as well on the morphologic as topographic levels (Fig.4). The Southern hemisphere lands are old fields craterized and of high altitude (Highlands), whereas the Northern Hemisphere lands are of smoother appearance and lower altitude (Lowlands). A third an ancient type of lands makes exception to this rule: they are large volcanic provinces of Tharsis and Elysium.

These three great zones have very different thicknesses of crust. The origin of this North-South crustal dichotomy remains an enigma. The scientists are divided between an internal assumption, dependent on the convection in the martian mantle, and an external assumption, dependent on the impacts. It is worth noting that in spite of this North-South dichotomy, Highlands of the South dominating Lowlands of the North of approximately 5 km, we find the lowest point of the martian sphere in the Southern hemisphere, at the bottom of the basin of impact of Hellas, and the highest point in the Northern hemisphere, with summon gigantic volcano shield of Olympus Mons. The volcanic buildings are numerous besides on the surface of Mars, and some of them reach proportions much more significant than those their modest terrestrial cousins. The development of these gigantic volcanos would have been supported by the absence of plate tectonics over Mars. Mars has also polar caps. The permanent caps are made of a layer of ice of water, which rests on laminated deposits (Polar Layered Deposits PLD), composed in alternation of layers of ice of water and dust. A seasonal cap of CO_2 ice settles over each year, by condensation since the atmosphere. Moreover, on the level of the Southern cap, it remains in any season a residual CO_2 cap. The existence of this residual cap is probably related to the characteristics of the martian orbit, which imply that the winter is longer in the Southern hemisphere than in the Northern hemisphere.

Figure 4. Martian topography and sites of failed and succeeded missions (NASA CREDIT)

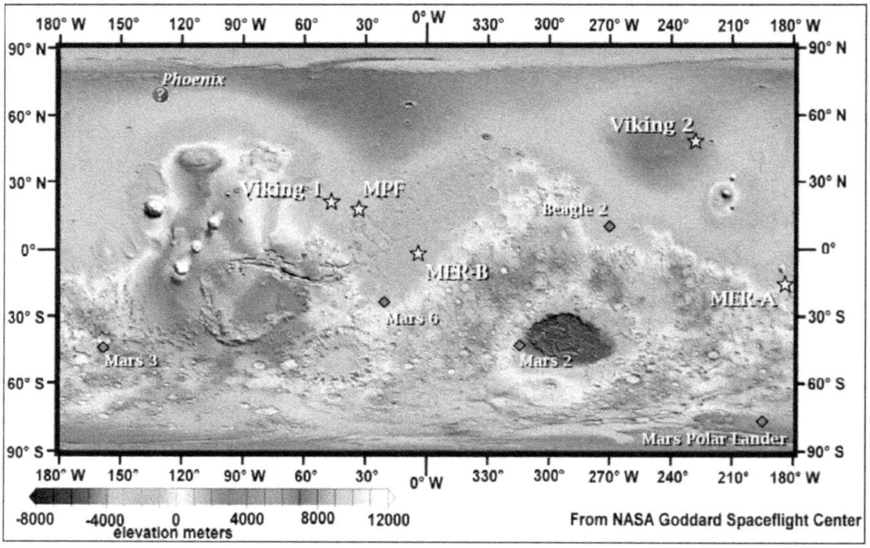

GEOLOGICAL HISTORY OF MARS

Martian Geological Eras Based on Mineralogy

On the basis of the mineralogical data collected since 2004 by the instrument OMEGA (Observatory for Mineralogy, Water, Ices and the Activity) of the mission Mars Express train, a new martian geological time scale was proposed (Bibring et al., 2006). It is divided into three major eras, with a phase of transition (Mars Global Change) between the two first. By chronological order, this alternative scale comprises Phyllosian, Theiikian and Siderikian. These ages are due to variable thermodynamic conditions.

Phyllosian

First coined by Bibring et al. (2006), the term "Phyllosian" represent an era during the early history of Mars in which near-neutral pH solutions produced phyllosilicates with alteration of basaltic martian crustal rocks. It is era of phyllosilicates thermodynamics (Chevrier et al., 2007). Originating from phyllos (Greek), it is the era of the hydrated phyllosilicates, and more precisely of clays. The zones where clays were detected correspond to old fields (Noachian). They were probably formed during a long period of aqueous deterioration. A more in-depth formation (e.g., hydrothermalism) however is not completely excluded.

Theiikian

Theiikian from *theiikos* (sulfate in Greek) is the era of sulphates (sulphate thermodynamics) (MARS, 2005). They would be formed on the surface in an acid environment. Their formation requires not only great quantities of water (it was already the case for Phyllosian) but also a significant source of sulphur. Volcanic degasification could provide the necessary sulphur contribution. These conditions allowed the deterioration of mafic minerals and the phyllosilicates formed during Phyllosian. A radical change of the climatic conditions on the surface of Mars (Mars Global Change) would thus have taken place between Phyllosian and Theiikian, undoubtedly related to a peak of volcanic activity. This episode of acid deterioration would also explain the absence of carbonates in great quantities on the surface of Mars (they were not detected to date by OMEGA, in spite of its capacity to detect them, even in small proportions), if they were ever formed. It is probable that Mars during Noachian had a dense atmosphere of CO_2 and liquid water circulated on its surface (Chassefière et al., 2007). The respective roles of the trapping in subsurface of water and CO_2 (in the form of carbonates) and the atmospheric exhaust (under the action of solar wind), to explain their disappearance, are not yet well-known. Let us note that this question is itself related to the history of the internal activity of Mars and to the stop of its dynamo, the magnetic field that induces limitation of the erosion of atmosphere (Chassefière et al., 2007).

Siderikian

Siderikian, according to *siderikos* (ferric in Greek), is the era of anhydrous ferric oxides thermodynamics (McLennan, 2012). They would have been formed by a slow weathering, on the surface, without water. This type of deterioration was to be already active during Phyllosian and of Theiikian, but its effectiveness was clearly exceeded by the processes which allowed the formation of the phyllosilicates and of sulphates.

Martian Geological Eras According to the Observation of the Impact Craters

Despite the uncertainty on the variations of the rate of cratering in the course of time (e.g., Hartmann, 2005), three geological ages may be distinguished based on catering (Scott and Carr, 1978). The meteoritic bombardment is linked to increasing disorder in the solar systems (Demina et al., 2021). That is to say, it is quite linked to the thermodynamics of the solar system. By chronological order, one finds Noachian (referring to Noachis Terra), Hesperian (referring to Hesperia Planum) and the Amazonian one (referring to Amazonis Planitia). Finer subdivisions (e.g., Tanaka, 1986) were subsequently carried out including: Early, Middle, and Late Noachian; Early and Late Hesperian; Early, Middle, and Late Amazonian. There is on the other hand an, and the opinions on the absolute dates diverge somewhat. According to recent publications (Hartmann & Neukum, 2001; Hartmann, 2005), the probable limits between the geological eras would be collected in Table 3.

Table 3. Absolute age estimates for the surface of Mars (Hartmann & Neukum, 2001)

Epoch Absolute	Age range (Gy)
Late Amazonian	0.6–0.3 to present
Middle Amazonian	2.1–1.4 to 0.6–0.3
Early Amazonian	3.1–2.9 to 2.1–1.4
Late Hesperian	3.6 to 3.1–2.9
Early Hesperian	3.7 to 3.6
Late Noachian	3.82 to 3.7
Middle Noachian	3.95 to 3.82
Early Noachian	> 3.95

Noachian Period

Though decreasing with time, the meteoritic bombardment remains however intense during the Noachian era. It is in Late Noachian that the principal basins of impact were formed, like those of Hellas and Argyre in the Southern hemisphere. Volcanic activity had already been playing a significant role during Noachian (Jakosky & Phillips, 2001). However noticed during late periods, the formation of the volcanic knob of Tharsis probably began during the Late to Middle Noachian (Jakosky & Phillips, 2001). During the Late to Middle Noachian, the planet would have been subjected to intense erosion, which would explain the presence of the famous ramified valleys observed in Highlands of the South. Erosion by more exotic fluids (liquid CO_2, liquid SO_2, and liquid hydrocarbons) was also considered. The magnetometer of Total Mars Surveyor (MGS) observed magnetic anomalies in Highlands of the South, dependent on the residual magnetization of the rocks in the first 50 km of the crust (Ness et al., 1999; Zuber et al., 2000). Certain scientists were thus tempted to explain this result with a martian plate tectonics, operating at the time where the martian dynamo (and thus the magnetic field) had been active active. Contexts of convergence (Connerney et al., 1999) and divergence (Fairen et al., 2002) zones were considered, but with no convincing geological or geodynamic arguments (Nimmo & Tanaka, 2005). If ever plate tectonics took place on early Mars, it should be related to the North-South dichotomy (e.g., Lenardic et al., 2004). The most distinguishing features of the Noachian are high rates of cratering, erosion, and valley formation, the accumulation of most of Tharsis, and surface conditions that enabled widespread production of weathering products such as phyllosilicates. The impacts would have caused hydrothermal activity affecting groundwater movement and storage.

Hesperian Period

Compared with the Noachian, we do not find any more ramified valleys in Hesperian lands, which suggest that the climatic conditions did not favour erosion. The formation of the network of canyons of Marineris Valles would be a tectonic consequence of the rising of the domes of the Tharsis (these canyons would have been subsequently increased by erosion). The Hesperian period is mainly characterized by continuous to episodic volcanism to form extensive lava plains and formation of the largest outñow channels and their terminal lakes or seas. Alteration results in phyllosilicates and local accumulation of sulfate-rich deposits, particularly in the western hemisphere. As for valleys and channels, complex outñow channels were cut by large ñoods of water. The abrupt start of outñow channels indicates that they formed not by surface drainage immediately following precipitation but by the rapid release of large volumes of stored water. The storage medium could be a subsurface groundwater aquifer, or a lake, or ice. However controversial and coinciding with volcanic contacts (Carr & Head, 2003), several possible shorelines have been tentatively identiðed in and around the northern plains (Parker et al., 1989, 1993; Clifford & Parker, 2001) & Hellas (Moore & Wilhelms, 2001) but they remain, and often coincide with volcanic contacts (Carr & Head, 2003). Supporting evidence for the presence of former bodies of water of Hesperian age in the northern plains are partly buried ridges and craters, interpreted as the result of burial by sediments carried by the large ñoods (Kreslavsky & Head, 2002).

Amazonian Period

The Amazonian Period extends from approximately 3 billion years ago, the middle of the terrestrial Archean, to the present, encompassing two thirds of the history of Mars. However less frequent than Hesperian, the catastrophic floods continued certainly at the Early Amazonian. The volcanic activity radically decreased. On the other hand, the aeolian activity is indeed the principal process which altered the decoration of the Mars surface, except the polar zones, during the Middle to Late Amazonian. Processes driven by obliquity variations are also more evident for this era, although such processes likely occurred throughout all ofmartian history (Laskar et al., 2004).

PALEO-HYDROGEOLOGICAL CYCLE AND GROUNDWATER INFLUENCE ON MARS SEDIMENTOLOGY

Elucidating the hydrogeological cycle of Mars is one of the main challenges in the exploration of the Solar System. Actually, several models were proposed to explain martian hydrology and hydrogeology and their connectivity (e.g., Fig. 5) Long-term aqueous activity on the surface of the planet is indicated by phased degradation of impact craters (e.g., Craddock & Maxwell, 1993); fluvial (e.g., Malin & Edgett, 2000) and lacustrine features (e.g., Cabrol & Grin, 2001); permafrost (e.g., Lucchitta, 1981); periglacial (e.g., Squyres, 1979) and glacial landforms (e.g., Kargel et al., 1995); and outflow channels. The outflow channels began to form as early as the Noachian, and their activity extended to the Early Amazonian (Dohm et al., 2001), and even to the latest Amazonian in Elysium Planitia (Burr et al., 2002), with recurrent flooding for some of the systems, including Kasei Valles (Scott, 1993) and Mangala Valles (Chapman & Tanaka, 1993). Dominated by an increasing dryness, the martian landscape dealt on groundwater upwelling to feed wet sedimentology of some aqueous species such as sulfates. Subsurface aquifers can account for the majority of observed gullies on Mars (Dickson &

Head, 2009). Mellon & Phillips (2001) argued that a groundwater reservoir must be confined between two aquicludes to prevent vertical transport of water, and by another barrier behind the aquifer, forcing release under pressure along the slope face. Dohm et al. (2008) discussed the significance of the genetic model ascribing the episodic formation of Oceanus Borealis to cataclysmic outburst flooding of the outflow channels. It was suggested that a mechanism whereby CO_2 clathrate in the martian permafrost zone is destabilized by episodes of very high heat flow, such that released CO_2 from the lower permafrost zone (2–3km depth) and dissolved CO_2 from the underlying groundwater explosively forces out pressurized slurries of water and fractured rock fragments in massive outbursts. Subsequent sediment-charged water enters the ocean as hyperpycnal flows, generating density flows that extend deposits across the northern plains. This hypothesis explains the long epochs (10^8 years), during which the Mars surface had extremely cold, dry conditions similar to those prevailing today, terminated by short duration (10^4–10^5 years) episodes of much warmer, wetter conditions associated with a transient greenhouse climate.

Figure 5. Schematic diagram showing the Mars Episodic Glacial Atmospheric Oceanic Upwelling by the rmotectonic FLood Outbursts hypothesis of Baker et al. (2000)

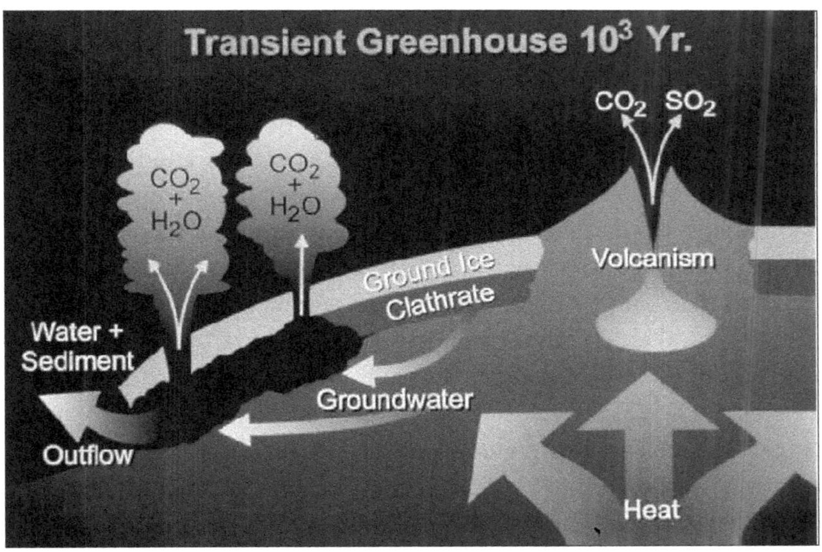

SEDIMENTOLOGY AND STRATIGRAPHY OF ARABIA TERRA, MERIDIANI PLANUM, TERBY CRATER, AND GALE CRATER IN RELATION WITH VARIABLE THERMODYNAMICS

The common key feature of these sites is the groundwater influence. They include the Vernal crater, the Meridiani Planum, the Terby crater, and the Gale crater.

Vernal Crater

Vernal crater is a 55 km diameter impact feature located at 6°N, 355.5°E, in southwestern Arabia Terra. Terra Arabia is one of the few equatorial regions on Mars where high abundance of near-surface hydrogen has been measured (Oehler & Allen, 2010). This abundance argues for the presence of shallow ice or hydrated minerals (Boynton et al., 2002; Feldman et al., 2002). As indicated by impact crater (Barlow & Perez, 2003) and elemental (e.g., Boynton et al., 2002) analyses, Arabia Terra province is one of the few visible water rich equatorial regions of Mars. This region records many unique characteristics. (1) The largest portion of heavily cratered terrain in the martian northern hemisphere. (2) High crater density indicates Noachian materials (e.g., Barlow, 1988). (3) A highland–lowland dichotomy boundary region distinctive from other boundary regions (Hynek & Phillips, 2001). (4) Some of the lowest topography displayed in the heavily cratered region (Smith et al., 2001). (5) A distinct center of tectonic activity (Anderson et al., 2006). (6) Compared to the rest of the cratered highlands, very few macrostructures are distinguished (Dohm et al., 2002). (7) The greatest extent of well-developed fretted terrain on Mars (Carr, 2001). (8) Outflow channels having no obvious source regions (Scott et al., 1995). (9) Highly concentrated multiple layer ejecta and central pit impact craters, suggesting a concentration of volatile rich materials (Barlow & Perez, 2003). (10) Higher albedo than the surrounding highland provinces (US Geological Survey, 1991). (11) Lower thermal inertia than its surroundings (Christensen et al., 2001). (12) Relatively to the rest of the cratered highlands, it has a free-air gravity signal (Yuan et al., 2001). (13) Presence of distinct magnetic anomalies (Arkani-Hamed, 2004). (14) Unique elemental signatures for H_2O and Cl (Feldman et al., 2002a). Vernal Crater is a Noachian impact structure that exhibits layered sediments, potential remnants of fluvio-lacustrine activity, and also indications of aeolian processes (Oehler & Allen, 2008, 2010). Putative spring mounds have been described at Vernal Crater (Oehler & Allen, 2010), probably formed as the result of subsurface fluid migration. Subsurface water would have sloped uniformly from the northwest rim down to the level of the springs, providing a potential hydraulic head, and therefore suggesting a hydraulic origin for the spring mounds. Water likely traveled along bedding planes, faults/fractures, or porous units, and the general flow could have been artesian and/or thermal. Seven geomorphic were mapped at Vernal Crater (Fig. 6) (Oehler & Allen, 2008). The northern half of the crater contains Noachian Plains units (NPuu & NPmd) dominated by features typical of aeolian processes (dunes and yardangs). The southern half may expose Noachian fluvio-lacustrine sediments; it contains few dunes or obvious yardangs but has possible meandering and braided stream units (NPmc & NPmm), layered deposits (NIlu) corresponding to the IR-bright feature, and a potential lake deposit (NIll) with associated shoreline like bedding (NIls) (Fig.6). Stratigraphically, the youngest units (in the northern half of Vernal Crater) appear to lie a minimum of four hundred meters below those investigated by Opportunity in Meridiani (Edgett, 2005); the older and deeper units to the south (the layered unit and possible lake/ shoreline deposits) would be 900 to 1200m (respectively) below the sediments being investigated by Opportunity.

Meridiani Planum

Hydrological modeling demonstrates that the Meridiani Planum and the surrounding Arabia Terra region would have been characterized by a shallow water table and sustained groundwater upwelling because of the unique topography of Arabia Terra (Andrews-Hanna et al., 2007). Imagery studies by the Opportunity rover at Meridiani Planum revealed sulfate rich sandstones that formed in a playa environment in the presence of a fluctuating water table (Grotzinger et al., 2005; Arvidson et al., 2006) (Fig.7).

On Meridiani Planum, mixtures of roughly equal amounts of altered siliciclastic debris of impure reworked evaporitic sandstones are preserved. They have a basaltic provenance (40±10% by mass), and chemical constituents, dominated by evaporitic minerals (jarosite, Mg-, Ca-sulfates, ±chlorides, ±Fe-, Na-sulfates), hematite, and secondary silica (60± 10%) (McLennan et al., 2005). These chemical constituents have been reworked by aeolian and subaqueous transport. Eventually, they are formed by evaporation of acidic waters derived from interaction with olivine-bearing basalts and subsequent diagenetic alteration. These rocks at Meridiani Planum experienced an extended diagenetic history, with at least two and up to four distinct episodes of cementation (McLennan et al., 2005). Diagenetic features are consistent with formation during later diagenesis in the phreatic (fluid saturated) zone or capillary fringe of a groundwater table under near isotropic hydrological conditions such as those expected during periodic groundwater recharge. This paragenetic sequence is consistent with an extended history of syndepositional through post-depositional diagenesis in the presence of a slowly fluctuating, chemically evolving, but persistently high ionic strength groundwater system (McLennan et al., 2005). Fig.8 shows a schematic model of the diagenetic history and the role of groundwater on the Burns formation at Meridiani Planum (McLennan et al., 2005). Diagenesis took place

Figure 6. Geomorphic units in Vernal Crater (Oehler & Allen, 2008)

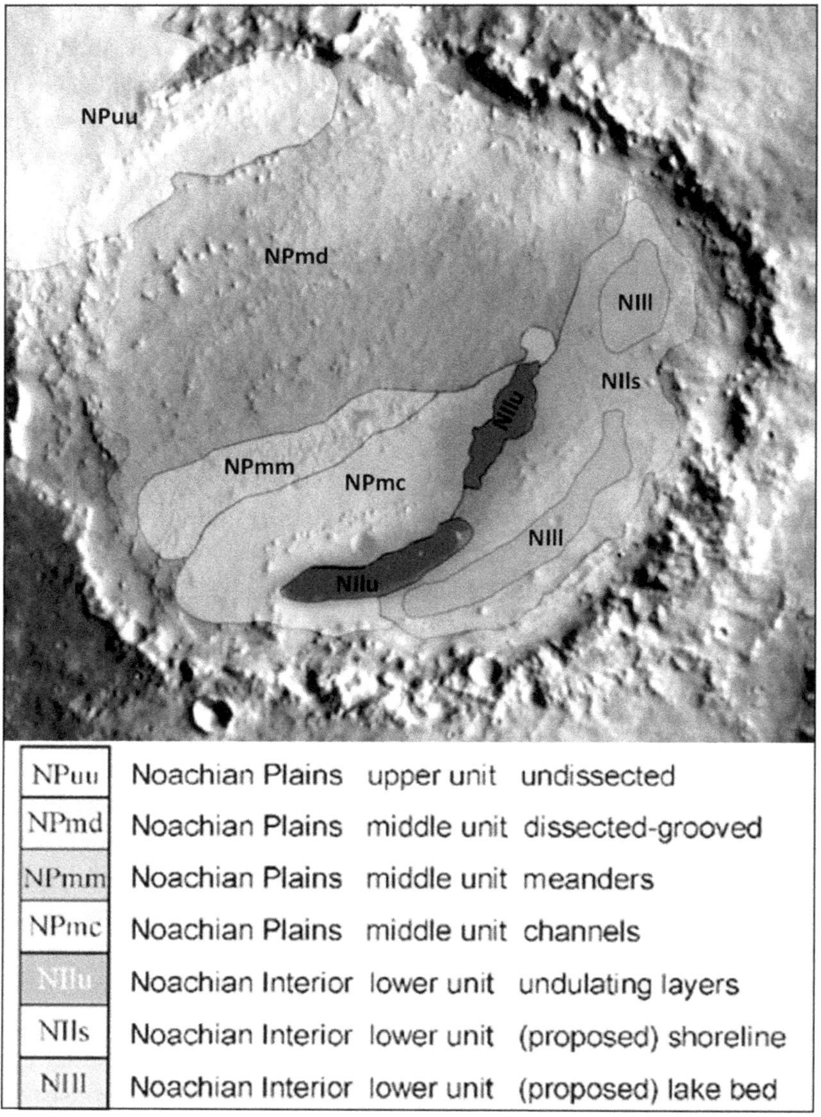

NPuu	Noachian Plains	upper unit	undissected
NPmd	Noachian Plains	middle unit	dissected-grooved
NPmm	Noachian Plains	middle unit	meanders
NPmc	Noachian Plains	middle unit	channels
NIlu	Noachian Interior	lower unit	undulating layers
NIls	Noachian Interior	lower unit	(proposed) shoreline
NIll	Noachian Interior	lower unit	(proposed) lake bed

according different stages. During the Early Diagenesis, evaporation of near surface groundwater table or capillary fringe of groundwater table results in intrasediment formation of mm-scale euhedral crystals of highly soluble evaporite mineral. At about the same time, early pore-filling cements form by evaporative processes. At the limit Early/Late Diagenesis, slow recharge of chemically distinct groundwater (higher pH and/or more oxidizing than pre-existing groundwater conditions) results

Figure 7. Map of Meridiani Planum region also showing the location of Opportunity landing site (Andrews-Hanna et al., 2007)

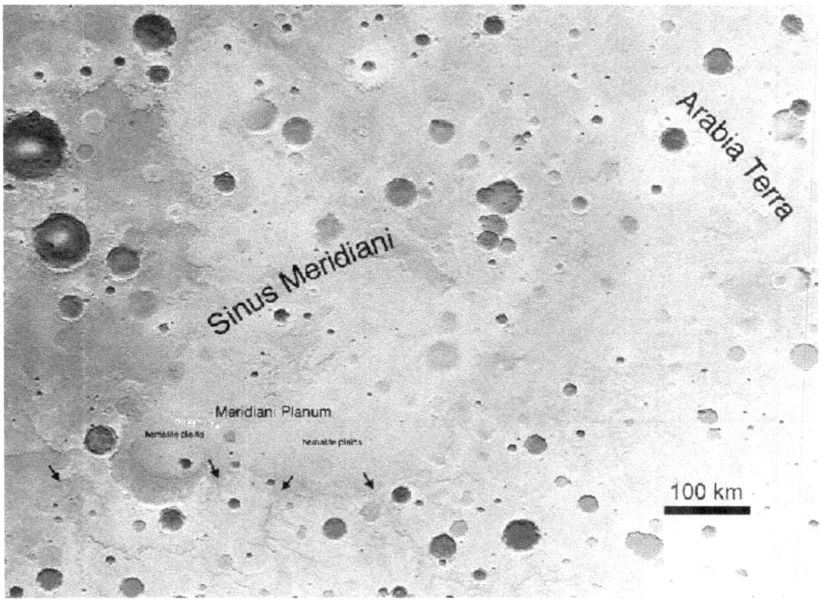

in breakdown of jarosite or other Fesulfate (such as melanterite) to form hematitic concretions. This process is likely very rapid and thus marks a convenient boundary between early and late diagenesis. During Late Diagenesis, we notice the formation of secondary crystal moldic porosity due to dissolution of syndepositional mm-scale euhedral evaporite crystals and secondary sheet-like vug porosity due to dissolution of relatively soluble porefilling cements.

Terby Crater

The 174-km-diameter Terby impact crater (28.0°S; 74.1°E), located on the northern rim of the Hellas basin, displays an anomalous inner morphology (e.g., De Hon, 1992; Cabrol & Grin, 1999) including a flat floor and light-toned layered deposits (Malin & Edgett, 2000; Ansan & Mangold, 2004) (Fig.9). The geometry obtained is consistent with that of clastic sediments that settled mainly in a sub-aqueous environment during the Noachian period (Ansan et al., 2011). To the north, the thickest sedimentary sequences observed are fan deltas, as identified by 100 m to 1 km long clinoforms and further defined by horizontal beds passing to foreset beds dipping by 6°-10° toward the center of the Terby crater (Ansan et al., 2011). The identification of distinct subaqueous fan sequences, separated by unconformities and

Figure 8. Diagenetic stages and groundwater influence at Meridiani Planum (McLennan et al., 2005)

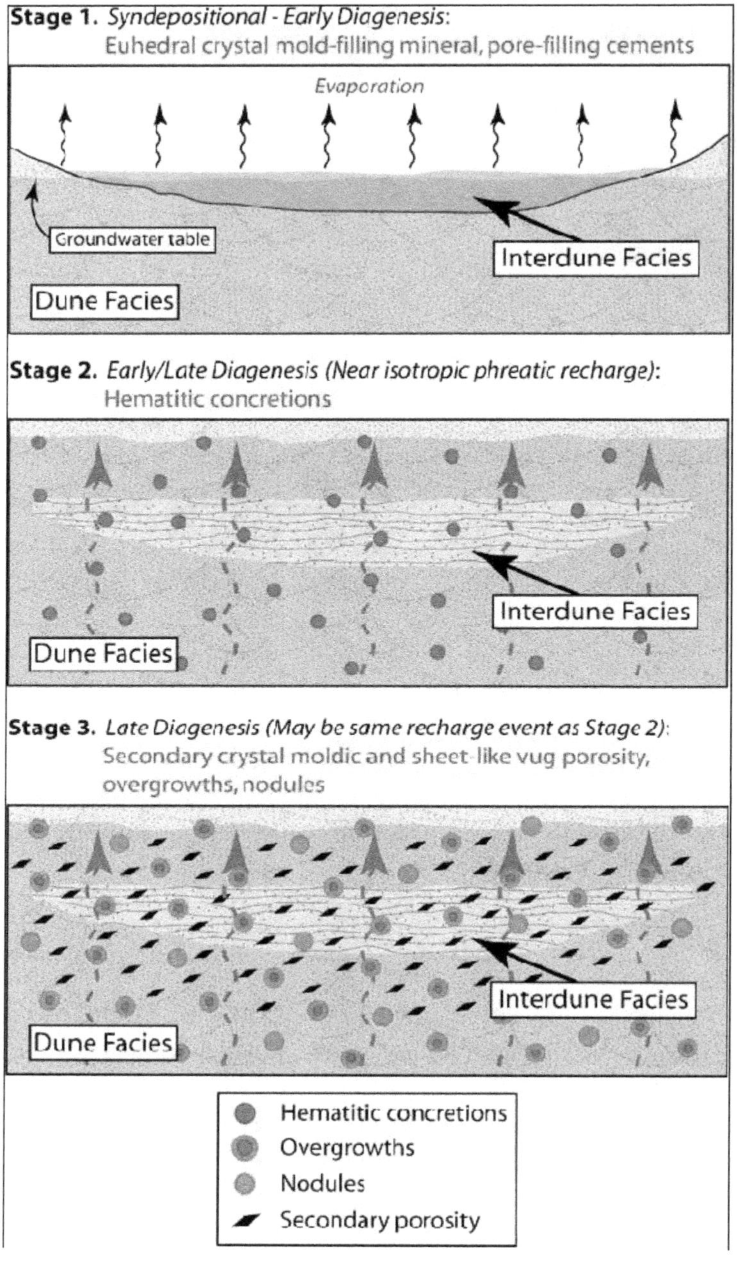

local wedges, suggest the accumulation of sediments from prograding/onlapping depositional sequences, due to lake level and/or sediment supply variations. Fig. 9a shows the location of Terby impact crater to the north of HellasBasin (white box) on the MOLA altimetry map. The flat inner Terby impact crater stands at -4 km in elevation, except in its northern area where closed depressions border high flat appendices named mesas.

Gale Crater

Presumed to be one of the oldest terrains on Mars, thedensely cratered and heavily eroded Arabia Terra is a large upland region in the north of Mars. It covers as much as 4500 kilometers at its longest extent, centered roughly atWikiMiniAtlas19°47¢N 30°00¢E / 19.79°N 30°E / 19.79; 30Coordinates: Click the blue globe to open an interactive map. 19°47¢N 30°00¢E / 19.79°N 30°E / 19.79; 30 with its eastern and southern regions rising 4 kilometers above the north-west (Fig.10a). Gale crater is a 152 km diameter impact crater centered at 5.3 S latitude, 222.3 W longitude in the Aeolis Mensae region (Milliken et al., 2010). A slightly off-centered mound of layered material, up to 5 km in height and nearly 100 km wide (Malin & Edgett, 2000), occupies the central area of Gale. Thick sequences of layered deposits are apparent in the lower reaches of the central mound (Milliken et al., 2010). The characteristics of such layers vary notably within the crater due to multiple possible formation mechanisms (including aeolian activity, volcanism, lacustrine deposition and spring-related activity). The mixture of lacustrine and aeolian sediments makes it challenging to understand the origin of the surface materials on Gale crater from the available imagery to date. On the one hand, a lake within Gale could have been originated by two separate events (Cabrol et al., 1999): (1) the south transgression of a northern large body of water, possibly a large basin, and/or (2) the influx of a fluvial valley network entering the crater from the south. On the other hand, the presence of wide extensions of localized dune fields, some of them occupying an extension up to 300 km^2, indicate that aeolian processes have played an important role in shaping some of the intracrater surfaces, and such processes could still be active. It was anticipated that a lander mission to Gale crater could provide a rich science return, answering multiple questions about Mars and its history, and particularly resolving questions related to interpreting remote-sensing data. That was the rationale to send the Curiosity rover to Gale crater, culminating in its successful landing in August 2012.

Milliken et al. (2010) carried out a stratigraphic subdivision of the outcropping sediment at Gale Crater. Two formations were identified: the Upper and the Lower formations (Figure 10b) (formation is with lower case because the study is carried out by remote sensing tools and not field expeditions. The lower formation itself is also subdivided into three members (Figure 10a).

Figure 9. (a) the location of Terby impact (b) MOLA topography of Terby impact crater with height contour lines with 100 m spacing in sinusoidal projection centered at 75°E of longitude(c) N-S topographic profile. (d) E-W topographic profile (Ansan et al., 2011)

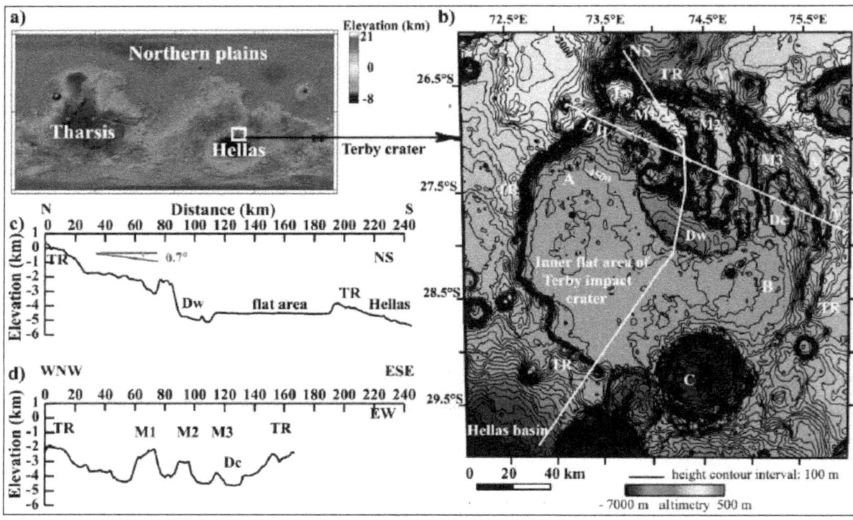

The Dynamic Albedo of Neutrons (DAN) (see section methods), provided information about the content of water at Gale Crater (Figure 11). The most likely hydrogen in Gale Crater outcrops is in hydrated minerals. These new data proved a groundwater intervention and wetter past conditions.

The comparison of the elemental composition of typical soils at three landing regions on Mars: Gusev Crater, Meridiani Planum, and Gale Crater shows their similarity (Fig. 12). It is worth noting also that concentrations of silicon dioxide and iron oxide were divided by 10, and nickel, zinc and bromine levels were multiplied by 100.

CONCLUSION

Actually, dealing with all aspect of thermodynamic processes shaping martian geology in this short overview is out of reach. Nevertheless, the major conclusion found out from this literature review is the importance of groundwater in shaping the martian land. To sum up, Mars oscillates between two extreme conditions according to its climatic conditions. If humid conditions prevail, water flows and temporary saline environments evade the planet. Otherwise, the red planet is a huge desert with aeolian

Figure 10. Lithostratgigraphic Units at Gale Crater (Milliken et al., 2010)

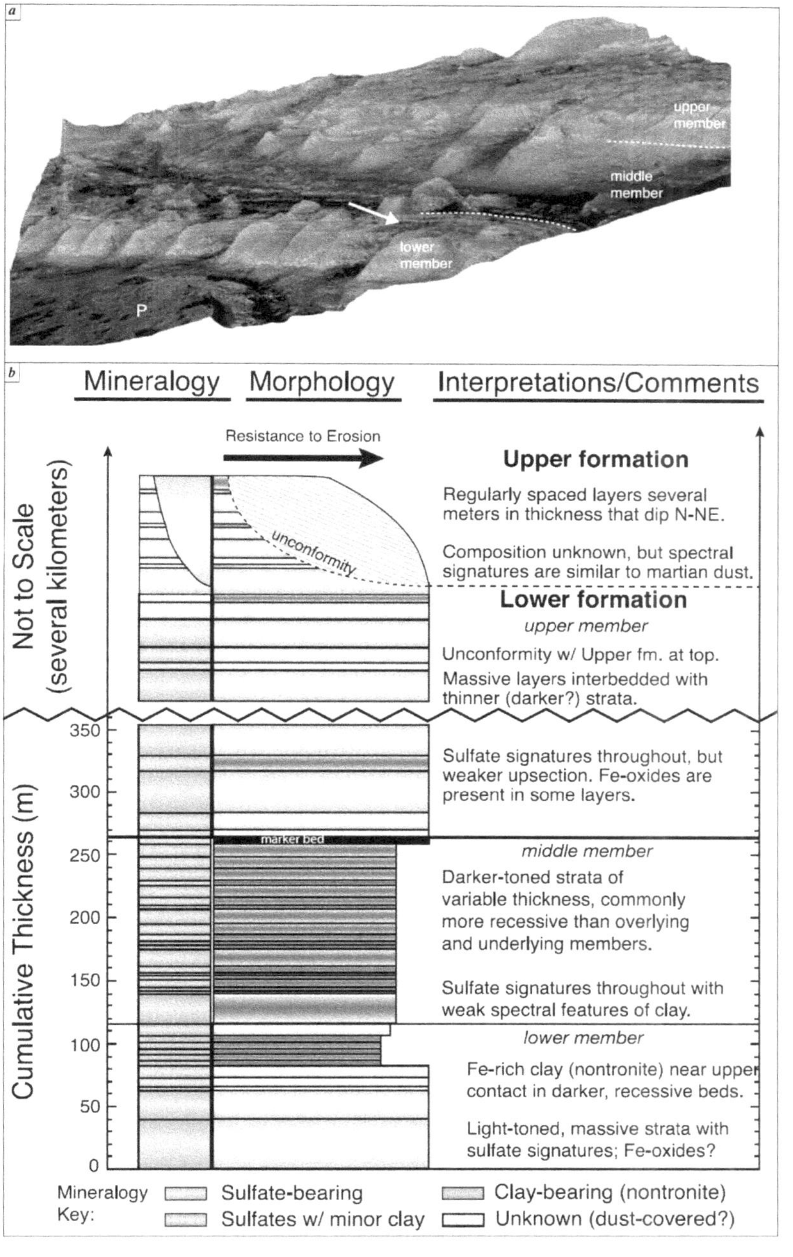

sedimentation. Subsequently, water dives under the ground and progressively feeds the surface through seepage or upwelling mechanisms. Previous studies showed that martian sites have similar conditions to those of terrestrial analogues. Consequently, they represent serious promise to satisfy the analogy with Tunisian sites.Aqueous conditions are common feature of all sites. The groundwater influence is twofold: (1) it may originate spring mounds; then, it controls their development through the setting of an aeolian wet sedimentology; (2) it is a determinant factor in sequence stratigraphy through controlling the repartition, progradation, and retrogradation of different types of aeolian sediment.

Figure 11. Comparative curve between a standard site and Mars showing the abundance of hydrate minerals and wetter conditions. Image Credit: NASA/JPL-Caltech/Russian Space Research Institute

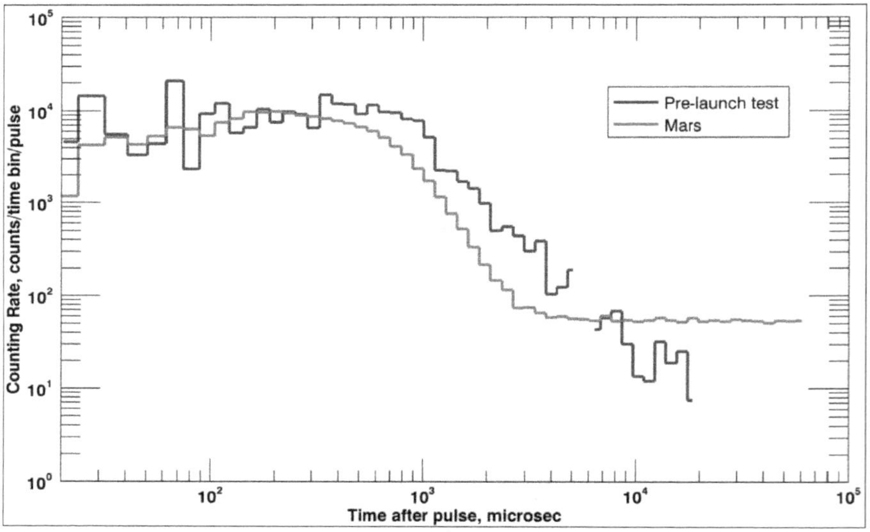

Figure 12. Similarity between different landing sites on Mars. Image Credit: NASA/ JPL-Caltech/University of Guelph

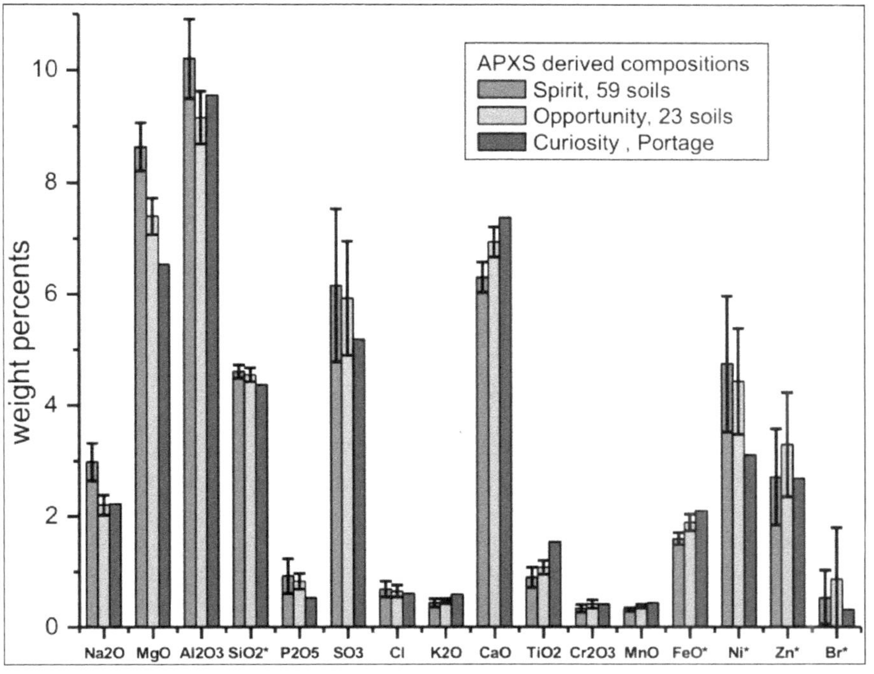

REFERENCES

Agee, C. B., Wilson, N. V., McCubbin, F. M., Ziegler, K., Polyak, V. J., Sharp, Z. D., Asmerom, Y., Nunn, M. H., Shaheen, R., Thiemens, M. H., Steele, A., Fogel, M. L., Bowden, R., Glamoclija, M., Zhang, Z., & Elardo, S. M. (2013). Unique Meteorite from Early Amazonian Mars: Water-Rich Basaltic Breccia Northwest Africa 7034. *Science*, *339*(6121), 780–785. Advance online publication. doi:10.1126cience.1228858 PMID:23287721

Ahrens, C., Cataldo, V., & Leone, G. (2021). Volcanic Eruptions on Mars, Lava Flow Morphology, and Thermodynamics. In *Mars: A Volcanic World* (pp. 71–94). Springer. doi:10.1007/978-3-030-84103-4_4

Anderson, R.C., Dohm, J.M., Haldemann, A.F.C., Pounders, E., Golombek, M., & Castano, A. (2006). Centers of tectonic activity for the eastern hemisphere. *Icarus*. (submitted for publication)

Andrews-Hanna, J., Phillips, R., & Zuber, M. (2007). Meridiani Planum and the global hydrology of Mars. *Nature*, *446*(7132), 163–168. doi:10.1038/nature05594 PMID:17344848

Ansan, V., Loizeau, D., Mangold, N., Le Mouélic, S., Carter, J., Poulet, F., Dromart, G., Lucas, A., Bibring, J.-P., Gendrin, A., Gondet, B., Langevin, Y., Masson, P., Murchie, S., Mustard, J. F., & Neukum, G. (2011). Stratigraphy, mineralogy, and origin of layered deposits inside Terby crater, Mars. *Icarus*, *211*(1), 273–304. doi:10.1016/j.icarus.2010.09.011

Ansan, V., & Mangold, N. (2004). Impact crater paleolakes in Hellas and Thaumasia areas, Mars. In Early Mars Conference, Jackson Hole, USA.

Appelbaum, J., & Flood, D. J. (1990). Solar radiation on Mars. *Solar Energy*, *45*(6), 353–363. doi:10.1016/0038-092X(90)90156-7

Arkani-Hamed, J. (2004). Timing of the martian core dynamo. *Journal of Geophysical Research*, *109*(E3), E03006. Advance online publication. doi:10.1029/2003JE002195

Arvidson, R. E., Poulet, F., Morris, R. V., Bibring, J.-P., Bell, J. F. III, Squyres, S. W., Christensen, P. R., Bellucci, G., Gondet, B., Ehlmann, B. L., Farrand, W. H., Fergason, R. L., Golombek, M., Griffes, J. L., Grotzinger, J., Guinness, E. A., Herkenhoff, K. E., Johnson, J. R., Klingelhöfer, G., ... Wolff, M. (2006). Nature and origin of the hematitebearing plains of Terra Meridiani based on analyses of orbital and Mars Exploration Rover data sets. *Journal of Geophysical Research*, *111*(E12). Advance online publication. doi:10.1029/2006JE002728

Baker, L. L., Agenbroad, D. J., & Wood, S. A. (2000). Experimental hydrothermal alteration of a martian analog basalt: Implications for martian meteorites. *Meteoritics & Planetary Science*, *35*(1), 31–38. doi:10.1111/j.1945-5100.2000.tb01971.x

Barbieri, R. (2012). *Continental evaporites from arid systems and their microfossils: a promising target in the search for traces of Martian life*. 3rd Conference on Terrestrial Mars Analogues, Marrakech, Morocco.

Barbieri, R., & Stivaletta, N. (2011). Continental evaporites and the search for evidence of life on Mars. *Geological Journal*, *46*(6), 513–524. doi:10.1002/gj.1326

Barbieri, R., & Stivaletta, N. (2012). Halophiles, continental evaporites and the search for biosignatures in environmental analogues for Mars. In Life on Earth and other Planetary Bodies. Springer. doi:10.1007/978-94-007-4966-5_3

Barlow, N. G. (1988). Crater size-frequency distributions and a revised Martian relative chronology. *Icarus*, *75*(2), 285–305. doi:10.1016/0019-1035(88)90006-1

Barlow, N. G., & Perez, C. B. (2003). Martian impact crater ejecta morphologies as indicators of the distribution of subsurface volatiles. *Journal of Geophysical Research, 108*(E8), 5085. Advance online publication. doi:10.1029/2002JE002036

Battler, M. M., Osinski, G. R., & Neil, B. R. (2012). Mineralogy of saline perennial cold springs on Axel Heiberg Island, Nunavut, Canada and implications for spring deposits on Mars. *Icarus*.

Bibring, J. P., Langevin, Y., Mustard, J. F., Poulet, F., Arvidson, R., Gendrin, A., Gondet, B., Mangold, N., Pinet, P., Forget, F., Berthé, M., Bibring, J.-P., Gendrin, A., Gomez, C., Gondet, B., Jouglet, D., Poulet, F., Soufflot, A., Vincendon, M., ... Neukum, G. (2006). Global Mineralogical and Aqueous Mars History Derived from OMEGA/ Mars Express Data. *Science, 312*(5772), 400–404. doi:10.1126cience.1122659 PMID:16627738

Bishop J. L., Zent A. P., & Pieters, C. M. (2002). A model for formation of dust, soil, and rock coatings on Mars: Physical and chemical processes on the Martian surface. *Journal of Geophysical Research, 107*(5097). . doi:10.1029/2001JE001581

Boynton, W. V., Feldman, W. C., Squyres, S. W., Prettyman, T. H., Brückner, J., Evans, L. G., Reedy, R. C., Starr, R., Arnold, J. R., Drake, D. M., Englert, P. A. J., Metzger, A. E., Mitrofanov, I., Trombka, J. I., d'Uston, C., Wänke, H., Gasnault, O., Hamara, D. K., Janes, D. M., ... Shinohara, C. (2002). Distribution of hydrogen in the near surface of Mars: Evidence for subsurface ice deposits. *Science, 297*(5578), 81–85. doi:10.1126cience.1073722 PMID:12040090

Brasser, R., & Walsh, K. J. (2011). Stability analysis of the martian obliquity during the Noachian era. *Icarus, 213*(1), 423–427. doi:10.1016/j.icarus.2011.02.024

Burr, D. M., Grier, J. A., McEwen, A. S., & Keszthelyi, L. P. (2002a). Repeated Aqueous Flooding from the Cerberus Fossae: Evidence for Very Recently Extant, Deep Groundwater on Mars. *Icarus, 159*(1), 53–73. doi:10.1006/icar.2002.6921

Cabrol, N. A., & Grin, E. A. (2001). The evolution of lacustrine environments on Mars: Is Mars only hydrologically dormant? *Icarus, 149*(2), 291–328. doi:10.1006/icar.2000.6530

Cabrol, N. A., Grin, E. A., Newsom, H. E., Landheim, R., & McKay, C. P. (1999). Hydrogeologic Evolution of Gale Crater and Its Relevance to the Exobiological Exploration of Mars. *Icarus, 139*(2), 235–245. doi:10.1006/icar.1999.6099

Cabrol, N. A., Grin, E. A., Newsom, H. E., Landheim, R., & McKay, C. P. (1999). Hydrogeologic Evolution of Gale Crater and Its Relevance to the Exobiological Exploration of Mars. *Icarus, 139*(2), 235–245. doi:10.1006/icar.1999.6099

Carr, M. (2001). Mars Global Surveyor observations of martian fretted terrain. *Journal of Geophysical Research, 106*(E10), 23571–23594. doi:10.1029/2000JE001316

Carr, M. H., & Head, J. W. III. (2003). Basal melting of snow on early Mars: A possible origin of some valley networks. *Geophysical Research Letters, 30*(24). Advance online publication. doi:10.1029/2003GL018575

Carrier, B. L., Beaty, D. W., Meyer, M. A., Blank, J. G., Chou, L., DasSarma, S., ... Xu, J. (2020). *Mars extant life: What's next?* Conference Report.

Cassata, W. S., Shuster, D. L., Renne, P. R., & Weiss, B. P. (2012, September–October). Trapped Ar isotopes in meteorite ALH 84001 indicate Mars did not have a thick ancient atmosphere. *Icarus, 221*(1), 461–465. doi:10.1016/j.icarus.2012.05.005

Chapman, M. G., & Tanaka, K. L. (1993). *Geologic map of theMTM-05152 and-10152 quadrangles, Mangala Valles region of Mars.* USGS Misc. Inv. Ser. Map I-2294 (1:500,000).

Chassefière, E., Leblanc, F., & Langlais, B. (2007). The combined effects of escape and magnetic field histories at Mars. *Planetary and Space Science, 55*(3), 343–357. doi:10.1016/j.pss.2006.02.003

Chevrier, V., Poulet, F., & Bibring, J. P. (2007). Early geochemical environment of Mars as determined from thermodynamics of phyllosilicates. *Nature, 448*(7149), 60–63. doi:10.1038/nature05961 PMID:17611538

Christensen, P. R., Bandfield, J. L., Hamilton, V. E., Ruff, S. W., Kieffer, H. H., Titus, T. N., Malin, M. C., Morris, R. V., Lane, M. D., Clark, R. L., Jakosky, B. M., Mellon, M. T., Pearl, J. C., Conrath, B. J., Smith, M. D., Clancy, R. T., Kuzmin, R. O., Roush, T., Mehall, G. L., ... Greenfield, M. (2001). The Mars Global Surveyor Thermal Emission Spectrometer experiment: Investigation description and surface science results. *Journal of Geophysical Research, 106*(E10), 23823–23871. doi:10.1029/2000JE001370

Clark, R. N., Carlson, R., Grundy, W., & Noll, K. (2013). Observed Ices in the Solar System. *The Science of Solar System Ices Astrophysics and Space Science Library, 356*, 3–46. doi:10.1007/978-1-4614-3076-6_1

Clifford, S. M., & Parker, T. J. (2001). The evolution of the Martian hydrosphere: Implications for the fate of a primordial ocean and the current state of the northern plains. *Icarus, 154*(1), 40–79. doi:10.1006/icar.2001.6671

Connerney & Acuna. (1999). Magnetic lineations in the ancient crust of Mars. *Science, 284*, 794.

Connerney, J. E. P., Acuña, M. H., Wasilewski, P. J., Kletetschka, G., Ness, N. F., Rème, H., Lin, R. P., & Mitchell, D. L. (2001). The global magnetic field of Mars and implications for crustal evolution. *Geophysical Research Letters*, *28*(21), 4015–4018. doi:10.1029/2001GL013619

Craddock, R. A., & Maxwell, T. A. (1993). Geomorphic evolution of the martian highlands through ancient fluvial processes. *Journal of Geophysical Research*, *98*(E2), 3453–3468. doi:10.1029/92JE02508

De Hon, R. A. (1992). Martian lake basins and lacustrine plains. *Earth, Moon, and Planets*, *56*(2), 95–122. doi:10.1007/BF00056352

de Pablo Hernández, M. Á., & Carrillo, J. D. C. (2012). *Geomorphological map of the lower NW flank of the Hecates Tholus volcano, Mars (scale 1:100,000)*. doi:10.1080/17445647.2012.703902

De Vera, J. P., Boettger, U., Noetzel, R. T., Sánchez, F. J., Grunow, D., Schmitz, N., Lange, C., Hübers, H.-W., Billi, D., Baqué, M., Rettberg, P., Rabbow, E., Reitz, G., Berger, T., Möller, R., Bohmeier, M., Horneck, G., Westall, F., Jänchen, J., ... Spohn, T. (2012). Supporting Mars exploration: BIOMEX in Low Earth Orbit and further astrobiological studies on the Moon using Raman and PanCam technology. *Planetary and Space Science*, *74*(1), 103–110. doi:10.1016/j.pss.2012.06.010

Demina, N. Y., Andreev, A. O., & Nefedyev, Y. A. (2021, November). Meteorite hazard model for a space mission to Mars. *Journal of Physics: Conference Series*, *2103*(1), 012031. doi:10.1088/1742-6596/2103/1/012031

Dickson, J. L., & Head, J. W. (2009). The formation and evolution of youthful gullies on Mars: Gullies as the late-stage phase of Mars' most recent ice age. *Icarus*, *204*(1), 63–86. doi:10.1016/j.icarus.2009.06.018

Dohm, J. M., Anderson, R. C., Baker, V. R., Ferris, J. C., Rudd, L. P., Hare, T. M., Rice, J. W. Jr, Casavant, R. R., Strom, R. G., Zimbelman, J. R., & Scott, D. H. (2001). Latent outflow activity for western Tharsis, Mars: Significant flood record exposed. *Journal of Geophysical Research*, *106*(E6), 12301–12314. doi:10.1029/2000JE001352

Dohm, J. M., Maruyama, S., Baker, V. R., Anderson, R. C., Ferris, J. C., & Hare, T. M. (2002). *Plate tectonism on early Mars: Diverse geological and geophysical evidence*. Academic Press.

Edgett, K.S. (2005). The sedimentary rocks of Sinus Meridiani: Five key observations from data acquired by the Mars Global Surveyor and Mars Odyssey orbiters. *Mars*, *1*, 5–58.

Ehlmann, B. L., Berger, G., Mangold, N., Michalski, J. R., Catling, D. C., Ruff, S. W., Chassefière, E., Niles, P. B., Chevrier, V., & Poulet, F. (2013, January). Geochemical Consequences of Widespread Clay Mineral Formation in Mars' Ancient Crust. *Space Science Reviews*, *174*(1-4), 329–364. doi:10.100711214-012-9930-0

El Maarry, M. R., Kodikara, J., Wijessoriya, S., Markiewicz, W. J., & Thomas, N. (2012, March). Desiccation mechanism for formation of giant polygons on Earth and intermediate-sized polygons on Mars: Results from a pre-fracture model. *Earth and Planetary Science Letters*, *323–324*(15), 19–26. doi:10.1016/j.epsl.2012.01.016

Essefi, E. (2009). *Multidisciplinary study of Sidi El Hani Saline Environment: the History and the Climatic Variability* (Master thesis). Faculty of sciences of Sfax, University of Sfax.

Essefi, E. (2013). *Wet Aeolian Sedimentology and Sequence Stratigraphy within the Terrestrial Analogues in Eastern Tunisia: Implications for Wet Aeolian Sedimentology and Sequence Stratigraphy on Mars*. National Engineering School of Sfax.

Essefi, E., Komatsu, G., Fairén, A. G., Chan, M. A., & Yaich, C. (2013). Alignment of fault spring mounds at El-Guetiate, Southeast Tunisia: Terrestrial analogue implications for martian tectonics. *Lunar and Planetary Science*, 1229.

Essefi, E., Komatsu, G., Fairén, A. G., Chan, M. A., & Yaich, C. (2014). Groundwater influence on the aeolian sequence stratigraphy of the Mechertate–Chrita–Sidi El Hani system, Tunisian Sahel: Analogies to the wet–dry aeolian sequence stratigraphy at Meridiani Planum, Terby crater, and Gale crater, Mars. *Planetary and Space Science*, *95*, 56–78. doi:10.1016/j.pss.2013.05.010

Feldman, W. C., Boynton, W. V., Tokar, R. L., Prettyman, T. H., Gasnault, O., Squyres, S. W., Elphic, R. C., Lawrence, D. J., Lawson, S. L., Maurice, S., McKinney, G. W., Moore, K. R., & Reedy, R. C. (2002). Global distribution of neutrons from Mars: Results from Mars Odyssey. *Science*, *297*(5578), 75–78. doi:10.1126cience.1073541 PMID:12040088

Fernández-Remolar, D. C. (2012). *Carbonate precipitation under bulk acidic conditions as a potential biosignature for searching life on Mars Earth and Planetary Science Letters*. Academic Press.

Foley, B. J., Bercovici, D., & Landuyt, W. (2012, May). The conditions for plate tectonics on super-Earths: Inferences from convection models with damage. *Earth and Planetary Science Letters*, *331–332*(15), 281–290. doi:10.1016/j.epsl.2012.03.028

Gainey, S. R., Hausrath, E. M., & Hurowitz, J. A. (2022). Thermodynamic and kinetic analysis of transitions in clay mineral chemistry on Mars. *Icarus*, *372*, 114733. doi:10.1016/j.icarus.2021.114733

Goody, R. M. (2012). The Atmosphere of Mars. Weather, 12(1), 3–15. doi:10.1002/j.1477-8696.1957.tb00381.x

Greeley, R., & Guest, J.E. (1987). *Geologic map of the eastern equatorial region of Mars*. USGS Misc. Inv. Ser. Map I-1802B (1:15,000,000).

Gross, M. (2012). The search for life on Earth and other planets. *Current Biology, 22*(7). . doi:10.1016/j.cub.2012.03.040

Gudipati, M. S., & Cooper, P. D. (2013). *Chemistry in Water Ices: From Fundamentals to Planetary Applications Astrophysics and Space Science Library*. Academic Press.

Haberle, R. M., Forget, F., Head, J., Kahre, M. A., Kreslavsky, M., & Owen, S. J. (2013). Summary of the Mars recent climate change workshop NASA/Ames Research Center. *Icarus, 222*(1), 415–418.

Hargitai, H., & Gide, M. (2009). Three virtual globes of Mars: topographic, albedo and a historic globe. *EPSC Abstracts, 4*.

Harry, Y., & McSween, J. (2012). *What we have learned about Mars from SNC meteorites*. doi:10.1111/j.1945-5100.1994.tb01092.x

Hartmann, W. K. (2005, April). Martian cratering 8: Isochron refinement and the chronology of Mars. *Icarus, 174*(2), 294–320. doi:10.1016/j.icarus.2004.11.023

Hartmann, W. K., & Neukum, G. (2001). Cratering chronology and the evolution of Mars. In *Chronology and evolution of Mars* (pp. 165–194). Springer. doi:10.1007/978-94-017-1035-0_6

Hirt, C., Claessens, S.J., Kuhn, M., & Featherstone, W.E. (2012). Kilometer-resolution gravity field of Mars: MGM2011. *Planetary and Space Science, 67*(1), 147–154. doi:10.1016/j.pss.2012.02.006

Hynek, B. M., & Phillips, R. J. (2001). Evidence for extensive denundation of the martian highlands. *Geology, 29*(5), 407–410. doi:10.1130/0091-7613(2001)029<0407:EFEDOT>2.0.CO;2

Jakosky, B., & Phillips, R. J. (2001). Mars' volatile and climate history. *Nature, 412*(6843), 237–244. doi:10.1038/35084184 PMID:11449285

Jakosky, B. M., Lin, R. P., Grebowsky, J. M., Luhmann, J. G., Mitchell, D. F., Beutelschies, G., Priser, T., Acuna, M., Andersson, L., Baird, D., Baker, D., Bartlett, R., Benna, M., Bougher, S., Brain, D., Carson, D., Cauffman, S., Chamberlin, P., Chaufray, J.-Y., ... Zurek, R. (2015). The Mars atmosphere and volatile evolution (MAVEN) mission. *Space Science Reviews, 195*(1), 3–48. doi:10.100711214-015-0139-x

Johnston, D. H., & Toksöz, M.N. (1977). *Internal structure and properties of Mars.* Academic Press.

Jouannic, G., Gargani, J., Costard, F., Ori, G. G., Marmo, C., Schmidt, F., & Lucas, A. (2012). Morphological and mechanical characterization of gullies in a periglacial environment: The case of the Russell crater dune (Mars). *Planetary and Space Science, 71*(1), 38–54. doi:10.1016/j.pss.2012.07.005

Kaltenegger, L. (2013). *Planetary Atmospheres and Chemical Markers for Extraterrestrial Life*. Astrochemistry and Astrobiology Physical Chemistry in Action. doi:10.1007/978-3-642-31730-9_5

Kargel, J. S., Baker, V. R., Beget, J. E., Lockwood, J. F., Pewe, T. L., Shaw, J. S., & Strom, R. G. (1995). Evidence of ancient continental glaciation in the martian northern plains. *Journal of Geophysical Research, 100*(E3), 5351–5368. doi:10.1029/94JE02447

Katterfel'd, G. N., Benesh, K., Khain, V. Ye., & Khodak, Yu. A. (1968). Problems of comparative planetologyInternational. *Dizhi Lunping, 10*(9), 989–1017. doi:10.1080/00206816809474964

Khatib, T., Abunajeeb, I., & Heneni, Z. (2020). Determination of Mars Solar-Belt by Modeling of Solar Radiation Using Artificial Neural Networks. *Journal of Solar Energy Engineering, 142*(1), 011007. doi:10.1115/1.4044304

Klima, R. (2013). The Evolution of the Lunar Highlands Crust: A Complicated History. *Second Conference on the Lunar Highlands Crust, 94*(2), 21–28. doi:10.1002/2013EO020009

Klose, M., & Shao, Y. (2013). Large-eddy simulation of turbulent dust emission. *Aeolian Research, 8*, 49–58. doi:10.1016/j.aeolia.2012.10.010

Kreslavsky, M. A., & Head, J. W. III. (2002). Mars: Nature and evolution of young latitude-dependent water–ice-rich mantle. *Geophysical Research Letters, 29*(15), 14–21. doi:10.1029/2002GL015392

Laskar, J., Robutel, P., Joutel, F., Gastineau, M., Correia, A., & Levrard, B. (2004). A long term numerical solution for the insolation quantities of the Earth. Astron. *Astrophys.* http://hal.ccsd.cnrs.fr/ccsd-00001603

Lenardic, A., Nimmo, F., & Moresi, L. (2004). Growth of the hemispheric dichotomy and the cessation of plate tectonics on Mars. *Journal of Geophysical Research, 109*(E2), E02003. doi:10.1029/2003JE002172

Limaye, A. B. S., Aharonson, O., & Perron, J. T. (2012). Detailed stratigraphy and bed thickness of the Mars north and south polar layered deposits. *J. Geophys. Res., 117*. doi:10.1029/2011JE003961

Lin, S., & Chen, N. (2021). DEM Based Study on Shielded Astronomical Solar Radiation and Possible Sunshine Duration under Terrain Influences on Mars by Using Spectral Methods. *ISPRS International Journal of Geo-Information, 10*(2), 56. doi:10.3390/ijgi10020056

Lu, X., & Kieffer, S. W. (2009). Thermodynamics and mass transport in multicomponent, multiphase H2O systems of planetary interest. *Annual Review of Earth and Planetary Sciences, 37*(1), 449–477. doi:10.1146/annurev.earth.031208.100109

Lucchitta, B. K. (1981). Mars and Earth: Comparison of cold-climate features. *Icarus, 45*(2), 264–303. doi:10.1016/0019-1035(81)90035-X

Malin, M. C., & Edgett, K. S. (2000). Sedimentary rocks of Early Mars. *Science, 290*(5498), 1927–1937. doi:10.1126cience.290.5498.1927 PMID:11110654

Mars, C. (2005). Sulfate formation and its relevence to environmental conditions on early. *Science, 307*, 1594-1597.

Mayhew, K. W. (2020). New Thermodynamics: Inelastic Collisions and Cosmology. *European Journal of Applied Physiology, 2*(6).

McLennan, S. M. (2012). *Geochemistry of sedimentary processes on Mars*. Academic Press.

McLennan, S. M., Bell, J. F. III, Calvin, W. M., Christensen, P. R., Clark, B. C., de Souza, P. A., Farmer, J., Farrand, W. H., Fike, D. A., Gellert, R., Ghosh, A., Glotch, T. D., Grotzinger, J. P., Hahn, B., Herkenhoff, K. E., Hurowitz, J. A., Johnson, J. R., Johnson, S. S., Jolliff, B., ... Yen, A. (2005). Provenance and diagenesis of the evaporite-bearing Burns formation, Meridiani Planum, Mars. *Earth and Planetary Science Letters, 240*(1), 95–121. doi:10.1016/j.epsl.2005.09.041

McMahon, S., & Cosmidis, J. (2022). False biosignatures on Mars: Anticipating ambiguity. *Journal of the Geological Society, 179*(2), jgs2021-050. doi:10.1144/jgs2021-050

Mendoza, V. M., Mendoza, B., Garduño, R., Cordero, G., Pazos, M., Cervantes, S., & Cervantes, K. (2021). Thermodynamic simulation of the seasonal cycle of temperature, pressure and ice caps on Mars. *Atmosfera, 34*(1), 1–23.

Milliken, R. E., Grotzinger, J. P., & Thomson, B. J. (2010). Paleoclimate of Mars as captured by the stratigraphic record in Gale Crater. *Geophysical Research Letters, 37*. doi:10.1029/2009GL041870

Mischna, M. A., Baker, V., Milliken, R., Richardson, M., & Lee, C. (2013). Effects of obliquity and water vapor/trace gas greenhouses in the early martian climate. *Journal of Geophysical Research. Planets, 118*(3), 560–576. doi:10.1002/jgre.20054

Moore, J. M., & Wilhelms, D. E. (2001). Hellas as a possible site of ancient ice-covered lakes on Mars. *Icarus, 154*(2), 258–276. doi:10.1006/icar.2001.6736

Moores, J. E., Brown, R. H., Lauretta, D. S., & Smith P. H. (2012). Experimental and theoretical simulation of sublimating dusty water ice with implications for D/H ratios of water ice on Comets and Mars. *Planetary Science, 1*(2).

Ness, N. F., Acuña, M. H., Connerney, J., Wasilewski, P., Mazelle, C., Sauvaud, J., Vignes, D., d'Uston, C., Reme, H., Lin, R., Mitchell, D. L., McFadden, J., Curtis, D., Cloutier, P., & Bauer, S. J. (1999). 14 colleagues MGS magnetic fields and electron reflectometer investigation: Discovery of paleomagnetic fields due to crustal remanence. *Advances in Space Research, 23*(11), 1879–1886. doi:10.1016/S0273-1177(99)00271-9

Nicolaou, G., Wicks, R. T., Owen, C. J., Kataria, D. O., Chandrasekhar, A., Lewis, G. R., ... Bruno, R. (2021). Deriving the bulk properties of solar wind electrons observed by Solar Orbiter: A preliminary study of electron plasma thermodynamics. *Astronomy & Astrophysics, 656*, A10. doi:10.1073/iti0213110

Niles, P., Catling, D. C., Berger, G., Chassefière, E., Ehlmann, B. L., Michalski, J. R., Morris, R., Ruff, S. W., & Sutter, B. (2013, January). B. & 8 colleagues, 2013. Geochemistry of Carbonates on Mars: Implications for Climate History and Nature of Aqueous Environments. *Space Science Reviews, 174*(1-4), 301–328. doi:10.100711214-012-9940-y

Nimmo, F., & Tanaka, K. L. (2005). Early crustal evolution of Mars. *Annual Review of Earth and Planetary Sciences, 33*(1), 133–161. doi:10.1146/annurev.earth.33.092203.122637

Oehler, D. Z., & Allen, C. C. (2008). Ancient hydrothermal springs in Arabia Terra, Mars. *Lunar and Planetary Science*, 39.

Oehler, D. Z., & Allen, C. C. (2010). Evidence for pervasive mud volcanism in Acidalia Planitia, Mars. *Icarus, 208*(2), 636–657. doi:10.1016/j.icarus.2010.03.031

Oehler, D. Z., & Allen, C. C. (2012, June). Giant Polygons and Mounds in the Lowlands of Mars: Signatures of an Ancient Ocean? *Astrobiology*, *12*(6), 601–615. doi:10.1089/ast.2011.0803 PMID:22731685

Oehler, J. F., Lénat, J. F., & Labazuy, P. (2008). Growth and collapse of the Reunion Island volcanoes. *Bull. Volcanol.*, *70*(6), 717–742. doi:10.100700445-007-0163-0

Parker, T. J., Gorsline, D. S., Saunders, R. S., Pieri, D. C., & Schneeberger, D. M. (1993). Schneeberger D.M., Coastal geomorphology of the martian northern plains. *Journal of Geophysical Research*, *98*(E6), 11061–11078. doi:10.1029/93JE00618

Parker, T. J., Saunders, R. S., & Schneeberger, D. M. (1989). Transitional morphology in the west Deuteronilus Mensae region of Mars: Implications for modification of the lowland/upland boundary. *Icarus*, *82*(1), 111–145. doi:10.1016/0019-1035(89)90027-4

Pasckert, J. H., Hiesinger, H., & Reiss, D. (2012). Rheologies and ages of lava flows on Elysium Mons, Mars. *Icarus*, *219*(1), 443–457. doi:10.1016/j.icarus.2012.03.014

Phillips, R. J., Zuber, M. T., Solomon, S. C., Golombek, M. P., Jakosky, B. M., Banerdt, W. B., Smith, D. E., Williams, R. M. E., Hynek, B. M., Aharonson, O., & Hauck, S. A. II. (2001). Ancient geodynamics and global-scale hydrology on Mars. *Science*, *291*(5513), 2587–2591. doi:10.1126cience.1058701 PMID:11283367

Portyankina, G., Pommerol, A., Aye, K.-M., Hansen, C. J., & Thomas, N. (2012). Polygonal cracks in the seasonal semi-translucent CO2 ice layer in Martian polar areas. *Journal of Geophysical Research*, *117*(E2), E02006. Advance online publication. doi:10.1029/2011JE003917

Righter, K., Schönbächler, M., Pando, K., Rowland, I. I. R. II, Righter, M., & Lapen, T. (2020). Ag isotopic and chalcophile element evolution of the terrestrial and martian mantles during accretion: New constraints from Bi and Ag metal-silicate partitioning. *Earth and Planetary Science Letters*, *552*, 116590. doi:10.1016/j.epsl.2020.116590

Robertson, K., & Bish, D. (2012). Constraints on the distribution of $CaSO_4{\cdot}nH_2O$ phases on Mars and implications for their contribution to the hydrological cycle. *Icarus*.

Rodriguez-Manfredi, J. A., de la Torre Juárez, M., Alonso, A., Apéstigue, V., Arruego, I., Atienza, T., Banfield, D., Boland, J., Carrera, M. A., Castañer, L., Ceballos, J., Chen-Chen, H., Cobos, A., Conrad, P. G., Cordoba, E., del Río-Gaztelurrutia, T., de Vicente-Retortillo, A., Domínguez-Pumar, M., Espejo, S., ... Zurita, S. (2021). The Mars Environmental Dynamics Analyzer, MEDA. A suite of environmentalsensors for the Mars 2020 mission. *Space Science Reviews*, *217*(3), 1–86. doi:10.100711214-021-00816-9 PMID:34776548

Scott, D. H. (1993). *Geologic map of the MTM 25057 and 25052 quadrangles, Kasei Valles region of Mars*. USGS Misc. Inv. Ser. I-Map 2208 (1:500,000).

Scott, D. H., & Carr, M. H. (1978). *Geologic map of Mars*. U.S. Geol. Survey, Misc. Inv. Map I-1083.

Scott, D. H., Dohm, J. M., & Rice, J. W., Jr. (1995). *Map of Mars showing channels and possible paleolake basins*. USGS Misc. Inv. Ser. Map I-2461 (1:30,000,000).

Sefton-Nash, E., Teanby, N. A., Montabone, L., Irwin, P. G. J., Hurley, J., & Calcutt, S. B. (2013, January). Climatology and first-order composition estimates of mesospheric clouds from Mars Climate Sounder limb spectra. *Icarus*, *222*(1), 342–356. doi:10.1016/j.icarus.2012.11.012

Selco, J. I. (1995). Chemical Thermodynamics on Mars. *Journal of Chemical Education*, *72*(7), 599. doi:10.1021/ed072p599

Smith, D. E., Zuber, M. T., Frey, H. V., Garvin, J. B., Head, J. W., Muhleman, D. O., Pettengill, G. H., Phillips, R. J., Solomon, S. C., Zwally, H. J., Banerdt, W. B., Duxbury, T. C., Golombek, M. P., Lemoine, F. G., Neumann, G. A., Rowlands, D. D., Aharonson, O., Ford, P. G., Ivanov, A. B., ... Sun, X. (2001). Mars Orbiter Laser Altimeter: Experimental summary after the first year of global mapping of Mars. *Journal of Geophysical Research*, *106*(E10), 23689–23722. doi:10.1029/2000JE001364

Smith, L., Forster, C. B., & Evans, J. P. (1990). Interaction of fault zones, fluid flow, and heat transfer at the basin scale. Hydrogeology of Permeability Environments. *International Association of Hydrogeologists Symposium, 28th Int. Geol. Congress*, 2, 41-67.

Solomon, S. C., & Qian, L. (2012). The Interaction of Solar and Geomagnetic Activity with ClimateChange in the Thermosphere. EGU General Assembly 2012, Vienna, Austria.

Squyres, S. W. (1979). The distribution of lobate debris aprons and similar flows on Mars. *Journal of Geophysical Research*, *84*(B14), 8087–8096. doi:10.1029/JB084iB14p08087

Šrámek, O., & Zhong, S. (2012). Martian crustal dichotomy and Tharsis formation by partial melting coupled to early plume migration. *Journal of Geophysical Research*, *117*(E1), E01005. Advance online publication. doi:10.1029/2011JE003867

Stanley, B. D., Schaub, D. R., & Hirschmann, M. M. (2012, November-December). CO_2 solubility in primitive martian basalts similar to Yamato 980459, the effect of composition on CO_2 solubility of basalts, and the evolution of the martian atmosphere. *The American Mineralogist, 97*(11-12), 1841–1848. doi:10.2138/am.2012.4141

Stivaletta, N., & Barbieri, R. (2009). Endolithic microorganisms from spring mound evaporite deposits (southern Tunisia). *Journal of Arid Environments, 73*(1), 33–39. doi:10.1016/j.jaridenv.2008.09.024

Tanaka, K. L. (1986). The stratigraphy of Mars. *Journal of Geophysical Research, 91*(B13), E139–E158. doi:10.1029/JB091iB13p0E139

Vaz, D. A. (2012). *Tectonic lineament mapping of the Thaumasia Plateau, Mars: Comparing results from photointerpretation and a semi automatic approach.* Academic Press.

Villanueva, G. L., Mumma, M. J., Novak, R. E., Radeva, Y. L., Käufl, H. U., Smette, A., Tokunaga, A., Khayat, A., Encrenaz, T., & Hartogh, P. (2013, March). A sensitive search for organics (CH4, CH3OH, H2CO, C2H6, C2H2, C2H4), hydroperoxyl (HO2), nitrogen compounds (N2O, NH3, HCN) and chlorine species (HCl, CH3Cl) on Mars using ground-based high-resolution infrared spectroscopy. *Icarus, 223*(1), 11–27. doi:10.1016/j.icarus.2012.11.013

Wakabayashi, J., & Shervais, J. W. (2012). Initiation and Termination of Subduction: Rock Record, Geodynamic Models, and Modern Plate Boundaries. *Lithosphere, 4*(6), 467–468. doi:10.1130/LINT1.1

Williamson, M. C. Johnson, C. L., & Sylvester, P. J. (2013). Introduction to Special Issue of. Canadian Journal of Earth Sciences: Canadian contributions to planetary geoscience. *Revue canadienne des sciences de la Terre.*. doi:10.1139/cjes-2012-0164

Yin, A. (2012). An episodic slab-rollback model for the origin of the Tharsis rise on Mars: Implications for initiation of local plate subduction and final unification of a kinematically linked global plate-tectonic network on Earth. *Lithosphere, 4*(6), 553-593. doi:10.1130/L195.1

Yuan, D. N., Sjorgren, W. L., Konopliv, A. S., & Kucinskas, A. B. (2001). Gravity field of Mars: A 75th degree and order model. *Journal of Geophysical Research, 106*(E10), 23377–23401. doi:10.1029/2000JE001302

Zuber, M. T. (2000). Internal Structure and Early Thermal Evolution of Mars from Mars Global Surveyor Topography and Gravity. Science, 287(5459), 1788-1793. Doi:10.1126cience.287.5459.1788

Chapter 2
Numerical Simulation on Renewable Solar Energy

Prasad G.
https://orcid.org/0000-0002-5709-9182
Dayananda Sagar University, Bangalore, India

Reshma G.
Dayananda Sagar University, Bangalore, India

Nikitha K. N.
Dayananda Sagar University, Bangalore, India

Pooja Reddy B.
Dayananda Sagar University, Bangalore, India

Gourav Nand Tiwary M.
Dayananda Sagar University, Bangalore, India

ABSTRACT

Due to the significance in renewable energy and rising energy demand, solar energy is becoming increasingly important. Although many academics are interested in the development of photovoltaic panels, there hasn't been enough research done to identify the loads operating on these systems' supporting structures. In this chapter, computational fluid dynamics (CFD) analysis is largely used to simulate, analyze, and grasp the impacts of wind forces on solar panels utilizing high-speed computer capabilities. Furthermore, unlike previous studies in this field, this study focuses on the sequential arrangement of ground-mounted solar panels and the effects of the sheltering effect. The steady state SST k-omega turbulence model is used in the CFD analysis. 3D studies are carried out, and it is discovered that for the solar panel configurations investigated in this chapter, they give accurate findings with robust modelling and solutions in terms of computing time.

DOI: 10.4018/978-1-7998-8801-7.ch002

INTRODUCTION

The 21st century has made us realize the importance of the need for the development and use of renewable energy sources. The most significant amount of research and development has been carried out on solar and wind energy sources, since they are cost efficient and reliable. The advancement in the technology on the solar energy front resulted in the development of Photovoltaic (PV) panels which are cheaper and energy efficient. While building the solar farm, researchers and engineers should also pay attention to the designing of the structural supporting system for solar panels in the first design stage. These supporting structures should be designed in such a way that they are strong enough to protect PV panels and which allows them to generate uninterrupted electricity throughout their service life (Assmus, M., & Koehl, M. 2012).

Advance level load calculations have to be carried out with regards to the wing forces acting on several engineering structures, since the major load on the PV panel has been found to be caused by wind. Wing loads acting on the solar panel and their effect on analysis and design of steel supporting structures are inadequate due to lack of experience and knowledge. Even though the private companies have their own different analysis approach and load calculations with regards to loads acting on solar panels, they are kept confidential (Banks 2010).

Considering the wing loads in the design phase of a structure, ASCE 7-10 (ASCE,2010) and Euro code (CEN, 2004) sufficiently provide an estimation of wind loads acting on mono-slope free roofs and canopies. The wind tunnel experiments and Computational Fluid Dynamics (CFD) analysis are the main approaches used by the solar companies to calculate the wind loads acting on consecutively placed panels in solar farms. When the wind flows on the ground mounted solar panels placed on the open terrains, wind flow has been blocked by the first row of the panels in the solar farm and wind loads acting on the other rows exhibit different trends. This has been caused due to the variation in the direction and speed of wind after facing all the rows of panels.

In this article analysis on the consecutively placed ground mounted solar panels in solar farms, where flat-plate PV panels were used. CFD by using ANSYS FLUENT was employed to model, analyze and to know the effects of wind force acting on solar panels (Schellenberg et al 2013). The sheltering effect was investigated with the placement of 1 and 2 consecutive panels using 3D CFD analysis approach.

Background

Overall, the global solar sector had a good year in 2019, growing at a low two-digit rate after essentially halting the previous year. Beyond the recently installed solar power around the world, solar development prospects appear to be even brighter. Despite China's sharp decline in demand last year, global solar installations increased by double digits, indicating that the world's appetite for solar power has been diversifying, with an expanding number of countries turning to the sun. Solar, in any way, remained the most appealing power generation option added in 2019. Not only was more solar PV built this year than all fossil fuel and nuclear power generation capacity combined, but it was also nearly twice as much as wind and more than all renewables combined. In addition to these excellent accomplishments, solar's proportion of total power output climbed to 48 percent in 2019, up from 42 percent in 2018.

However, these encouraging developments demonstrating solar's dominance of yearly worldwide power generation capacity expansions must be viewed in context. When it comes to solar's overall market share, it's still quite little, accounting for only 8.5 percent by the end of 2019. In terms of actual output, all solar PV systems together produced only 2.6 percent of world power. This has been also true in compared to renewables as a whole, which in 2019 accounted for nearly a third of overall generation capacity and 23% of global power output. The good news has been that solar has a huge market potential, and its cost-competitiveness has been improving all the time, allowing it to capture a larger share of the market. Solar's triumph over other technologies has been due to a number of factors, one of which has been its tremendous cost decrease over the previous decade, which has now propelled solar to the top of the cost leaderboard. The numerical modelling and analysis techniques introduced in the preceding chapter will be validated using literature data. The numerical analysis undertaken by (Jubayer & Hangan, 2012) and the verification study investigation (Warsido et al., 2014) is taken into account.

MAIN FOCUS OF THE CHAPTER

Theoretical background and general information on computational fluid dynamics (CFD) analysis, modelling, and analysis will be presented. Instead of conducting wind tunnel testing, CFD analytical tools have become increasingly popular for solving engineering challenges. ANSYS FLUENT has been used in this article to simulate the effects of wind forces on solar panels. When it comes to representing the physical situation at hand, three-dimensional (3D) modelling has been conceivable. A solar panel array with the following dimensions was used to describe the problem. The dimensions are Single solar panel are as follows Length=150mm, Height=90mm, Extrusion =10mm, PV-cell- Length=75mm; Width=2.7m. The dimensions are double solar panel are as follows Length=300mm, Height=180 mm, Extrusion =20 mm, PV-cell- Length=150mm; Width=5.4m.

Figure 1. Grid geometry of single cell solar panel and double cell solar panel

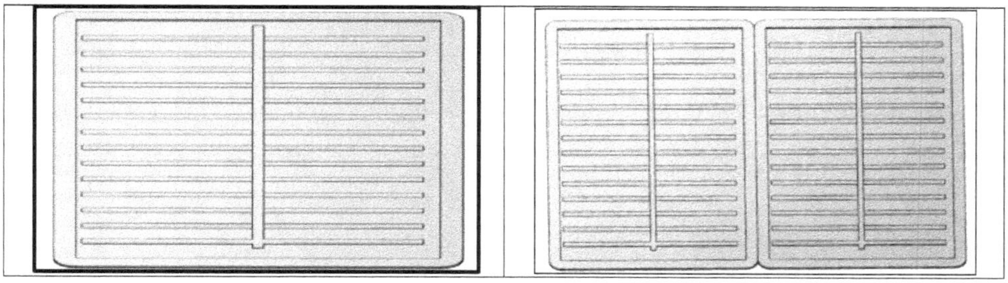

Figure 1 illustrates the Grid Geometry of Single cell solar panel and double cell solar panel A volume of fluid, which in this case will be air, must be specified in order to undertake CFD analysis. This volume will serve as the flow domain for the fluid. The domain boundary, as well as the boundary between the fluid flow and the solar panels, should be presented to the CFD analysis tool with great care.

SOLUTIONS AND RECOMMENDATIONS

Figure 2. Grid generation on single cell solar panel and double cell solar panel

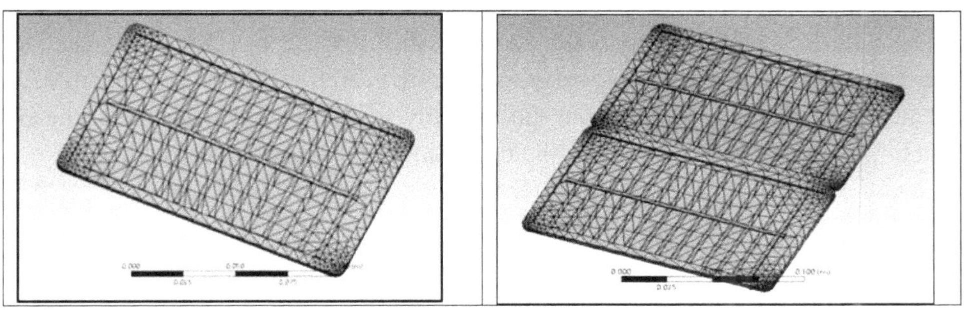

It's also important to capture the fairly turbulent airflow near the wall portions accurately. Significant mesh refining has been required towards the bottom wall region and around the panels to achieve this. Figure 2 shows the Grid Generation on Single cell solar panel and double cell solar panel. In order to assess mesh quality, three stages of mesh refinement are evaluated, as indicated by cost Guidelines, with coarse, medium, and fine meshes. It's crucial to ensure that the model has been mesh independent in terms of mesh refinement, which means that the difference between the results produced from medium and fine meshing should be small enough that fine meshing can be considered as the convergent solution in terms of mesh refinement.

Figure 3. Boundary conditions for single cell solar panel and double cell solar panel

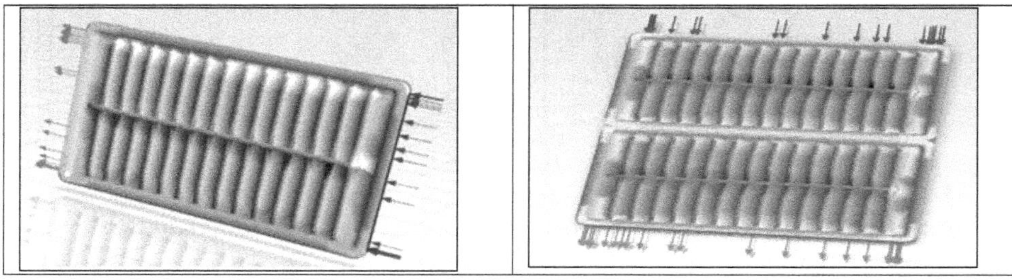

The top and side limits of the air domain are given as symmetry to increase the flow space near these boundaries while also eliminating the requirement for mesh refinement. Figure 3 depicts the Boundary Conditions for Single cell solar panel and double cell solar panel. The rough wall boundary condition, which requires the specification of roughness height and constant, has been applied to the bottom section of the air domain. It's also crucial to specify the boundary condition between the air and the solar panel, as well as the panel's supporting structural system (if one exists), which has been chosen as a smooth wall in this article. The air domain's front face has been designated as a velocity intake, while the back face has been designated as a pressure exit. Within the air domain, no pressure difference has been mentioned. Because the panel volume has been substantially smaller than the volume of the air domain, the element sizes through the panel thickness are much refined.

Figure 4. Temperature distribution contours for single cell solar panel and double cell solar panel

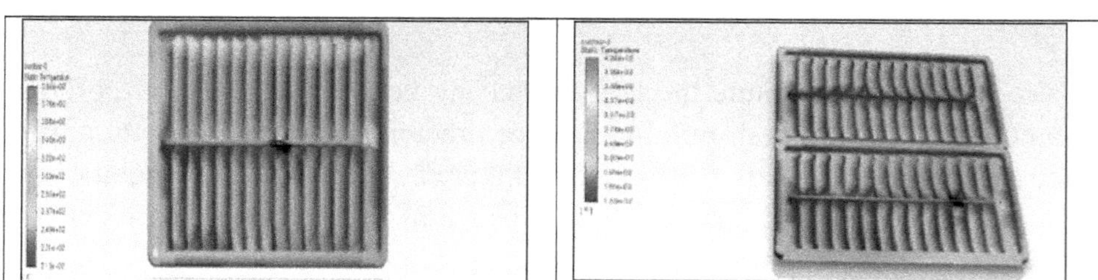

In this project, the FLUENT Module in ANSYS R 2021 has been used to do CFD analysis. Double precision was chosen for the analysis. Figure 4 shows the Temperature distribution contours for Single cell solar panel and double cell solar panel. A pressure-based solver has been preferred for the solver. The absolute velocity formulation has been chosen. In terms of getting the dominating wind loads operating on the solar panels, the steady state method has been thought to be adequate. The SST k- omega model was proposed for the viscous model (Menter, Kuntz, & Langtry, 2003).

The turbulent viscosity ratio has been set at ten. The gauge pressure has been set to zero in the pressure outlet section, and the intensity and viscosity ratio technique has been chosen. (Chung, et al 2011). Despite the fact that the distance between the panel and the outlet has been adequate, the turbulence level in the outlet region has been projected to be higher than the inlet face due to the meeting of obstacles (solar panel). As a result, the backflow turbulence intensity has been reduced to 5%, while the backflow turbulent viscosity ratio remains at 10. Because the solar panel in the domain has been stationary, no dynamic meshing has been required, and the panel's structural vibrations are ignored in this study (Essefi, 2014).

Figure 5. Iteration and convergence for single cell solar panel and double cell solar panel

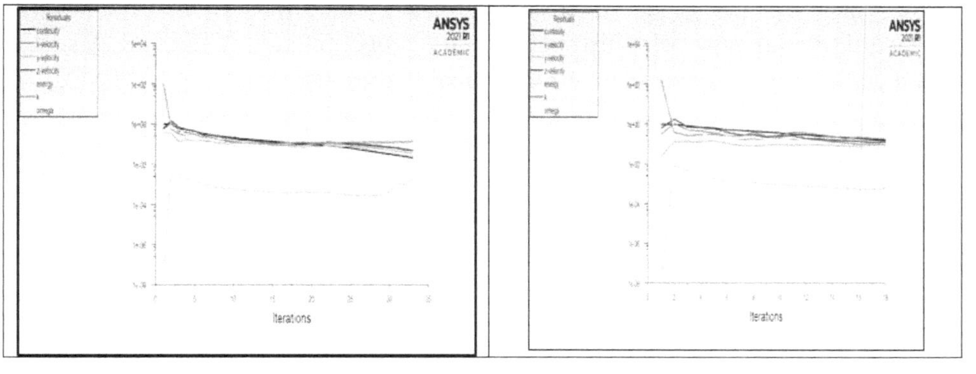

For the solution approaches, the pressure-velocity coupling scheme has been chosen. As a result, the solver will solve both the pressure and velocity equations at the same time in each step.

Figure 5 shows the Iteration and convergence for Single cell solar panel and double cell solar panel. Solving these two equations together will help to achieve substantially faster convergence by utilizing roughly twice as much computing resources. The numerical analysis section of this work makes use of ANSYS FLUENT. Double accuracy was required for the analysis. The pressure-based solver has been chosen as the solver. The absolute velocity formulation has been required. Only the steady state solution has been taken into account. SST k- omega without low Reynolds corrections has been chosen for the viscous model. The pressure-velocity coupled solution has been chosen as the equations. Convergence of the solutions has been determined in the calculation portion by selecting acceptable tolerance levels and examining the convergence charts of drag, lift, and moment coefficients.

More complicated models in 2D and 3D geometries can be physically robust, but their application in a regulation control system becomes challenging and unsuitable due to their processing costs. These complicated methodologies, on the other hand, are well suited to extensive investigation of heat transport behavior within the solar collector in order to optimize component design and construction. In this context, CFD research has shown that it can be utilized to improve the efficiency of solar collectors.

FUTURE RESEARCH DIRECTIONS

Three fourth of the top 20 countries have bright political support prospects through 2024, with all of them anticipated to have double-digit annual growth rates. Our prognosis has been hazy for three countries, two of which are likely to grow at single-digit rates. The government of Japan has done little to change the general downward trend in new solar installations - the reorganization of the energy sector has taken far too long, and solar systems remain prohibitively expensive. Instead of utilizing solar's low-cost potential, the government has been attempting to resurrect nuclear against popular will for years, and has recently began to turn to coal at a time when most countries are abandoning this CO_2-intensive fuel.

Italy has been another country with bleak prospects. For years, the industry has remained at a low installation level, despite initiatives to harness the country's abundant sunshine - the most recent example being the failure of technology neutral renewables auctions, in which solar received just minor volumes. Australia, the third country in this group, has been unique in that it has the world's highest solar housing density (2.3 million) and has been growing for years. However, for utility-scale assistance, the national Renewable Energy Target (RET), a carbon price certificate model mechanism based on the offsetting of emissions in the grid by PV output, will terminate in 2020. The current federal government has no plans to prolong the programme, putting a large pipeline of utility-scale solar projects in jeopardy. Turkey and Mexico are the two markets with a bleak outlook.

While Turkey has experienced another year of decline since the climate for solar changed two years ago due to a financial crisis and dwindling political support, effectively abandoning the previously successful feed-in tariff scheme, a change in government in Mexico has led to a new path in energy politics, with a full focus on centralized fossil fuel power plants while challenging the policy.

CONCLUSION

In this research, the wind loads acting on single and consecutively placed ground mounted flat plate solar panels are examined. Computational fluid dynamic analysis has been primarily used to model, analyze, and comprehend the impacts of wind forces operating on the panels, particularly the protection provided by the first row of solar panels in terms of drag, lift, and moment coefficients to the succeeding panels. When compared to 3D CFD analyses with supporting columns, 3D CFD analyses on solar panels without supporting structures yield robust and on the safe side wind loads. When the width of solar panels has been large enough, 2D CFD analyses produce more robust and safe wind loads than 3D CFD analyses, as long as the angle of attack of the wind has been not oblique.

It has been interesting to examine optimization studies on wind loads operating on solar panels and supporting structures in terms of solar panel tilt, spacing factor, and panel length as variable parameters. Actually, the best way to optimize these parameters has been to maximize the amount of energy produced by a solar farm based on the latitude of the location on the planet. While the cost of the structural supporting system has been minor in comparison to the cost of photovoltaic panels, the percentage of the cost related to the supporting system's use and construction might be decreased further while maintaining the solar farm's maximum energy output. It's important to note that the research study described in this article was limited to steady state analysis. Transient effects generated by wind flow through consecutively installed solar panels can be determined using unsteady analysis methods.

The first examples of solar support as part of COVID-19 economic stimulus packages can already be observed all around the world. As part of its recovery measures, the Malaysian government announced a new tender for 1 GWAC utility scale solar capacity; the Swiss government approved 46 million CHF (48.5 million USD) to support the expansion of solar PV systems this year; and Japan included an economic stimulus package worth nearly 1 billion USD to support corporate PPAs to facilitate the development of solar PV systems. The European Commission proposed a two-year 750 billion EUR COVID-19 recovery instrument, 'Next Generation EU,' with the European Green Deal at its core – this has been expected to roll out solar energy projects across member states and launch a massive renovation of the EU's building stock and infrastructure, both of which will benefit solar. Bremen,

Germany's smallest state, made solar mandatory on all new dwellings, including public buildings, in early June. But this could only be the beginning.

While we predict a substantial 34 percent growth rate to 150 GW in 2021 in our Medium Scenario, which does require large levels of state recovery support, this capacity would still be 6 percent behind last year's 2021 forecast (Revathy et al 2021):. To get back on track, it would take until 2022 to reach 169 GW. The virus's effects aren't likely to be totally forgotten until 2024. However, if the right policy support measures are taken to accelerate the deployment of the lowest-cost clean power generation sources, such as solar and wind, as well as large-scale production of renewable hydrogen to help decarbonize our society before 2050, the 2020s could indeed evolve into a solar decade, fully unleashing the sun's power.

REFERENCES

ANSYS Inc. (2021). ANSYS Fluent Theory Guide. ANSYS Inc.

Assmus, M., & Koehl, M. (2012). Experimental investigation of the mechanical behavior of photovoltaic modules at defined inflow conditions. *Journal of Photonics for Energy*, *2*(1), 022002. doi:10.1117/1.JPE.2.022002

Banks, D. (2010). *How Wind Load Studies Will Impact the Solar Industry*. Academic Press.

Chung, K., Chang, K., & Chou, C. (2011). Wind loads on residential and large-scale solar collector models. *Journal of Wind Engineering and Industrial Aerodynamics*, *99*(1), 59–64. doi:10.1016/j.jweia.2010.10.008

Essefi, E., Mefteh, S., Medhioub, M., & Yaich, C. (2014). Magnetic Study of the Heated and Unheated Sedimentary Fillings of Sebkha Mhabeul, Southeast Tunisia: A Geophysical Method for Paleoclimatic Investigation and Tephrochronological Dating. *International Journal of Geophysics*, *2014*, 1–7. doi:10.1155/2014/908395

Jubayer, C. M., & Hangan, H. (2012). Numerical Simulation of Wind Loading on Photovoltaic Panels. *Structures Congress*, 1180–1189. 10.1061/9780784412367.106

Menter, F., Kuntz, M., & Langtry, R. (2003). Ten years of industrial experience with the SST turbulence model. *Turbulence, Heat and Mass Transfer.*

Revathy, S. M., Rangaraj, A. G., Srinath, Y., Boopathi, K., Shobana Devi, A., Balaraman, K., & Prasad, D. M. R. (2021). Impact on solar radiation parameters in India during COVID-19 lockdown: A case study. *International Journal of Sustainable Energy*, *40*(8), 806–820. doi:10.1080/14786451.2021.1893726

Schellenberg, A., Maffei, J., Telleen, K., & Ward, R. (2013). Structural analysis and application of wind loads to solar arrays. *Journal of Wind Engineering and Industrial Aerodynamics Journal*, *123*, 261–272. doi:10.1016/j.jweia.2013.06.011

Warsido, W., Bitsuamlak, G., Barata, J., & Gan Chowdhury, A. (2014). Influence of spacing parameters on the wind loading of solar array. *Journal of Fluids and Structures*, *48*, 295–315. doi:10.1016/j.jfluidstructs.2014.03.005

Chapter 3
Comparative Thermodynamic Theoretical Modeling of Brines From Chott Djerid, Southeastern Tunisia and Sebkha Bazer, Northeast Algeria

Sana Bedoui
Higher Institute of Water Sciences and Techniques of Gabes, University of Gabes, Tunisia

Elhoucine Essefi
University of Gabes, Tunisia

Younes Hamed
University of Gafsa, Tunisia

ABSTRACT

The thermodynamic theoretical modeling of brines from the Chott Djerid, Southeastern Tunisia and Sebkha Bazer, Northeast Algeria between the initial solution and evaporation of 100% shows different geochemical evolutions and variable precipitated mineral species. This is due to different initial geochemical compositions, which are basically related to different geological and climatic contexts. A rigorous thermodynamic model has been presented for determining the crystallization sequence during the different stages of the evaporation process.

DOI: 10.4018/978-1-7998-8801-7.ch003

A geochemical software program (PHREEQC) was used. Based on the analyses, PHREEQC appears the precipitation of a sequence of minerals. These results are confirmed by the mineralogical analyses (XRD results). The gradual evaporation process of brine showed that thenardite, gypsum, halite, and huntite are the feasible salts that could be extracted. Halite is the most abundant mineral along with the evaporation experiment of the two sebkhas.

INTRODUCTION

Saline systems are widely distributed in the world. Several studies have treated mineral/solution interactions as well as the geochemical evolution of salts and brines in the lakes, chotts, and sebkhas ecosystems (Geuddari et al., 1984; Rettig et al., 1980; Jones & Deocamp, 2003; Hacini et al., 2008). The thermodynamic calculations based on the Pitzer (1973) equations are often used to analyze the geochemical evolution of saline lake waters during evaporation. Harvie & Weare (1980) and Eugster et al., (1980) adopted these equations for the evaporation of natural brines in closed systems, and they were successfully applied to the interpretation of a number of evaporites systems (Eugster & Maglione, 1979; Gac et al., 1979; Gueddari et al., 1984). The initial brine composition and the equilibrium constants of evaporite minerals entirely control the chemical evolution of evaporative systems in closed basins. For brine systems, thermodynamic models are typically used to describe phase equilibrium. The exploitation of brine resources can be governed by the phase equilibrium of saltwater systems (Pengsheng, 2016; Mianping et al., 2016). The crystallization sequence during evaporation of natural brine at the Chott Djerid and Sebkha Bazer concentration ponds reflected the findings of both Hardie and Eugster's evolutionary model and the PHREEQC simulated model. The evaporation simulation was performed using the phase diagram and Pitzer's model for the quaternary system. The main goal of this chapter is to develop a functional model of reactivity for the analysis of complex systems and to provide a better understanding of the crystallization sequence during the evaporation process, especially the water-mineral chemical interaction, using geochemical speciation modeling software. Consequently, the calculation of the activity coefficients of various species is therefore very important. In fact, the Pitzer model was chosen because it had already been successfully applied to complex salt systems (Lach, 2015; Essefi et al., 2020; Essefi et al., 2021a,b).

THERMODYNAMIC AND GEOCHEMICAL DESCRIPTION OF BRINES

Macroscopic Properties of Brines

According to Kharaka and Hanor's definition, brines are salty solutions containing more than 35 g/L of solids (Kharaka & Hanor, 2003). When brine is exposed to evaporation, it forms salt crystals from the dissolved elements. In fact, the physicochemical characteristics of brines are determined by the nature of the numerous dissolved minerals. Thus, the thermodynamics of solutions is used to describe the behavior of brines. The physicochemical properties of some dissolved minerals are conferred on brines by their natural state (Thadée, 2017).

According to Truong-Meyer, there are two methods of representing an aqueous solution which are apparent and true species (Truong-Meyer, 2012). The apparent species composition of a saline solution is the amount of salts dissolved to make the solution. For example, if sodium chloride is dissolved in water, the apparent composition can be determined based on the amounts of H_2O and NaCl (Thadée, 2017). Thus, apparent species seem to be either neutral or considered as non-dissociated. True species represent all the constituents of the system; it means all neutral molecules and ions, in other words, the dissociated species actually present in the aqueous solution. If the same saline solution is used, the true species will show the amounts of ions (H^+, OH^-, Na^+, and Cl^-) as well as neutral molecules (H_2O and NaCl) (Thadée, 2017). In this chapter, we used the true species to describe the studied systems.

Classification of Hydrochemistry

Salt lake brines can be classified into five varieties based on their chemical compositions: chloride, sulfate, carbonate, nitrite, and borate types among the world's salt lake resources (Gao et al., 2007). Salt lake brines can be described using the complex salt-water multicomponent system (Li – Na – K – Ca – Mg – H – Cl – SO_4 – B_4O_7 – OH- HCO_3 – CO_3 – H_2O)(Deng, 2012).

1. **Carbonate type:** the principal precipitated minerals are thermonatrite ($Na_2CO_310H_2O$), baking soda ($NaHCO_3$), natron ($Na_2CO_310H_2O$), glauber salt, and halite, it belongs to the system (Na – K – Cl – CO_3 – SO_4 – H_2O).
2. **Chloride type:** the brine composition of this type of saline system can be gathered as the chloride type system (Na – K – Mg – Cl – H_2O), with halite (NaCl), sylvite (KCl), carnallite ($KClMgCl_2H_2O$), and bischofite ($MgCl_2H_2O$) as the major precipitation of salts.

3. **Sulfate type:** this type of salt lake resource can be classified into two types: sodium sulfate and magnesium sulphate types. The main precipitation of salts is halite, glauber salt, blodite ($Na_2SO_4.MgSO_4.7H_2O$), and epsom salt ($MgSO_4.7H_2O$). The system of this kind of salt lake is the same as the seawater system ($Na - K - Mg - Cl - SO_4 - H_2O$).
4. **Nitrite type:** it belongs to ($Na - K - Mg - Cl - NO_3 - SO_4 - H_2O$). The natratime saltier ($NaNO_3$), the niter (KNO_3), the darapskite ($NaNO_3.Na_2SO_4{\cdot}H_2O$), the potassium darapskite ($KNO_3.K_2SO_4.H_2O$), the humberstonite ($NaNO_3.Na_2SO_4.2MgSO_4.6H_2O$) were the main precipitation of salts.
5. **Borate type:** This kind of saline system can be divided into carbonate-borate hypotype and sulfate-borate hypotype.

Basic Concepts for Geochemical Modeling

The geochemistry principles are explained in this part, which includes an introduction to thermodynamics, geochemical processes, and the geochemical modeling. The first geochemical models were developed in the 1970s (Westall et al., 1976; Wolery, 1979). Indeed, these models have been used to solve complex geochemical problems like speciation, mineral saturation index determination.

Geochemical Modeling

The process of modeling the distribution and the reactivity of solute molecules in a given solution is known as geochemical modeling. There are two types of geochemical models which are:

1. **Geochemical Equilibrium Models:** these models are built on the concept that thermodynamic equilibrium may be obtained in a short period of time (no time factor is included in the calculation). Only equilibrium reactions are considered (Zou et al., 2021).
2. **Geochemical Kinetic Models:** It incorporates kinetic reactions as well as the time factor. The geochemical characterization data of the static and kinetic tests are used in the thermodynamic equilibrium modeling approach which is widely used (Declercq et al., 2017). Chemical speciation modeling of equilibrium is based on thermodynamic data. These data are accurate equilibrium constants for the dissolution and precipitation of solid phases of interest between aqueous and solid phases, as well as the production of dissolved species in aqueous solution (Fatah et al., 2022).

Limitations of Geochemical Modeling

All geochemical modeling programs encounter errors, the majority of which are caused by the quantity and the quality of thermodynamic data used to calculate equilibrium constants, complex dissociation constants, and mineral dissolution constants. The utility of the obtained results is also limited by the incomplete state and the reliability of the physical and chemical parameters reported and used as program inputs (Benlahcen, 1996). Moreover, the degree of representativeness of water sample, which can come from a single zone of the aquifer or from the whole aquifer (Hull et al., 1985).

The Basics of Thermodynamic Modeling

This part summarized the basic concepts of precipitation/dissolution reactions, surface complexation and solid solutions.

Precipitation/dissolution

The basic concepts of thermodynamic equilibrium modeling are well-explained in a several publications (Stumm and Morgan, 1996; Appelo and Postma, 2005). The minerals dissolve in solution when the aqueous phase of a mineral is under saturated. When the aqueous phase is oversaturated with regard to a mineral, it will precipitate or remain precipitated. Thermodynamic principles are used to calculate if a solution is under- or over-saturated.

Law of Mass Action

The law of mass action is one of the most basic equation for geochemical equilibrium modelling. According to this law, the ratio of the product of the activities of the reaction products to the product of the activities of reagents is constant. The law of mass action is written in the following way:

$$K = \frac{\sum_{i=1}^{np} \left[P_i\right]^{v_{p,i}}}{\sum_{i=1}^{nr} \left[R_i\right]^{v_{r,i}}}$$

- **K:** the equilibrium constant
- **[ ]:** the denotes activity
- **Ri and Pi:** the reactants and products (aqueous species, surface complexes, and minerals)
- **nr and np:** the number of reactants and products
- **Vr, i and Vp, i:** the stoichiometric coefficients of the reactants and products

The thermodynamic equilibrium theory is the most appropriate concept used for identifying numerous variables in determining the relationship between the solution and the minerals (Stumm and Morgan, 1970).Coudrain-Ribstein showed that models that calculate the speciation of aqueous solutions and minerals can be identified in the development of geochemical models (Coudrain-Ribstein, 1988).

Pitzer Model

To better understand the interactions between the elements contained in natural brines- salts, a reactivity model can be used. The modeling of aqueous electrolyte solutions has been reported in various studies such as; (Pitzer, 1973; Fraenkel, 2011; Hu et al., 2011, Li et al., 2011, Tian et al., 2012), but Pitzer's approach has been the most commonly used of the several approaches proposed (Pitzer, 1973). Pitzer's equations (Pitzer, 1991) allow a practical calculation of thermodynamic solutions in an aqueous solution. This theory has also attracted attention since Harvie & Weare (1980) and Harvie et al., (1984) demonstrated that it could be used to anticipate successive dissolution of salts derived from evaporation of seawater. Therefore, Pitzer equations have frequently used in aqueous geochemistry. It has been combined with a number of software packages, such as PHREEQC (Parkhurst & Appelo, 1999) and OLI Analyser (Reddy & Lewis, 2006). Pitzer's model is based on the Debye-Hucke theory.

Thermodynamic Modeling Through PHREEQC Software

PHREEQC is a geological solubility software that may be used for highly concentrated solutions (Plummer, 1988; Mariah et al., 2006). PHREEQC is a C and C^{++} application used to calculate the geochemistry of three-phase systems. It is a universally geochemical modeling software available from the USGS (Parkhurst, 1995). It is made to perform a diverse of aqueous geochemical calculations. The concentration of different species in the brine, as well as other requirements such as temperature and pH, are input parameters for PHREEQC. Various calculations were performed by this software such as:

- Thermodynamic characterization of electrolytic solutions;
- Thermodynamic equilibrium between liquid-vapor phases (activity coefficient, gas solubility, pH, etc.) and solid-liquid (solubility of salts in aqueous solutions);
- Balance and kinetics of chemical reactions;
- Speciation of ions in solution.

PHREEQC can be used also to estimate the sequence in which the salts precipitated as a function of brine concentration. Therefore, to test the applicability of these models, a simulation of the geochemical evolution of brine using mathematical models will be required. The geochemical model PHREEQC can simulate water-mineral equilibrium interactions, as well as ion exchangers and surface complexes (Charlton and Parkhurst, 2011).PHREEQC is based on a variety of geochemical equilibrium equations, such as aqueous solution interacting with minerals, gases, solid solutions, exchangers, and sorption surfaces (Heredia, 2017). The dissolution/ precipitation of minerals, cationic exchanges, and surface complexation are all examples of interactions between water androck. By default, to achieve thermodynamic equilibrium, the PHREEQC code assumes the chemical reactions that occur in real time. Moreover, in the scientific community, other geochemical calculation codes exist. This software is widely used. It has the advantages of being free, and including the Pitzer formalization for dealing with the geochemistry of hypersaline systems.

Every chemical equilibrium problem's PHREEQC formulation is derived from a set of equations (Parkhurst & Appelo, 1999):

PHREEQC main types of geochemical predefined equations

f_{alk}: Mole balance alkalinity
f_e: Mole balance for exchange site
f_g: Mole balance gas
f_H: Mole balance of hydrogen
f_{H2O}: Activity of water
f_m: Mole balance of master species except H+, e -, H_2O, and the alkalinity
f_O: Mole balance of oxygen
f_P: total Equilibrium with a fixed pressure multicomponent gas phase
f_p: Equilibrium with a pure phase
f_{pss}: Equilibrium with solid solution
f_{sk}: Mole balance for surface sites
f_z: Aqueous charge balance
$f_{z,s}$: Surface charge balance
f_{μ}: Ionic strength Surface charge potential

Functioning of the Geochemical Code PHREEQC v3: Database

The database is a keyword data block that connects the model to the user database. PHREEQC may access nine databases, each including the definitions for all elements, exchange processes, surface complexation reactions, mineral phases, gas components, and rate expressions. These databases come from various resources, and each one is tailored to a specific set of simulations. To achieve this model, the specific geochemical composition of the brines was extracted from the databases. The PhreeqC code requires an input file and a thermodynamic database to function (Database). It provides default output file (output) and customizes output files. The user specifies the contextual information of the study in the input file, such as the concentrations of the different elements of the simulated chemical system, the temperature, pressure settings, and the ph. When water-rock-gas interactions are expected, the different components (solid or gas) involved must be declared (Thadée, 2017). In the output file, PhreeqC provides the characteristics of the solutions such as:

- pH, temperature, density, speciation,
- Saturation indices of all minerals in the chemical system studied,
- The evolution of the mineralogical and/or gas composition when water-rock-gas interactions were declared as input.

Equilibrium Between Solid Phases and Dissolved Species

A complete set of solid phase formation reactions and solubility constant are the minimal thermodynamic data required to determine equilibrium between dissolved species and precipitated solid phases (Meeussen et al., 2009). Furthermore, the chemical equilibrium is used to calculate the formation constants of aqueous species and solid phases. The equilibrium among dissolved aqueous species frequently occurs (Meeussen et al., 2009).

RESULTS

Thermodynamic Modeling of Chott Djerid Brine

Thermodynamic analysis was used with the aim to predict which salts and in which order can crystallize from natural brine. In this study, PHREEQC program version 3.0 (Parkhurst & Appelo, 2013), a geochemical software was used to model the thermodynamics of Chott Djerid brine. Figure 1 shows the achieved results considering the salts that can be formed during the evaporation process.

The thermodynamic theoretical modeling of Chott Djerid brine between the initial solution and evaporation of 100% converges toward a stable solution with the precipitation of evaporitic minerals. According to the prediction of PHREEQC, different types of calcium carbonates and calcium sulfates are going to precipitate from the concentration of the natural brine when the corresponding salinity increases (Fig 1). The Djerid brines are highly alkaline and rich in Na^{+}, K^{+}, Cl^{-}, and SO_4^{2-}. The Djerid brines solution were concentrated along the evaporation experiment resulting the precipitation of minerals. At 20% of evaporation process, according to the geochemical composition determined by PHREEQC, the precipitate was likely to be either Gypsum ($CaSO_4.2H_2O$) or Halite (NaCl). The theoretical evolution of the geochemical elements of the Djerid brine is equivalent to the evolution values obtained during the progressive evaporation. Furthermore, some minor minerals such as mirabilite ($Na_2SO_4.10\ (H_2O)$), carnalite ($KMgCl_3.\ 6(H_2O)$), hexahidrite ($MgSO_4.\ 6(H_2O)$) are not detected, this is due to the absence of some minerals in the PHREEQC database. At the last stage of evaporation process 100%, we notice the precipitation of variable evaporite minerals such as sylvite (KCl).

Figure 1. Output of the thermodynamic modeling using as input the concentration of solid phases recuperated after each stage of evaporation process of Djerid brine

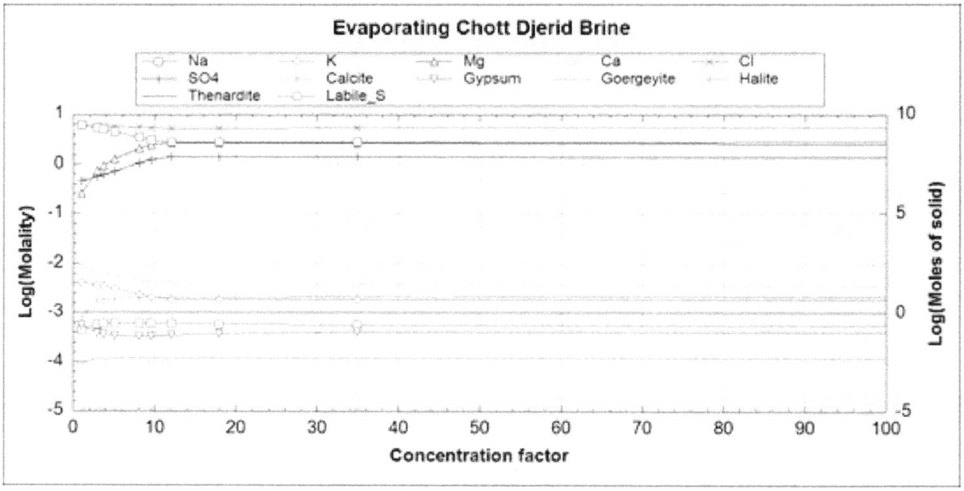

Thermodynamic Modeling of Sebkha Bazer Brine

Therefore, PHREEQC software can be used to estimate the sequence in which the salts precipitate from the brine (Macedonio et al., 2013). The precipitation of the minerals was first studied through the concentration of solid phases recuperated after each stage of the evaporation process of Bazer brine. Figure 2 shows the achieved results considering all salts that can be precipitated between the initial volume of the solution and the amount of water remaining after each evaporation step. A test of five brine samples was run in PHREEQC to predict the precipitated salts from brines. The results of the thermodynamic modeling from using the chemical composition of the evaporated brine showed that the majority of the minerals consist of halite, and thenardite because of the high concentration of Na, SO_4^{2-}, and Cl^- during the evaporation process. To have a better focus on the modeling of Bazer brines, the minerals considered for the thermodynamic model were gradually identified. The PHREEQC model of Bazer brine summarized the crystallization sequence of salts. The evolution values acquired during progressive evaporation are equivalent to the theoretical evolution of the Bazer geochemical elements. Thus, in the late evaporating process, the output of the thermodynamic model of multicomponent saltwater of Bazer brine system demonstrates the precipitation of the carbonate mineral such as huntite ($Ca\ Mg_3\ (CO_3)_4$) which is accompanied by the dramatic increase of Mg amount. The PHREEQC model of Bazer brine during the evaporation experiment is composed of a successive crystallization sequence corresponding to gypsum, halite, thenardite. It can be seen that the results of thermodynamic modeling of Bazer brine are generally consistent with the experimental data.

Figure 2. Output of the thermodynamic modeling using as input the concentration of solid phases recuperated after each stage of evaporation process of Bazer brine

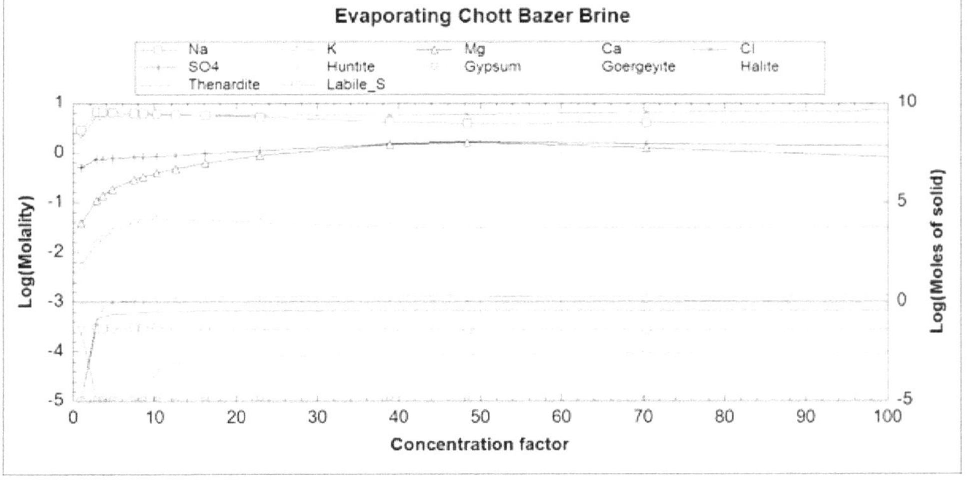

CONCLUSION

A rigorous thermodynamic model has been presented for determining the crystallization sequence during the different stages of the evaporation process. A geochemical software program (PHREEQC) was used. Based on our analyses PHREEQC appears the precipitation of a sequence of minerals. These results are confirmed by the mineralogical analyses (XRD). The Gradual evaporation process of brine showed that thenardite, gypsum, halite, and glaserite, huntite are the feasible salts that could be extracted. Halite is the most abundant mineral along the evaporation experiment of the two wetlands.

REFERENCES

Appelo, C. A. J., & Postma, D. (2005). *Geochemistry, groundwater and pollution* (2nd ed.). Balkema.

Appelo, C. A. J., & Postma, D. (2005). *Groundwater and pollution*. Balkema.

Bedoui, S., Essefi, E., Besser, H., Benour, A., Ayadi, Y., Khelifi, F., & Hamed, Y. (2022). Geochemical, Mineralogical and Isotopic Fracturing during the Progressive Evaporation of Chott Djerid Brine (Southern Tunisia). In International Journal of Engineering Research in Africa (Vol. 58, pp. 107-125). Trans Tech Publications Ltd.

Benlahcen, A. (1996). *Modélisation de saumures carbonatees par le traitement multivariable en hydrogéochimie à la mine Niobec de Saint-Honoré (Québec)* [Modeling of carbonate brines by multivariate treatment in hydrogeochemistry at the Niobec mine in Saint-Honoré, Quebec]. Université du Québec à Chicoutimi.

Charlton, S. R., & Parkhurst, D. L. (2011). Modules based on the geochemical model PHREEQC for use in scripting and programming languages. *Computers & Geosciences*, *37*(10), 1653–1663. doi:10.1016/j.cageo.2011.02.005

Declercq, J., Charles, J., Bowell, R., Warrender, R., & Barnes, A. (2017). Comparison of thermodynamic equilibrium and kinetic approach in the predictive evaluation of waste rock seepage quality in Northern Finland. Proceedings of International Mine Water Association, Mine Water and Circular Economy, 664-671.

Deng, T. (2012). Stable and Metastable Phase Equilibria in the Salt-Water Systems. *Advances in Crystallization Processes*, 399-430.

Essefi, E. (2021b). Geoeconomic interest versus environmental and health issues of the mineralogical assemblage of sebkha Oum El Khialate, southeastern Tunisia. *Arabian Journal for Science and Engineering*, *46*(6), 5835–5845. doi:10.100713369-020-05244-5

Essefi, E., Smida, N. B., Jandoubi, I., Othmani, M. A., & Tagorti, M. A. (2020). Progressive evaporation of brine of sebkhaMchiguig, central Tunisia: A geo-economical comparative study of salt and brine. *Carbonates and Evaporites*, *35*(2), 1–10. doi:10.100713146-020-00592-7

Essefi, E., & Tagorti, M. A. (2021a). Geoeconomic interest of minerals assemblage of sebkha El Melah, southeastern Tunisia. *Water Practice & Technology*, *16*(2), 633–647. doi:10.2166/wpt.2021.004

Eugster, H. P., Harvie, C. E., & Weare, J. H. (1980). Mineral equilibria in a six-component seawater system, Na-K-Mg-Ca-SO4-Cl-H2O, at 25° C. *GeochimicaetCosmochimicaActa*, *44*(9), 1335–1347. doi:10.1016/0016-7037(80)90093-9

Eugster, H. P., & Maglione, G. (1979). Brines and evaporites of the Lake Chad basin, Africa. *GeochimicaetCosmochimicaActa*, *43*(7), 973–981. doi:10.1016/0016-7037(79)90087-5

Fatah, A., Mahmud, H. B., Bennour, Z., Gholami, R., & Hossain, M. (2022). Geochemical modelling of CO_2 interactions with shale: Kinetics of mineral dissolution and precipitation on geological time scales. *Chemical Geology*, *592*, 120742. doi:10.1016/j.chemgeo.2022.120742

Fraenkel, D. (2011). Monoprotic mineral acids analyzed by the smaller-ion shell model of strong electrolyte solutions. *The Journal of Physical Chemistry B*, *115*(3), 557–568. doi:10.1021/jp108997f PMID:21192660

Gac, J. Y., Al-Droubi, A., Paquet, H., Fritz, B., & Tardy, Y. (1979). Chemical model for origin and distribution of elements in salts and brines during evaporation of waters. Application to some saline lakes of Tibesti, Chad. *Physics and Chemistry of the Earth*, *11*, 149–158. doi:10.1016/0079-1946(79)90018-1

Gao, S. Y., Song, P. S., Xia, S. P., & Zheng, M. P. (2007). *Salt lake chemistry-new type borate salt lake*. Science Press.

Gueddari, M., Monnin, C., Perret, D., Fritz, B., & Tardy, Y. (1983). Geochemistry of brines of the chott El Jerid in southern Tunesia—Application of Pitzer's equations. *Chemical Geology*, *39*(1-2), 165–178. doi:10.1016/0009-2541(83)90078-5

Hacini, M., Kherici, N., & Oelkers, E. H. (2008). Mineral precipitation rates during the complete evaporation of the MerouaneChott ephemeral lake. *GeochimicaetcosmochimicaActa, 72*(6), 1583–1597.

Harvie, C. E., & Weare, J. H. (1980). The prediction of mineral solubilities in natural waters: The Na K Mg Ca Cl SO4 H2O system from zero to high concentration at 25° C. *Geochimica et Cosmochimica Acta, 44*(7), 981–997. doi:10.1016/0016-7037(80)90287-2

Heredia, D. J. (2017). *Improvement of the numerical capacities of simulation tools for reactive transport modeling in porous media* (Doctoral dissertation). Université Rennes 1.

Hu, Y. F., Chu, H. D., Li, J. G., Liu, Z. C., Peng, X. M., Ling, S., & Zhang, J. Z. (2011). Extension of the simple equations for prediction of the properties of mixed electrolyte solutions to the mixed ionic liquid solutions. *Industrial & Engineering Chemistry Research, 50*(7), 4161–4165. doi:10.1021/ie1022496

Hull, R. W., Kharaka, Y. K., Maest, A. S., & Fries, T. L. (1985). Sampling and analysis of subsurface waters: A summary of current methodology. In *Proceedings First Canadian/American Conference on Hydrology*. National Water Well Association.

Jones, B. F., & Deocampo, D. M. (2003). Geochemistry of saline lakes. *Treatise on Geochemistry, 5*, 605.

Kharaka, Y. K., &Hanor, J. S. (2003). Deep fluids in the continents: I. Sedimentary basins. *Treatise on Geochemistry, 5*, 605.

Lach, A. (2015). *Modélisation thermodynamique des propriétés d'excès des saumures naturelles et industrielles* [Thermodynamic modelling of the excess properties of natural and industrial brines.] (Doctoral dissertation). Pau.

Li, M. Y., Wang, L. S., & Gmehling, J. (2011). Thermodynamics of phase equilibria in aqueous strong electrolyte systems. *Industrial & Engineering Chemistry Research, 50*(6), 3621–3631. doi:10.1021/ie101428j

Macedonio, F., Quist-Jensen, C. A., Al-Harbi, O., Alromaih, H., Al-Jlil, S. A., Al Shabouna, F., & Drioli, E. (2013). Thermodynamic modeling of brine and its use in membrane crystallizer. *Desalination, 323*, 83–92. doi:10.1016/j.desal.2013.02.009

Mariah, L., Buckley, C. A., Brouckaert, C. J., Curcio, E., Drioli, E., Jaganyi, D., & Ramjugernath, D. (2006). Membrane distillation of concentrated brines—Role of water activities in the evaluation of driving force. *Journal of Membrane Science, 280*(1-2), 937–947. doi:10.1016/j.memsci.2006.03.014

Meeussen, J. C. L., van der Sloot, H. A., Dijkstra, J. J., & Kosson, D. S. (2009). *Review of thermodynamic and adsorption databases*. Energy Research Centre of the Netherlands and Vanderbilt University/CRESP.

Mianping, Z., Yongsheng, Z., Xifang, L., Wen, Q. I., Fanjing, K., Zhen, N., & Yongjie, L. (2016). Progress and prospects of salt lake research in China. *ActaGeologicaSinica-English Edition*, *90*(4), 1195–1235. doi:10.1111/1755-6724.12767

Parkhurst, D. L. (1995). User's guide to phreeqc - a computer program for speciation, reaction-path, advective-transport, and inverse geochemical calculations. U.S. Geological Survey Water-Resources Investigations Report.

Parkhurst, D. L., & Appelo, C. A. J. (2013). *Description of input and examples for PHREEQC version 3: a computer program for speciation, batch-reaction, one-dimensional transport, and inverse geochemical calculations (No. 6-A43)*. US Geological Survey.

Pengsheng, S., (2016). The phase diagram of salt-water systems and utilization of salt lake resources. *Journal of Salt Lake Research*, 3.

Pitzer, K. S. (1973). Thermodynamics of electrolytes. I. Theoretical basis and general equations. *Journal of Physical Chemistry*, *77*(2), 268–277. doi:10.1021/j100621a026

Pitzer, K. S. (1991) Ionic Interaction Approach: Theory and Data Correlation. In Activity Coefficients in Electrolyte Solutions. CRC Press.

Pitzer, K. S., & Mayorga, G. (1973). Thermodynamics of electrolytes. II. Activity and osmotic coefficients for strong electrolytes with one or both ions univalent. *Journal of Physical Chemistry*, *77*(19), 2300–2308. doi:10.1021/j100638a009

Plummer, L. N. (1988). A computer program incorporating Pitzer's equations for calculation of geochemical reactions in brines (Vol. 88, No. 4153). Department of the Interior, US Geological Survey.

Reddy, S. T., & Lewis, A. E. (2006). Water recovery and salt recovery from brine solutions. *Proc. 13th Int. Workshop Industrial Crystallization Univ. Technol.*, 328.

Rettig, S. L., Jones, B. F., & Risacher, F. (1980). Geochemical evolution of brines in the Salar of Uyuni, Bolivia. *Chemical Geology*, *30*(1-2), 57–79. doi:10.1016/0009-2541(80)90116-3

Stumm, W., & Morgan, J. J. (1970). *Aquatic chemistry: An introduction emphasizing chemical equilibria in natural waters*. Academic Press.

Stumm, W., Morgan, J. J., & Drever, J. I. (1996). Aquatic chemistry. *Journal of Environmental Quality*, *25*(5), 1162. doi:10.2134/jeq1996.2551162x

Thadée, A. L. (2017). *Étude des saumures naturelles et industrielles: Approche expérimentale et par modélisation de l'extraction du lithium par évaporation* [Study of natural and industrial brines: Experimental and modelled approach to lithium extraction by evaporation] (Doctoral dissertation). Pau.

Tian, P., Ning, P., Cao, H., & Li, Z. (2012). Determination and Modeling of Solubility for CaSO4· 2H2O–NH4+–Cl—SO42—NO3—H2O System. *Journal of Chemical & Engineering Data*, *57*(12), 3664–3671. doi:10.1021/je300871p

Westall, J.C. (1976). *MINEQL: A computer program for the calculation of chemical equilibrium composition of aqueous systems*. Mass. Inst. Technol.

Wolery, T. J. (1979). *Calculation of chemical equilibrium between aqueous solution and minerals: the EQ3/6 software package (No. UCRL—52658)*. California University.

Zou, Y., Zheng, C., & Sheikhi, S. (2021). Role of ion exchange in the brine-rock interaction systems: A detailed geochemical modeling study. *Chemical Geology*, *559*, 119992. doi:10.1016/j.chemgeo.2020.119992

Chapter 4
Desorption Isotherms and Thermodynamic Properties of Prickly Pear Seeds

Samia Mtori
Higher Institute of Technological Studies of Zaghouan, Tunisia

Amira Touil
Laboratoire de Recherche Sciences et Technologies de l'EnvironnementTechnopole, Borj Cedria, Tunisia

Fethi Zagrouba
Laboratoire de Recherche Sciences et Technologies de l'EnvironnementTechnopole, Borj Cedria, Tunisia

ABSTRACT

Sorption isotherms of prickly pear seeds were determined by static gravimetric method at temperatures 45°, 60°, and 70°C, over a relative moisture range of 5-95%. Sorption isotherms are important to define dehydration limits of the product, estimate moisture content alterations under environment conditions, and to acquire moisture content values for safe storage. Four mathematical models were applied to analyze the experimental data. Equilibrium moisture contents of prickly pear seeds decreased with temperature increment at a constant value of relative humidity. The GAB model showed the best fitting to the experimental data. Isosteric heat and differential entropy, determined by applying the Clausius-Clapeyron and Gibbs-Helmholtz equations respectively, decreased strongly as the moisture content increased and could be well adjusted by an empirical exponential relationship. Enthalpy-entropy compensation theory is valid for the sorption of prickly pear seeds, in which the water sorption mechanism in seeds can be considered to be enthalpy controlled.

DOI: 10.4018/978-1-7998-8801-7.ch004

INTRODUCTION

Prickly pear *(Opuntia ficus-indica)* or nopal cactus belongs to the dicotyledonous angiosperm *Cactaceae* family, a family that includes about 1500 species. *O. ficus indica* is a tropical and subtropical plant; it can grow in arid and semi-arid climates with a geographical distribution encompassing Mexico, Latin America, South Africa and Mediterranean countries (Butera et al.., 2002). Tunisia cultivates extensively prickly pear '*Opuntia ficus-indica*' grown mainly in region of Sahel, Kasserine and Kairouan. The current production is estimated at more than 1200 000 t of fruit. The cactus pear fruit is an oval-shaped berry with an average weight of 100–200 g. The juicy pulp contributes 60–70% to the total fruit weight and contains many hard-coated seeds that contribute 10–15% to the pulp weight (Cantwell, 1995). These seeds are the waste materials of the prickly pear fruit processing while proper utilization of these waste products could lead to an important new source of good quality edible oil (Tlili et al., 2011; Chougui et al., 2013). Prickly pear seeds are characterized by their high humidity content which can be the most important cause of their contaminations. Drying these seeds at equilibrium moisture content is recommended to stabilize and store the product for further use such as oil extraction.

Therefore, the quality of the stored product depends on the amount of moisture, moisture migration and moisture adsorption through the food material during storage. Consequently, for proper storage, it is necessary to determine the sorption isotherms of the products. A sorption isotherm is the relationship between equilibrium moisture content and moisture content under constant temperature and pressure (Garbalinska et al., 2017). Moreover, sorption isotherms are very important in order to define dehydration limits of the product, estimate moisture content alterations under environment conditions and to acquire safe moisture content values. These values are critical in order to prevent unwanted microorganism activity during storage. In addition, sorption isotherms indicate the required energy in order to dry the product, represented by isosteric heat of sorption (Noshad et al., 2012).

From the isotherms it is possible to determine the latent heat of vaporization, enthalpy, entropy, isokinetic theory and Gibbs free energy. These properties provide essential information about sorption mechanism and allow for the estimation of the energy requirements of the drying process (Arslan and Togrul., 2006). Thermodynamic parameters provide also information that enables the calculation of energy in heaiting and mass transfer in biological systems, thus enabling greater understanding of the properties of water molecules (Goneli et al., 2010).

This study aims to determine experimentally isotherms desorption of prickly pear seeds at 45, 60 and 70°C; to fit the sorption curves by four model and determine the suitable model that well describes the sorption curves of prickly pear seeds. This work aims also to study thermodynamic properties (isosteric heat of desorption, differential entropy, enthalpy-entropy relationship and Gibbs free energy).

MATERIAL AND METHOD

Plant Material

Seeds used in the drying experiments were taken from the prickly fruit (*Opuntia ficus indica)* of the (*Ameclyae)* variety, grown in Knais, region of Sousse (Tunisia). This variety is a small and oval yellowish prickly pear, with a relatively large number of seeds (35% on a dry weight basis equivalent to an average of 230 seeds per fruit), rich in polyphenols, flavonoids, tannins, fatty acids, and vitamins (Morales et al., 2012). Mature fruit samples were harvested in August and taken to the laboratory the same day, where they were washed with running water to remove impurities and were then air-dried and manually peeled. The seeds were separated by pressing the whole edible pulp and rinsing the pulp, abundantly with distilled water.

Method

Equilibrium Moisture Content (EMC)

In order to determine the desorption isotherms of prickly pear seeds at 45, 60 and 70°C, the gravimetric static method with saturated saline solutions was used to maintain a fixed RH. The atmosphere surrounding the product has fixed air moisture content for every working temperature imposed by the salt solution. Seven saturated saline solutions (NaOH, KOH, $MgCl_2$, K_2CO_3, NaBr, NaCl and $BaCl_2$) with water activity varying from 0.055 to 0.873 at different temperatures were used (Greenspan., 1977). Each saturated solution (100 ml) was placed in a hermetically closed glass container. Seeds were placed in the containers and hermetically closed. Each closed container holding the seeds samples was then placed in an environmental controlled chamber set at 45, 60 and 70°C. The experimental apparatus is shown in Figure 1.

The samples were weighted every 2 days using a fourth digit precision analytical balance until a constant weight was observed, indicating equilibrium between the samples and the saline solution. Once the equilibrium was reached, the moisture content of the seeds was determined by a drying oven whose temperature is fixed at 105°C. This operation lasts four or five hours, until the masses do not vary more with time. The difference of mass before and after drying in the oven gives the moisture content X_{eq} of the product at hygroscopic equilibrium (Equation1).

Figure 1. Experimental apparatus for sorption isotherms measurement: (1) thermostated bath; (2) glass jar containing salt solution; (3) sample holder; (4) product; (5) saturated salt solution

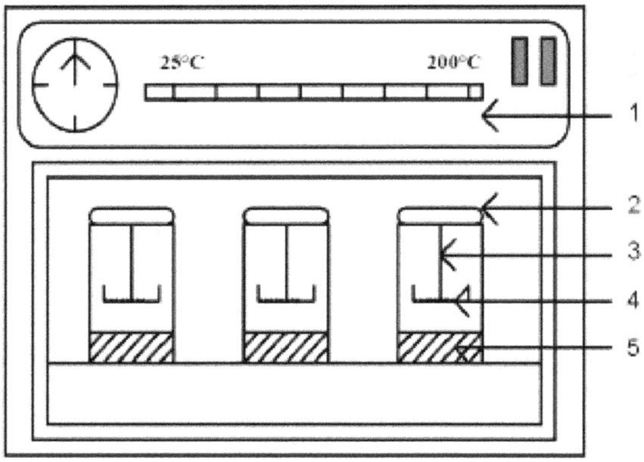

$$X_{eq} = \frac{X_w - X_d}{X_d} \tag{1}$$

Where X_{eq} is the equilibrium moisture content dry basis, expressed in kg water/kg dry matter, X_w is the equilibrium mass and X_d. The same experiment is reported at 45, 60 and 70°C.

Modeling Experimental Values by Sorption Models

Experimental moisture sorption data can be described by many sorption models. Four models were used to adjust the EMC experimental values (Table 1). These models are widely applied, in the scientific literature, to describe the moisture isotherms in food products. Model parameters were calculated by using non linear regression in computer program "CurveExpert Professional 1.5.0".

Statistical Analysis

The corresponding equations of mathematical models include two or three parameters. These parameters were identified by non-linear least square regression analysis, using CurvExpert 1.5 software. The goodness of fit of the tested mathematical models to the experimental data was evaluated from the correlation coefficient (r2) and the standard error (S). These parameters are deðned as following: Equation 6 and Equation 7.

Table 1. Models used to predict moisture desorption isotherms in prickly pear seeds

Model name	Model equation	Equation number
GAB (Van Den Berg 1981)	$Xeq = \frac{X_m CKaw}{(1-Kaw)(1-Kaw+CKaw)}$	(2)
Henderson (Henderson 1952)	$X_{eq} = X_m[-\ln(1-aw)]^n$	(3)
Oswin (Oswin 1946)	$X_{eq} = \text{Xm}\left(\frac{aw}{1-aw}\right)^n$	(4)
Halsey (Halsey 1948)	$X_{eq} = \left(\frac{-A}{\ln(aw)}\right)^{1/n}$	(5)

X_{eq}-Equilibrium moisture content, % db; X_m- Monolayer water content, % db; aw-Water activity, decimal; C and K are dimensionless parameters in the GAB equation; A is a dimensionless parameter related to the Henderson and Halsey models.

$$S = \sqrt{\frac{\sum_{i=1}^{n_{exp.data}} (Exp_i - Cali)^2}{n_{exp.data} - n_{param}}} \tag{6}$$

$$r = \sqrt{1 - \frac{\sum_{i=1}^{n_{exp.data}} (Exp_i - Cali)^2}{\sum_{i=1}^{n_{exp.data}} (Exp_i - Expi)^2}} \tag{7}$$

Where "Cal" is the value of the moisture ratio or of the drying rate (f) calculated by using the tested model, "Exp" is the experimental value of the moisture ratio or of the drying rate, n_{param} is the number of parameters of the particular model and n_{exp}. Data is the number of experimental points (Ghodake et al.2007).

Thermodynamic Properties

When water is removed from food, heat is absorbed, despite the increased osmotic pressure and water activity gradient. Sorption isosteric heat-differential-enthalpy is an indicator of the state of the water retained in the solid materials. Net sorption isosteric heat (ΔH_S) is the difference between total heat absorbed by food (ΔHd) and water vaporisation heat (ΔHvap) associated with the sorption process, which is calculated from experimental data using the Clausius-Clapeyron equation (Equation 8):

$$\left[\frac{d\left(lna_w\right)}{d(1/T)}\right] = -\frac{\Delta H_s}{R} \tag{8}$$

Where the isosteric heat of sorption is defined as molecules absorbed at a particular moisture content (Equation 9):

$$\Delta H_s = \Delta H_d - \Delta H_{vap} \tag{9}$$

By integrating and entering the boundary conditions (Equations 10 and 11):

$$\int_{a_{w1}}^{a_{w2}} d\left(lna_w\right) = -\frac{\Delta H_s}{R} \int_{T_1}^{T_2} d\left(1/T\right) \tag{10}$$

$$\ln\left(\frac{a_{w2}}{a_{w1}}\right) = \frac{\Delta H_s}{R}\left[\frac{1}{T_1} - \frac{1}{T_2}\right] \tag{11}$$

where: $aw1$, $aw2$ – water activity of $T1$ and $T2$; R – gas constant (8.314 kJ/mol·K) (Santanu-Basu et al., 2006).

The isosteric heat of sorption is an important parameter during drying. Its value in drying equilibrium should be considered at the end of the drying process, since it has a magnitude equal to the latent heat of vaporisation (Talla et al., 2005).

The differential enthalpies (ΔS_d) were calculated from isothermal data at different temperatures. In this way, the relationship between isosteric heat (ΔHd) and sorption differential enthalpy (ΔSd) is as follows:

$$d(lna_w) = (\Delta H_d / RT) - (\Delta S_d / R) \tag{12}$$

By plotting ln aw versus the inverse of temperature, the particular moisture content, W and ΔHd were obtained from the slope, and ΔSd was obtained from the intercept. Use of these equations at different moisture contents demonstrates the dependence of ΔHd and ΔSd on moisture content (Rouquerol et al., 2014).

The compensation theory proposes a linear relationship between ΔH_S and $\Delta S_{d:}$

$$\Delta H_S = T_\beta \cdot \Delta S_d + \Delta G_\beta \tag{13}$$

In this relation T_β is the isokinetic temperature and represents the temperature at which all reactions in the series proceed at the same rate.

Another test for validing the compensation theory is comparing the isokinetic temperature with the harmonic mean temperature T_{hm} (Krug et al., 1976a; Krug et al., 1976b), that is defined as:

$$T_{hm} = \frac{n}{\sum_{i=1}^{n}(1/T)} \tag{14}$$

Where n is the total number of isotherms used.

The isokinetics temperature (T_β) and constant ΔG_β were calculated using linear regression.

RESULTS AND DISCUSSION

Moisture Desorption Isotherms

The hygroscopic equilibrium of the prickly pear seeds is reached in 7 to 12 days. Figure 2 shows the experimental moisture desorption isotherms of prickly pear seeds samples kept at 45, 60 and 70°C, respectively. The observed isotherm patterns were classified as type III according to Brunauer et al. (1940) cited by Iguedjtal et al. (2008). Type III isotherms are characteristic of foods rich in soluble components. Prickly pear seeds (Hassini et al., 2015) grapes and apricots (Kaymak-Ertekin and Gedik, 2004), strawberries (Moraga et al., 2004) and kiwis (Moraga et al., 2006) were also reported as exhibiting type III isotherms. The desorption isotherms showed that with aw up to 0.55, fruits gained relatively low moisture. However, with aw values higher than 0.55, solids solubilisation and adsorption promoted a significant increase in moisture content (Hubinger et al., 1992).

For all isotherms, EMC increased as the a_w value increased, being more evident when aw values were higher than 0.55. This is a common pattern in food desorption processes. At constant aw, EMC increases with decreasing temperature. The behavior of desorption experiments is similar to many plants and food materials (Hassini et al., 2015; Lahsasni et al., 2002; Boudhrioua et al., 2008). This behavior may be explained by considering excitation states of water molecules. At increased temperatures, molecules are in an increased state of excitation, thus increasing their distance apart and decreasing the attractive forces between them. This leads to a decrease in the degree of water sorption at a given water activity with increasing temperature (García-Pérez J.V et al., 2008). Furthermore, at a constant temperature,

Figure 2. Influence of temperature on the desorption isotherms of seeds

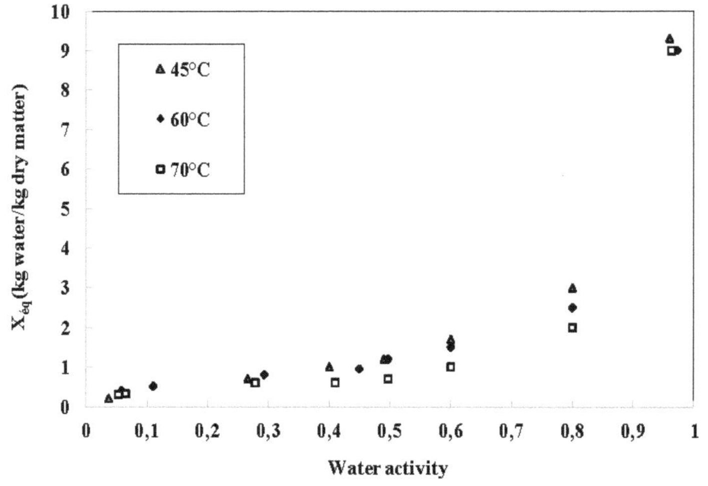

the equilibrium moisture content increases with increasing water activity. In a general way, both temperature and water activity have significant effect on experimental equilibrium moisture content. Similar trends have been observed in several other studies for agricultural and food products (Boudhrioua et al., 2008, Ait Mohamed et al., 2004).

Modeling Experimental Values of Sorption

The experimental data of the desorption curves of prickly pear seeds were fitted to four sorption model (Table 2). The results reveal the temperature dependance for the sorption behaviour, with an increase in temperature decreasing the sorption capacity. Activation of the water molecules due to the increase in temperature causes them to break away from water binding sites, thus lowering the equilibrium moisture content (Naji et al., 2010).

The sorption relationships detailled in Table 1 were fitted to the experimental data for prickly pear seeds. The results of nonlinear regression analysis of fitting the sorption equations to the experimental data are showen in Table 2.

The agreement between experimental and estimated data (Figure 3) indicates that GAB model is a good choice for simulating sorption isotherms of prickly pear seeds at the temperatures considered.

Table 2 shows the three GAB parameters obtained. The GAB model introduced a second well-differentiated sorption stage for water molecules and an additional energy constant, k. the obtained k values were, in all series, near to 1.0. The monolayer capacity is represented in one of the three GAB constants. At 45 and 60°C the monolayer moisture content (Xm) obtained by the GAB model was lower than that obtained by Henderson and Oswin models (Table 2). The third parameter, C, is also an energy constant known as the BET constant, but with slightly different physical meanings (Timmermann E.O et al., 2001).

Table 2. Different model for determination desorption isotherms of seeds

Model	Température	coefficients	r	s
GAB	45°C	X_m = 0,5949; C = 36,7164; K = 0,97418	0,99874	0,00113
	60°C	X_m = 0,6048; C = 5,22481; K = 0,95919	0,99802	0,00602
	70°C	X_m = 0,4119; C = 5,36874; K = 0,97726	0,99826	0,00321
Henderson	45°C	X_m = 1,4340; n = 1,59821	0,99425	0,01398
	60°C	X_m = 1,3831; n = 1,43877	0,99842	0,00580
	70°C	X_m = 0,8522; n = 1,77169	0,99748	0,00332
Oswin	45°C	X_m = 1,3216; n = 0,60285	0,99684	0,01037
	60°C	X_m = 1,3301; n = 0,51393	0,99771	0,00699
	70°C	X_m = 0,8402; n = 0,63237	0,98670	0,01573
Halsey	45°C	A = 4,72973; n = 2,11879	0,99687	0,01032
	60°C	A = 4,73373; n = 2,13694	0,99842	0,00580
	70°C	A = 4,81240; n = 2,25689	0,99811	0,00288

Figure 3. Isotherm curves experiment and predicted using four models at 45°C

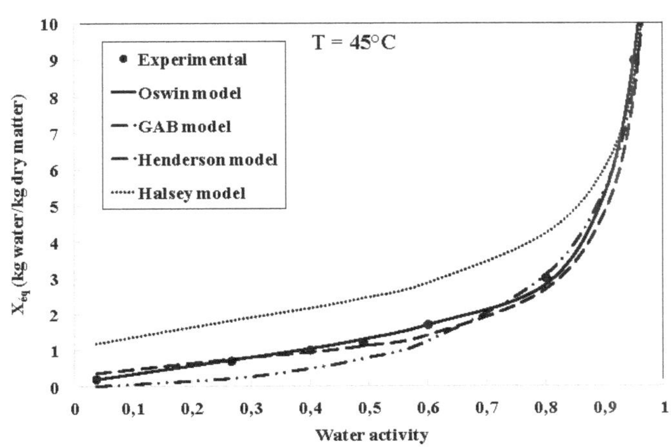

Isosteric Heat and Sorption Entropy

The sorption isosteric heat of seeds was obtained by applying the Clausius-Clapeyron equation to the experimental equilibrium isotherm data. The results indicated that the lower the moisture content of the sample, the more sorption isosteric heat was required (Figure 4). The net isosteric heat of desorption increases rapidly with deceasing equilibrium misture content, from 0.590 to 21.64 kJ/mol with a variation in the equilibrium water content ranging from 0.28 to 0.025 kg water/kg dry matter. Since in the early stages of adsorption, many polar active sites exist on the product surface, and as water molecules are attached to these sites as a monolayer, the energy required for removing water is very high. However, with increased moisture content the affinity of the molecules for the sample is reduced and the sorption isosteric heat is decreased (Moreira et al., 2008).

The variation of isosteric heat of sorption were similar to those observed bt other authors for many seeds: Karen et al. (2020) for sorption of pepper seeds; Vaquiro et al. (2011) for lime seeds; Correa et al. (2015) for cucumber seeds.

The strong dependence of differential entropy on moisture content lower than 0.05 kg.kg-1 d.b. can be observed in Figure 5, with an exponential trend similar to the exhibited for the isosteric heat of sorption. The differential entropy increases as the moisture content decreases (Figure 4), since the differential entropy, that measures the ordering change, is lower when the molecular movement is more restricted. Similar trends with respect to entropy were also observed in other seeds.

Figure 4. Desorption isosteric heat as a function of equilibrium moisture content

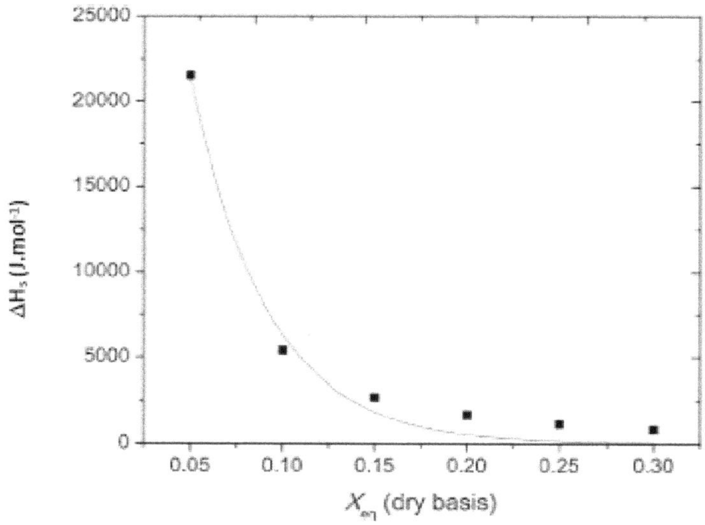

Figure 5. Desorption entropy as a function of equilibrium moisture content

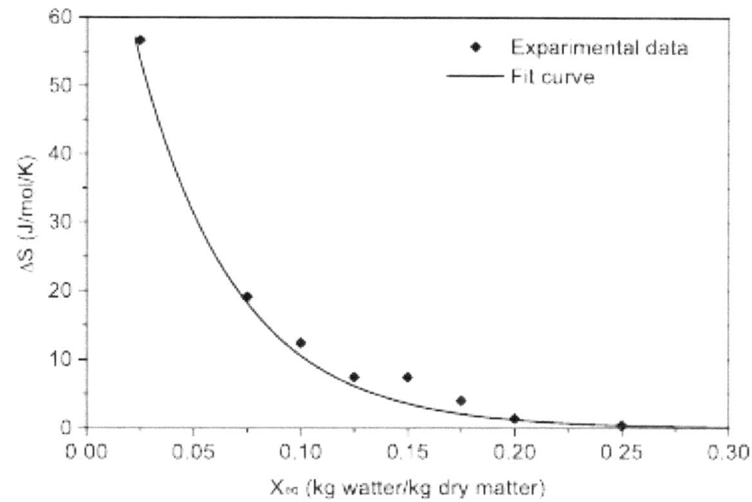

Enthalpy-Entropy Compensation Theory

Values of isosteric heat as a function of entropy values of desorption are presented at Figure 6. A good straight line related the net desorption isosteric heat to the differential desorption entropy for the prickly pear seeds, at all temperatures. This behavior confirms the enthalpy-entropy compensation hypothesis. It was considered that to a certain value of moisture content, isosteric heat and entropy are invariant with temperature (Aguerre et al., 1986).

The parameters T_β and ΔG_β were calculated from the data by linear regression.

T_β (K) =377,6 K and ΔG_β = 2609,6 J/mol.

Gibbs free energy is tied to the work necessary to make sorption sites available (Nkolo et al., 2008). For prickly pear seeds the positive value of ΔG_β indicates a non spontaneous desorption process.

The temperature at which all reaction in series occur at the same rate T_β was stablished at 377.6 K. if this temperature is different from harmonic temperature T_{hm} the second requirement to confirm a linear chemical compensation patter is evidenced (Krug et al., 1976a; Krug et al., 1976b). The harmonic temperature for prickly pear seeds was found to be T_{hm} = 345.4 K (calculated on the basis of Eq. (14)), which is lower than isokinetic temperature.

Figure 6. Enthalpy–entropy relationship for desorption process of prickly pear seeds

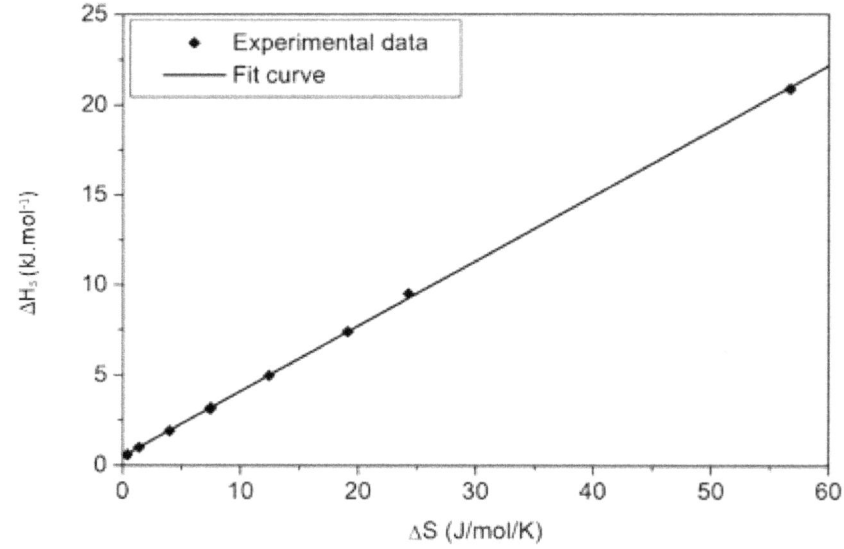

CONCLUSION

The sorption curves provide valuable information about the hygroscopic equilibrium of product. They give a clear idea on the stability domain of product after drying. These curves are indispensable in the food product, especially in the operation of storage and conservation of food product. The water sorption behavior of prickly pear seeds was evaluated in order to optimize drying kinetics and minimize the energy expenses involved in dehydration process.

The desorption isotherms of prickly pear seeds have been determined by experiment and then modeled by the GAB, Henderson, Oswin and Halsey equations. The observed isotherm patterns were classified as type III characteristic of foods rich in soluble components. Temperature affected the moisture sorption behavior where equilibrium moisture content decreased with increasing temperature at a constant water activity. The results show that GAB model seems to be the most suitable for describing the desorption isotherms of prickly pear seeds. By using the GAB model, the net isosteric heat of sorption and the differential entropy have been calculated as function of moisture content. The linear relationship betwwed isosteric heat and entropy showed that lower energy is necessary to remove a certain amount of water at higher moisture contents, which is associated to the decrease in the number of available sites to sorption followed by a reduction of interactions betweed prickly pear seeds and water molecules in the multilayer. This phenomenon makes the water removal process more spontaneous when prickly pear seeds presents higher moisture contents. The analysed properties were in agreement with the thermodynamic theory and were comparable with those reported in the literature for other agrofood materials.

ACKNOWLEDGMENT

I would like to thank Dr Ben Selma Zied, General director of "huilerie ben Selma", for supplying prickly pear seeds.

REFERENCES

Aguerre, R. J., Suarez, C., & Viollaz, P. E. (1986). Enthalpy-entropy compensation in sorption phenomena: Application to the prediction of the effect of temperature on food isotherms. *Journal of Food Science*, *51*(6), 1547–1549. doi:10.1111/j.1365-2621.1986.tb13856.x

Ait Mohamed, L., Kouhila, M., Lahsasni, S., Jamali, A., Idlimam, A., Rhazi, M., Aghfir, M., & Mahroue, M. (2004). Equilibrium moisture content and heat of sorption of prickly pear seeds. *Journal of Stored Products Research*, *41*, 199–209. doi:10.1016/j.jspr.2004.03.001

Arslan, N., & Togrul, H. (2006). The fitting of various models to water sorption isotherms of tea stored in a chamber under controlled temperature and humidity. *Journal of Stored Products Research*, *42*(2), 112–135. doi:10.1016/j.jspr.2005.01.001

Boudhrioua, N., Bahloul, N., Kouhila, M., & Kechaou, N. (2008). Sorption isotherms andisosteric heats of olive leaves (Chemlali variety): Experimental and mathematical investigations. *Food and Bioproducts Processing*, *86*(3), 167–175. doi:10.1016/j.fbp.2007.10.010

Brunauer, S., Emmett, P. H., & Teller, E. (1940). Adsorption of gases in multi-molecular layers. *Journal of the American Chemical Society*, *60*(2), 309–319. doi:10.1021/ja01269a023

Butera, D., Tesoriere, L., Gaudio, D. F., Bongiorno, A., Allegra, M., Pintaudi, A. M., Kohen, R., & Livrea, M. A. (2002). Antioxidant activities of Sicilian prickly pear (Opuntia ficus-indica) fruit extracts and reducing properties of its betalains: Betanin and indicaxanthin. *Journal of Agricultural and Food Chemistry*, *50*(23), 6895–6901. doi:10.1021/jf025696p PMID:12405794

Cantwell, M. (1995). Post-harvest management of fruits and vegetable stems. In Agroecology, Cultivation and Uses of Cactus Pear. FAO Plant Production and Protection Paper.

Chougui, N., Tamendjari, A., Hamidj, W., Hallal, S., Barras, A., Richard, T., & Larbat, R. (1995). Oil composition and characterisation of phenolic compounds of Opuntiaficus-indica seeds. *Food Chemistry*, *139*(1-4), 796–803. doi:10.1016/j.foodchem.2013.01.054 PMID:23561175

Correa, P. C., Goneli, A. L. D., Junior, P. C. A., De Oliveira, G. H. H., & Valente, D. S. M. (2015). Moisture sorption isotherms and isosteric heat of sorption of coffee in different processing levels. *International Journal of Food Science & Technology*, *45*(10), 2016–2022. doi:10.1111/j.1365-2621.2010.02373.x

Garbalinska, H., Bochnenek, M., Malorny, W., & Von Werder, J. (2017). Comparative analysis of the dynamic vapor sorption technique and the traditional method for sorption isotherms determination- Exemplified at autoclaved aerated concrete samples of four density classes. *Cement and Concrete Research*, *91*, 97–105. doi:10.1016/j.cemconres.2016.11.001

García-Pérez, J. V., Cárcel, J. A., Clemente, G., & Mulet, A. (2008). Water sorption isotherms for lemon peel at different temperatures and isosteric heats. *Lebensmittel-Wissenschaft + Technologie*, *41*(1), 18–25. doi:10.1016/j.lwt.2007.02.010

Ghodake, H. M., Goswami, T. K., & Chakraverty, A. (2007). Moisture sorption isotherms, heat of sorption and vaporization of xithered leaves, black and green tea. *Journal of Food Engineering*, *78*(3), 827–835. doi:10.1016/j.jfoodeng.2005.11.023

Goneli, A.L.D., Correa, P.C., Oliveira, G.H.H., & Botelho, F.M. (2010). Water sorption isotherms and thermodynamic properties of okra seeds. Trans. ASABE. *American Society of Agricultural and Biological Engineering*, *53*, 191-197.

Greenspan, L. (1977). Humidity fixed points of binary saturated aqueous solutions. *Journal of Research of the National Bureau of Standards-A. Physics and Chemistry*, *81*, 92-93.

Halsey, G. (1948). Physical adsorption on non-uniform surfaces. *The Journal of Chemical Physics*, *16*(10), 931–937. doi:10.1063/1.1746689

Hassini, L., Desmorieux, H., Torres, S. S., & Touil, A. (2015). Desorption isotherms and thermodynamic properties of prickly pear seeds. *Industrial Crops and Products*, *97*, 457–465. doi:10.1016/j.indcrop.2015.01.078

Henderson, S. M. (1952). A basic concept of equilibrium moisture. *Transactions of the ASAE. American Society of Agricultural Engineers*, *33*, 29–32.

Hubinger, M., Menegalli, F. C., Aguerre, R. J., & Suarez, C. (1992). Water vapor adsorption isotherms of guava, mango and pineapple. *Journal of Food Science*, *57*(6), 1405–1407. doi:10.1111/j.1365-2621.1992.tb06869.x

Iguedjtal, T., Louka, N., & Allaf, K. (2008). Sorption isotherms of potato slices dried and texturized by controlled sudden decompression. *Journal of Food Engineering*, *85*(2), 180–190. doi:10.1016/j.jfoodeng.2007.06.028

Karen, C. R., Hellismar, W., Isneider, L., Samuel, G. F., Daniel, P., & Renato, S. (2020). Isotherms and thermodynamic properties of water adsorption in Cumari pepper seeds. *Revista Brasileira de Engenharia Agrícola e Ambiental*, *24*(4), 280–285. doi:10.1590/1807-1929/agriambi.v24n4p280-285

Kaymak-Ertekin, F., & Gedik, A. (2004). Sorption isotherms and isosteric heat of sorption for grapes, apricots, apples and potatoes. *Lebensmittel-Wissenschaft + Technologie*, *37*(4), 429–438. doi:10.1016/j.lwt.2003.10.012

Krug, R. R., Hunter, W. G., & Grieger, R. A. (1976a). Enthalpy entropy compensation. 1- some fundamental statistical problems associated with the analysis of Van't Hoff and Arrhenius data. *Journal of Physical Chemistry*, *80*(21), 2335–2341. doi:10.1021/j100562a006

Krug, R. R., Hunter, W. G., & Grieger, R. A. (1976b). Enthalpy entropy compensation. 2- Separation of the chemical from the statistical effect. *Journal of Physical Chemistry*, *80*(21), 2341–2351. doi:10.1021/j100562a007

Lahsasni, S., Kouhila, M., Mahrouz, M., & Fliyou, M. (2002). Moisture adsorption-desorption isotherms of prickly pear cladode at different temperatures. *Energy Conversion and Management*, *44*, 923–936. doi:10.1016/S0196-8904(02)00094-8

Moraga, G., Martinez, N., & Chiralt, A. (2004). Water sorption and phase transitions instrawberries: Influence of pre-treatment. *Journal of Food Engineering*, *62*(4), 315–321. doi:10.1016/S0260-8774(03)00245-0

Moraga, G., Martinez, N., & Chiralt, A. (2006). Water sorption and phase transitions inkiwi fruit: Influence of pre-treatment. *Journal of Food Engineering*, *72*, 156–174. doi:10.1016/j.jfoodeng.2004.11.031

Morales, P., Ramírez-Moreno, E., Sanchez-Mata, M. C., Carvalho, A. M., & Ferreira, I. C. F. R. (2012). Nutritional and antioxidant properties of pulp and seeds of twoxoconostle cultivars (Opuntia joconostle FAC Weber ex Diguet and Opuntia matudae Scheinvar) of high consumption in Mexico. *Food Research International*, *46*(1), 279–285. doi:10.1016/j.foodres.2011.12.031

Moreira, R., Chenlo, F., Torre, M. D., & Vallejo, N. (2008). Thermodynamic analysis of experimental sorption isotherms of loquat and quince fruits. *Journal of Food Engineering*, *88*(4), 514–521. doi:10.1016/j.jfoodeng.2008.03.011

Naji, A., Idlimam, A., & Kouhila, M. (2010). Sorption isotherms and thermodynamic properties of powdered milk. *Chemical Engineering Communications*, *197*(8), 1109–1125. doi:10.1080/00986440903412936

Noshad, M., Shahidi, F., Mohebbi, M., & Mortazav, S. (2012). Desorption isotherms and thermodynamic properties of fresh and osmotic ultrasonic dehydrated quince. *Journal of Food Processing and Preservation*, *37*, 1–12.

Oswin, C. R. (1946). The kinetics of package life. III. The isotherm. *Journal of the Society of Chemical Industry*, *6*(12), 419–421. doi:10.1002/jctb.5000651216

Rouquerol, J., Rouquerol, F., Llewellyn, P., Maurin, G., & Sing, K. S. (2014). *Adsorption by Powders and Porous Solids: Principles, Methodology and Applications* (2nd ed.). Academic press.

Santanu-Basu, U. S., & Shivhare Mujumdar, A. S. (2006). Models for Sorption Isotherms for Foods: A Review. *Drying Technology*, *24*(8), 917–930. doi:10.1080/07373930600775979

Talla, A., Jannot, Y., Nkeng, G. E., & Puiggali, J. R. (2005). Experimental Determination and Modelingof Sorption Isotherms of Tropical Fruits: Banana, Mango, and Pineapple. *Drying Technology*, *23*(7), 1477–1498. doi:10.1081/DRT-200063530

Timmermann, E. O., Chirife, J., & Iglesias, H. A. (2001). Water sorption isotherms of foods and foodstuffs: BET or GAB parameters. *Journal of Food Engineering*, *48*(1), 19–31. doi:10.1016/S0260-8774(00)00139-4

Tlili, N., Bargougui, A., Elfalleh, W., Triki, S., & Nasri, N. (2011). Phenolic compounds, protein, lipid content and fatty acids compositions of cactus seeds. *Journal of Medicinal Plants Research*, *5*(18), 4519–4524.

Van Den Berg, C., & Bruim, S. (1981). Water activity and its estimation in food systems. In *Influence of food quality*. Academic Press.

Vaquiro, H. A., Simal, S., Reis de Carvalho, G., & Telis-Romero, J. (2011). Moisture desorption isotherms and thermodynamic properties of lime seeds. *European Drying Conference*, 26-28.

Chapter 5
Energy and Exergy Analysis of Desiccant Cooling System Under Hot Dry Climate

Sarra Belguith
Institut Supérieur des Sciences Appliquées et Technologies de Gabès, Tunisia

Zina Meddeb
Institut Supérieur des Sciences Appliquées et Technologies de Gabès, Tunisia

Bechir Chaouachi
National Engineering School of Gabes, Tunisia

ABSTRACT

In this chapter, in order to find the optimum coefficient of performance (COP) and exergy performance (COPex) of desiccant cooling system operating on ventilation and recirculation cycles under hot and dry climate, a simple theoretical model has been developed based on the first and second laws of thermodynamics. Then, the model was implemented in the MATLAB software. The obtained theoretical results were compared with those of the literature and showed a good agreement. Moreover, results showed that COP of ventilation and recirculation cycles are 1.89 and 1.13 respectively, greater than the corresponding COPex, which is equal to 0.7 and 0.38 respectively. In addition, the maximum destruction exergy percentages are provided by the desiccant wheel and the heat source, which are respectively 57% and 24.67% in ventilation cycle and 33.08% and 38.83% in recirculation cycle. Finally, the sensitivity of exergy destruction desiccant wheel and heat source with reference (dead-state) were explored.

DOI: 10.4018/978-1-7998-8801-7.ch005

INTRODUCTION

The desiccant cooling system is a cooling technique that could be respect the environment and require low grade energy like waste heat and solar energy. Because this, system works with environmentally friendly refrigerants and uses low temperature heat. In desiccant cooling system tow air streams, the first air stream is the fresh air dehumidified in the desiccant wheel and sensibly could in the heat exchanger and evaporatively cooled in indirect evaporative cooler before being deliver to conditioned space. The second stream is the regeneration air used to reactive desiccant wheel that remove moisture from the process air. Among this cooling technology is attractive solution for environmental and energy problems in the world. Many researchers are focused to the studies of desiccant cooling technologies

Demis et al. (2018) established a numerical study of novel desiccant cooling based on multi –stage cooling with Maisotsenko cycle and reported that the proposed system appears to be more efficient in moderate climates. Belguith et al. introduced a wave analyse of different cycle of desiccant cooling system under hot and dry climates and showed that the ventilation cycle have the best coefficient of performance and it strongly influenced by the outside conditions (Belguith et al., 2020). There are few study in the literature interested on exergy analysis of desiccant cooling system . Abbassi et al performed a comparative study between different configurations of desiccant cooling system, their studies related that the ventilation and recirculation cycle in single stage have higher performance than those of a double-stage system (Abbassi et al.,2017) .The energy analysis of desiccant cooling system which takes into account only the first law of thermodynamics is not sufficient to evaluate the performance of this air conditioning system therefore, it is necessary to evaluate the system from an exergetic point of view which is based on the second law of thermodynamics. Zhu et al, (2014) performed energy and exergy analysis of recirculation mode of desiccant cooling system. It was found that the desiccant wheel responsible on the major part of exergy losses in the system. Enteria et al., (2015) worked on exergoeconomic analysis to the desiccant cooling system. They were determined the exergy efficiency, exergy destruction ratio, cost rate and exergoeconomic factors. In addition it was discovered that the maximum exergoeconomic factors were related to the exit air fan, outdoor air fan and secondary heat exchanger. Xiong et al., (2010) carried an exergy analysis of two stage liquid desiccant dehumidification used calcium chloride ($CaCl_2$) solution. They conculuded that the thermally driven energy coefficient of performance of the overall proposed system increased from 0.24 to 0.73 and the corresponding exergy coefficient of performance increased from 6.8% to 23%. Carpinlioglu, (2015) evaluated the performance of desiccant cooling system by using entransy concept. It was found that the maximuim exergy destruction caused by the desiccant wheel.

In the present work. In order to provide an upper limit for the thermal COP and exergy efficiency, COPex values and to improuve the desiccant cooling system in ventilation and recirculation modes, the overall energy and exergy performance of all components and DCS installation are evaluated on the basis of thermodynamic analysis and energy analysis. Also, the components with higher losses of exergy are identifed. The sensitivity of desiccant wheel and heat source exergy analysis with reference (dead-state) temperature and humidity ratio is discussed.

SYSTEM DESCRIPTION

Ventilation Cycle

The ventilation cycle of desiccant cooling system is illustrated in Figure 1. Fresh air (stream 1) is supplied to desiccant wheel (DW) where its moisture is removed and it is warmed up due to the adsorption heat effect. Then, the hot and dry air (stream 2) flows through a sensible heat exchanger where its cooled and finally it(stream 3) is humidified and cooled in the humidifier (evaporative cooler 1) before being blown into the room (stream 4). On the regeneration side, the return air (stream 5) is cooled and humidified in the second humidifier (evaporative cooler 2). Then the air (stream 6) is preheated by the incoming air (stream 2) in the sensible heat exchanger .The warm air stream (stream 7) is then further heated by heat source (stream 8)to regenerate the wheel by allowing the desorption of water. Finally, it is exhausted to ambient (stream 9) (Ali et al., 2018, 2015, Hurdogen et al., 2010, Bourdoukan et al., 2010).

Figure 1. Desiccant cooling system in ventilation cycle (Belguith et al., 2020)

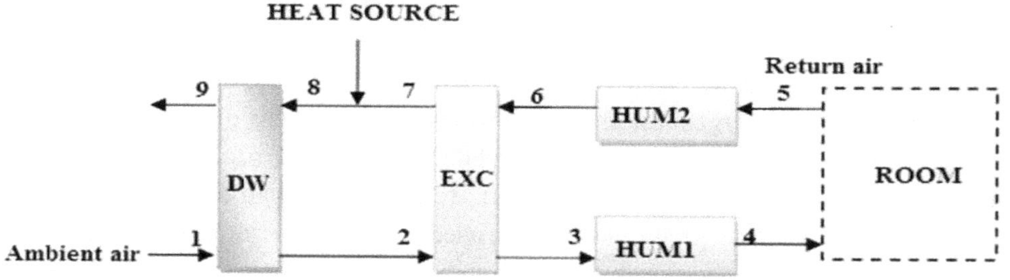

Recirculation Cycle

In the recirculation cycle of desiccant cooling, the air undergoes the following process (see Figure 2). In the process side the return air is supplied to desiccant wheel (DW) (1-2) where its moisture is eliminated and it is warmed up due to the adsorption heat effect. Then, then it is refreshed in the heat exchanger(2-3), and finally it is humidified and cooled in the humidifier (evaporative coller 1) (3-4) before being blown into the conditioned space. On the regeneration side, the process air is cooled and humidified in the second humidifier (evaporative cooler 2) (5-6). Then the air is preheated by the incoming air in the sensible heat exchanger (6-7). The warm air stream is then further heated by the heat source(7-8) to renovate the rotary desiccant by allowing the desorption of water(9-10). Finally, it is exhausted to ambient (Dezfouli et al. 2014, Belguith et al., 2020).

Figure 2. Desiccant cooling system in recirculation cycle (Belguith et al., 2020)

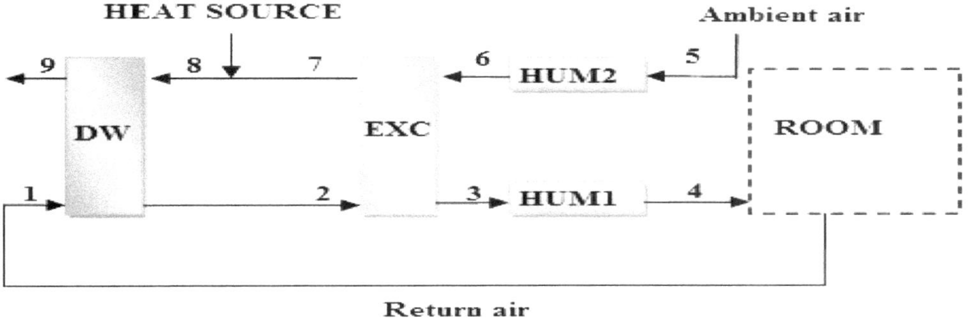

ENERGY AND EXERGY INVESTIGATION

Energy Analysis

The desiccant wheel characterized by two efficiency parameters ε_1 and ε_2 that represent the degree of approximation to the adiabatic state and the degree of dehumidification respectively (Panaras et al., 2010; Maclaine–Cross et al., 1972; Jurinak et al., 1982).

$$\varepsilon_1 = \frac{F_1(T_2, w_2) - F_1(T_1, w_1)}{F_1(T_8, w_8) - F_1(T_1, w_1)} \tag{1}$$

$$\varepsilon_2 = \frac{F_2(T_2, w_2) - F_2(T_1, w_1)}{F_2(T_8, w_8) - F_2(T_1, w_1)} \tag{2}$$

$$F_1 = -\frac{2865}{(T + 273.15)^{1.49}} + 4,344w^{0,8624} \tag{3}$$

$$F_2 = \frac{(273.15 + T)^{1.49}}{6360} - 1.127w^{0.07969} \tag{4}$$

The effectiveness of the heat exchanger and the humidifier are written as fellow in (4) and (5) (Sphaier et al., 2012, Threlkeld et al., 1998).

$$\varepsilon_{ech} = \frac{T_2 - T_3}{T_2 - T_6} \tag{5}$$

$$\varepsilon_{Hum} = \frac{T_3 - T_4}{T_3 - T_{sat}} \tag{6}$$

The coefficient of performance of desiccant cooling system is defined as follows (Daou et al.,2006):

$$COP = \frac{Q_{COOL}}{Q_{Regeneration}} = \frac{h_5 - h_4}{h_8 - h_7} \tag{7}$$

Exergy Analysis

The total system exergetic balance is written as follows (Caliskan et al., 2019)

$$\sum \dot{E}x_{in} = \sum \dot{E}x_{out} + \sum \dot{E}x_{dest} \tag{8}$$

The exergy of humid air has three types of exergy, chemical, mechanical and thermal exergy . For the desiccant cooling system, shown as above, the mechanical exergy is negligible. The formula that describes exergy is:

$$\dot{E}x_{air} = \dot{E}x_{ch,air} + \dot{E}x_{th,air} \tag{9}$$

$$\dot{E}x_{ch,air} = \dot{m}_{air}[R_a T_0[(1+1.608w)\ln(\frac{1+1.608w_0}{1+1.608w}) + 1.608w\ln(\frac{1.608w}{1.608w_0})]] \tag{10}$$

$$\dot{E}x_{th,air} = \dot{m}_{air}[(\mathrm{cp}_{air} + w.cp_v).(\mathrm{T} - \mathrm{T}_0 - T_0 \ln(\frac{T}{T_0}))] \tag{11}$$

Where Ra is ideal gas constant which equal to (0.287 kJ/kg K),

- **The exergy at the level of desiccant wheel**

The exergetic balance at the desiccant wheel is written as follows:

$$\dot{E}x_1 + \dot{E}x_8 = \dot{E}x_2 + \dot{E}x_9 + \dot{E}x_{Loss,Dw} + \dot{E}x_{dest.Dw} \tag{12}$$

$$\dot{E}x_{Loss,Dw} = \dot{Q}_{Loss}(1 - \frac{T_0}{T_{Dw}}) \tag{13}$$

T_{Dw} is the average temperature of the desiccant

$$T_{Dw} = \frac{T_1 + T_2 + T_8 + T_9}{4} \tag{14}$$

The exergetic balance at the desiccant wheel is written as follows:

$$\dot{m}_{air}h_1 + \dot{m}_{air}h_8 = \dot{m}_{air}h_2 + \dot{m}_{air}h_9 + \dot{Q}_{Loss} \tag{15}$$

The exergy efficiency of the desiccant wheel is determined by the following equation:

$$\varepsilon_{Dw} = \frac{\dot{E}x_2 - \dot{E}x_1}{\dot{E}x_8 - \dot{E}x_9} \tag{16}$$

- **The exergy at the level of Heat exchanger**

The exergetic balance at the heat exchanger is written as follows:

$$\dot{E}x_2 + \dot{E}x_6 = \dot{E}x_3 + \dot{E}x_7 + \dot{E}x_{Loss.exch} + \dot{E}x_{dest.exch} \quad (17)$$

$$\dot{E}x_{Loss,\mathrm{exch}} = \dot{Q}_{Loss}(1 - \frac{T_0}{T_{exch}}) \quad (18)$$

TDw is the average temperature of the desiccant wheel

$$T_{exch} = \frac{T_2 + T_3 + T_6 + T_7}{4} \quad (19)$$

The exergetic balance at the desiccant wheel is written as follows:

$$\dot{m}_{air} h_2 + \dot{m}_{air} h_6 = \dot{m}_{air} h_3 + \dot{m}_{air} h_7 + \dot{Q}_{Loss.exch} \quad (20)$$

The exergy efficiency of the heat exchanger is determined by the following equation:

$$\varepsilon_{ex,\mathrm{exch}} = \frac{\dot{E}x_7 - \dot{E}x_6}{\dot{E}x_2 - \dot{E}x_3} \quad (21)$$

- **The exergy at the level of the first Humidifier (Evaporative cooler 1)** (Tavakol et al.,2018)

$$\dot{E}x_{dest,Hum1} = \dot{E}x_3 - \dot{E}x_4 + \dot{E}x_{water} \quad (22)$$

$$\dot{E}x_{water} = \dot{m}_{air}(w_4 - w_3).[\mathrm{cp}_{water}(T - T_0) - T_0.\mathrm{cp}_{water}\ln(\frac{T}{T_0})] - R_v.T_0\ln(\varphi_0)] \quad (23)$$

The exergy efficiency of the evaporative cooler 1 is determined by the following equation:

$$\varepsilon_{ex,\mathrm{Hum1}} = \frac{\dot{E}x_4}{\dot{E}x_3 + \dot{E}x_{water}} \quad (24)$$

- **The exergy at the level of the second humidifier (Evaporative cooler 2)** (Tavakol et al.,2018)

$$\dot{E}x_{dest,Hum2} = \dot{E}x_5 - \dot{E}x_6 + \dot{E}x_{water} \quad (25)$$

$$\dot{E}x_{water} = \dot{m}_{air}(w_6 - w_5).[\mathrm{cp}_{water}(T - T_0) - T_0.\mathrm{cp}_{water}\ln(\frac{T}{T_0})] - R_v.T_0\ln(\varphi_0)] \quad (26)$$

The exergy efficiency of the evaporative cooler 2 is determined by the following equation:

$$\varepsilon_{ex,\mathrm{Hum2}} = \frac{\dot{E}x_6}{\dot{E}x_5 + \dot{E}x_{water}} \quad (27)$$

- **The exergy at the level of heat source** (Tavakol et al.,2018)

$$\dot{E}x_{dest,\mathrm{hs}} = \dot{E}x_7 - \dot{E}x_8 + \dot{Q}_{req}.(1 - \frac{T_0}{T_{reg}}) \quad (28)$$

$$\dot{Q}_{reg} = \dot{m}_{air}.(\mathrm{H}_8 - \mathrm{H}_7) \quad (29)$$

The exergy efficiency of the heat source is determined by the following equation:

$$\varepsilon_{ex.Hs} = \frac{\dot{E}x_8 - \dot{E}x_7}{\dot{Q}_{req}(1 - \frac{T_0}{T_{reg}})} \quad (30)$$

The thermal exergy of the overall system is determined by the following equation (Tavakol et al.,2018, kanoglu et al ., 2004))

$$COP_{ex,th} = \frac{\dot{E}x_{cooling}}{\dot{E}x_{heat}} \quad (31)$$

$$\dot{E}x_{cooling} = \dot{E}x_4 - \dot{E}x_1 \quad (32)$$

$$\dot{E}x_{heat} = \dot{E}x_8 - \dot{E}x_7 \quad (33)$$

Case Study

The energy and exergy analyses of desiccant cooling system on ventilation and recirculation cycles has been evaluated for the weather of Gabes city which is located in the south-eastern part of Tunisia. The outside air conditions are: Temperature 30.9 °C and relative humidity 71%. The return air conditions are 16g/kg humidity ratio and temperature 26°C. A constant mass flow of the process and the regeneration air which are equal to 0.125 kg /s is used. The temperature and humidity ratio of dead state air are the ambient air conditions .The values of ε_1 and ε_2 of desiccant wheel considered 0.05 and 0.95 respectively (Joudi et al ., 2001, Scheridan and Mitchell 1985). Moreover, the efficiency of heat exchanger is 1, also the efficiency of humidifiers is 1. Additionally, the regeneration temperature T8 is constant which equal to 60°C.

RESULTS AND DISCUSSION

The results are presented as curves and tables. The values determined by the model was compared with experimentally results found by konglu et al., (2014). Figure 3 shows that the obtained results by MATLAB software are in good agreement with the experimental data.

The propriety of air in different states in ventilation and recirculation cycles has been displayed in Table 1 and 2 respectively. Also, the psychrometric chart of ventilation and recirculation cycles solid desiccant air conditioning system has been presented in Figures 4 and 5 respectively.

The thermodynamic property of the desiccant cooling system is well understood when we know the types of destruction exergy which have classified into three categories.The first category is the external energy which is provided by the difference temperature and humidity ratio between the air conditioning system and the external environment. The second category is due to the large temperature difference between the two limits of the sensible heat exchanger and of the heat source. The third category includes the exergy destruction provided by the desiccant wheel and evaporative coolers.

Figure 3. Comparison between the results obtained by the model and experimental results of Konglu et al. (2004)

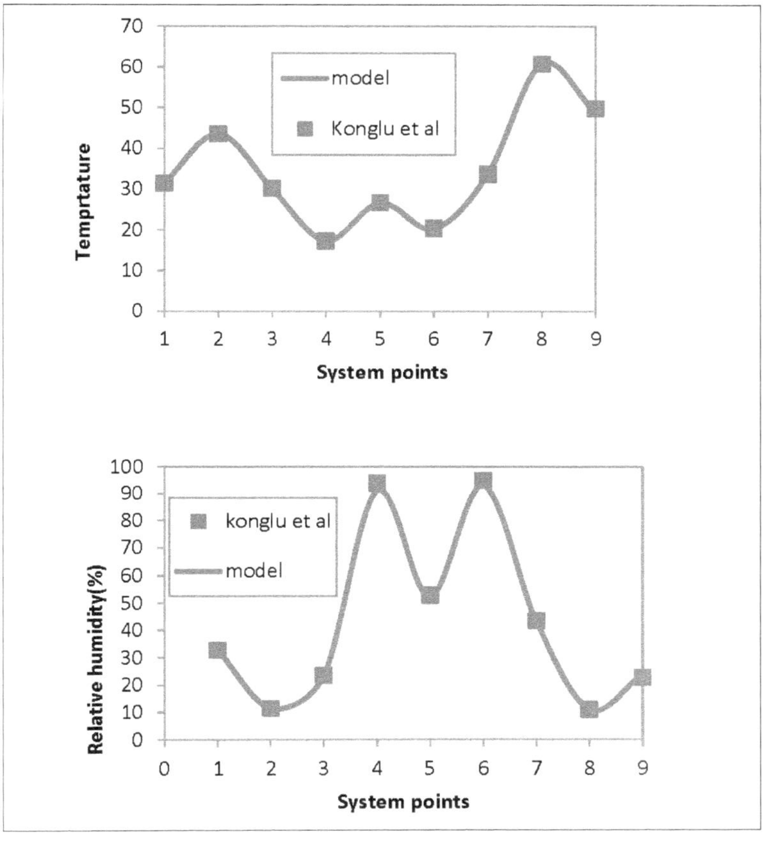

Table 1. proprieties of air in different states in ventilation cycle

POINT	T(°C)	w(kg/kg)	h (kJ/kg)	$\dot{E}x$ (kw)
1	30.9	0.02	82.21	0
2	52.72	0.0117	83.51	0.1265
3	22.62	0.0117	52.58	0.0466
4	18.48	0.0134	52.58	0.0521
5	26	0.016	66.91	0.0125
6	22.62	0.0174	66,91	0.0179
7	52.72	0.0174	98.14	0.0986
8	60	0.0174	105.69	0. 173
9	38.28	0.0256	104.40	0.0233

Table 2. Proprieties of air in different states in recirculation cycle

POINT	T(°C)	w(kg/kg)	h(kJ/kg)	$\dot{E}x$ (kw)
1	26	0.016	66.91	0.0125
2	45.74	0.0091	69.53	0.1011
3	26.59	0.0091	49.95	0.0641
4	17.76	0.0127	49.95	0.0613
5	30.9	0.02	82.21	0
6	26.59	0.0218	82.21	0.0053
7	45.74	0.0218	102.24	0.0431
8	60	0.0218	117.14	0.1725
9	40.51	0.287	114.52	0.0487

Figure 4. Psychrometric chart of ventilation cycle (Belguith et al., 2020)

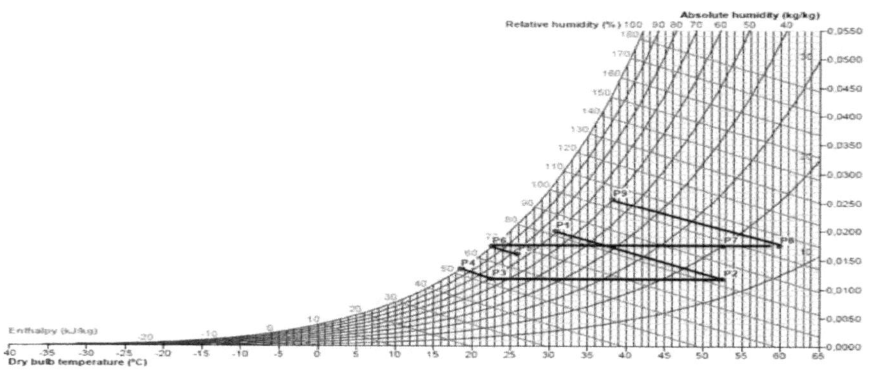

Figure 6 gives the destruction exergy distribution of different component of solid desiccant air conditioning system in ventilation cycle. It can be seen that the desiccant wheel have the highest which equal to 57%. The desiccant wheel has bigger temperature difference between the process and regeneration air. The irreversibility's in the rotary desiccant can be explained by the causes mentioned in the following. The inlet and the outlet temperature on the process air side (1-2) are 30.9.4°C and 52.72°C, in the regeneration side, the air enters and leaves the desiccant dehumidifier at a temperature of 60°C and 38.28 °C, respectively. Also, The difference of humidity ratio between the process and regeneration stream due to the major difference of vapor pressure between the process air flow and the desiccant material affect the desiccant wheel exergetic performance . The using of

Figure 5. Psychrometric chart of recirculation cycle (Belguith et al., 2020)

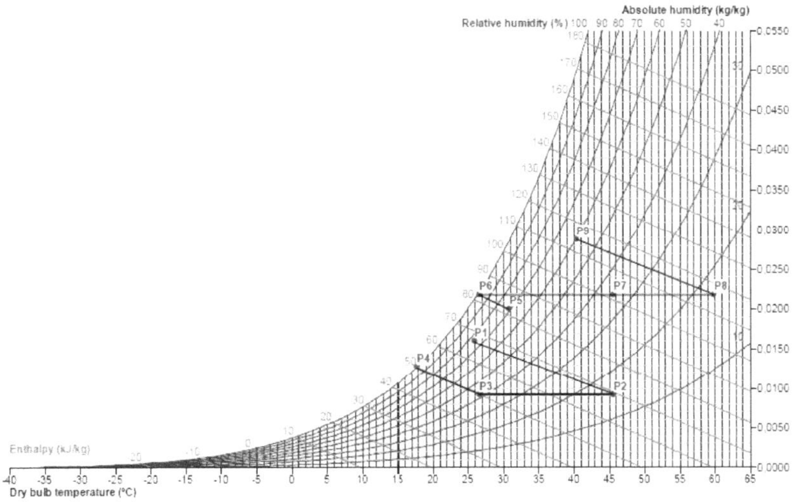

desiccant bed instead of desiccant wheel represent alternative solution to overcome the exergy problem caused by rotary desiccant Moreover, when the desiccant wheel is operated with adsorbent material a low temperature, the destruction exergy may be decreased and the exergy efficiency of the system improved.

The exergy destructions occurring in the evaporative cooler 1 (10.85%) is even more than those in the evaporative cooler 2 (7.4%). The air enters and leaves the evaporative cooler1 (3 -4) at a temperature and humidity ratio 22.62°C, 0.0134 kg/kg and 18.48° C, 0.0117 kg /kg respectively. In the evaporative cooler 2, Therefore, larger temperature and humidity ratios difference between the process inlet and outlet evaporative cooler air streams and the greater evaporation rate are the main causes for irreversibilities. Improvement in the exergy performance of the evaporative cooler 1 can be achieved by decreasing the size and capacity of the evaporative cooler 1.

The exergy destruction coefficient of the third category which includes the desiccant wheel and the evaporatives coolers is 75.25% (figure 7).

The heat source is the second component with important exergy destruction which equal to 24.67%. The causes of irreversibility for heat source are explained by high temperature difference between inlet and outlet of heat source. The air enters and leaves the heat source (7-8) at a temperature of 52.72°C and 60 °C, respectively. Basically, during the dehumidification process the outside air through the desiccant wheel where its moisture is removed and its temperature increased by the adsorption heat effect. The dehumidification process essentially due to the major difference of vapor pressure between the process and regeneration streams that depends of regeneration temperature. Therefore, for further moisture removal from the rotary desiccant the regeneration temperature should be higher. The exergy destruction in the heat source usually depends on the method adopted to regenerate the desiccant wheel and the regeneration requirement.

The heat exchanger has the lowest exergy destruction (0.06%). There are serval reasons for this. For the heat exchanger, the inlet and the outlet temperature on the process air side (2-3) are 52.72°C and 22.62°C, whereas those on the regeneration air side (6-7) are 22.62°C and 52.72°C . The heat exchanger can completely recover the heat the tow streams of air. This may seem expected since in the present analysis the heat exchanger effectiveness is 1 an exceedingly high value for a heat exchanger effectiveness. Nevertheless, this is justified the exergy destruction in the heat exchanger is directly determined by its effectiveness and approaches reversible at high effectiveness (Abbassi et al.,2017)

The exergy destruction coefficient of the second category which includes the heat source and the heat exchanger is 24.73% (figure 7).

Figure 8 gives the destruction exergy distribution of different component of solid desiccant air conditioning system in recirculation cycle. The heat source has found with the highest exergy destruction (38.83%). The causes of irreversibility for heat source the large difference temperature between the two terminals of the heat source. The temperature in the heat source towing terminals (7-8) are 45.74 ° C and 60 ° C, respectively. Therefore, to minimize the destruction of exergy in the heat source, you should use a desiccant material with a low regeneration temperature.

The desiccant wheel is second component with percentage of exergy destruction achieved 33.08%. The causes of irreversibility for the rotary desiccant are the difference temperature and humidity ratio between the process and the regeneration streams. These results are consistent with the experimental result of Rafique et al. (2016) who found that the desiccant wheel and heat source are the main sources of entropy generation.

The evaporative cooler 1 (first humidifier) is the second last with exergy destruction 22.9% more important than at evaporative cooler 2 (5.18%). The evaporative cooler has important temperature and humidity ratios difference between the process inlet and outlet humidifier air streams are the main causes for irreversibilities which due also because of the greater evaporation rate. The heat exchanger is the last with exergy destruction 0.014%. The exergy destruction coefficient of the third category which includes the desiccant wheel and the evaporatives coolers is 61.6% while the second category which includes the heat source and the heat exchanger is 38.84% (figure 9).

It can be seen from thermodynamic analyses of desiccant cooling cycle the absence of the first category of exergy destruction this can be explain by the fact that the present work interested by the exergy destructions due to irreversibilities during the internal processes (Figure 7 and 9).

Figure 6. Percentage of exergy destruction of different component of ventilation cycle

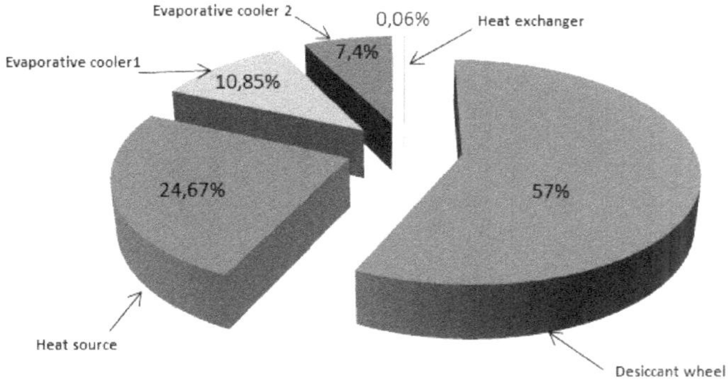

Figure 7. Percentage of exergy destruction of different categories in the ventilation cycle

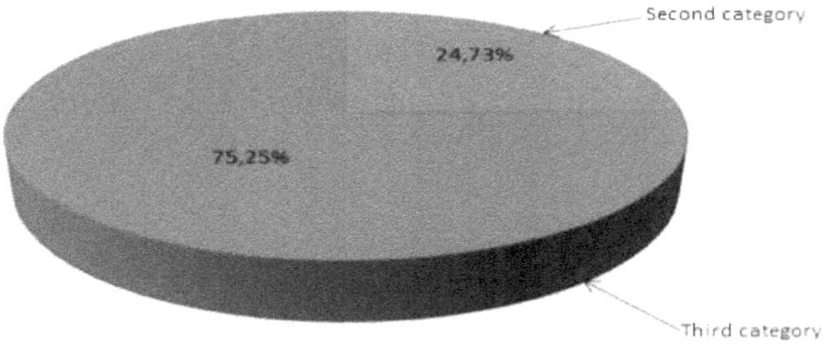

Figure 8. Percentage of exergy destruction of different component of recirculation cycle

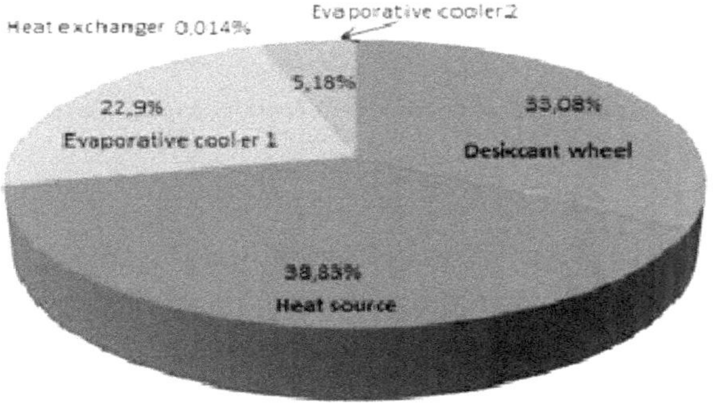

Figure 9. Percentage of exergy destruction of different categories in the recirculation cycle

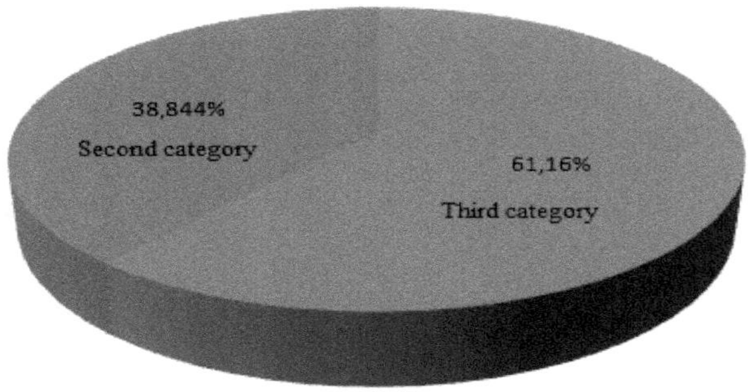

The exergy efficiency of desiccant wheel, heat exchanger, evaporative cooler 1, evaporative cooler 2 and heat source of ventilation and recirculation cycles are illustrated in Figure 10 and 11 respectively.

Figure 10. Exergy efficiency relative to each component of ventilation cycle

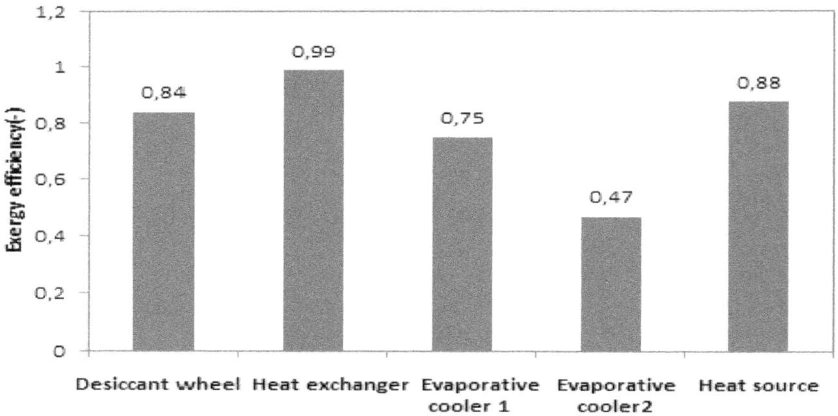

Figure 12 and 13 show the energetic and exergitic performance of desiccant cooling system in ventilation and recirculation cycles. Concerning the ventilation cycle, the energetic and exergetic performance are 1.89 and 0.7 respectively. Concerning the recirculation cycle, the energetic and exergetic performance are 1.13 and 0.38 respectively. These results can be explain by the fact, in the present analysis. The effectiveness of components is selected based on high efficient system in order to provide an upper limit for the COP and COP_{ex} values, which could be obtained with ideal components.

Figure 11. Exergy efficiency relative to each component of recirculation cycle

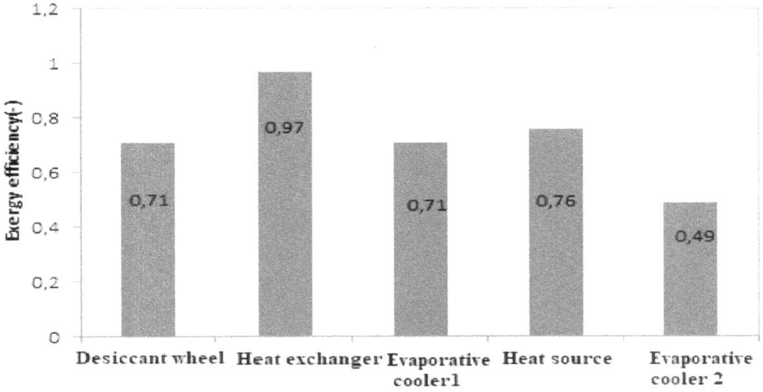

Figure 12. Energy and exergy performance of ventilation cycle

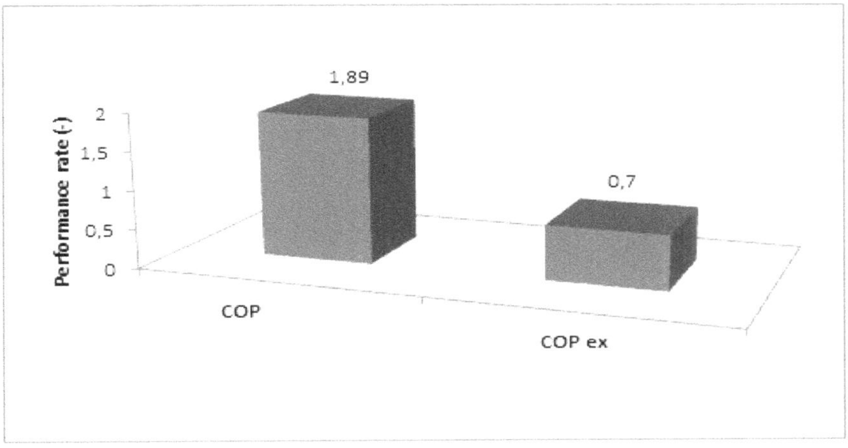

SENSITIVITY OF EXERGY DESTRUCTION IN DESICCANT WHEEL AND HEAT SOURCE

The dead-state point is a very significant parameter for the thermodynamic analysis (Goncalves et al.,2014).The sensitivity to energy destruction of the desiccant wheel and the heat exchanger with the reference temperature (dead state) and the humidity ratio was explored .As a result, the exergy destruction decreases when the reference temperature is increased from 30 °C to 40 °C and the humidity ratio increases from 10 g / kg to 20 g / kg. Therefore, when the outside conditions air (temperature and humidity ratio) increases, the potential of improving of the desiccant wheel and heat source decreases. Meaning the actual exergy destruction in desiccant wheel and heat source is less than the theoretical exergy destruction.

Figure 13. Energy and exergy performance of recirculation cycle

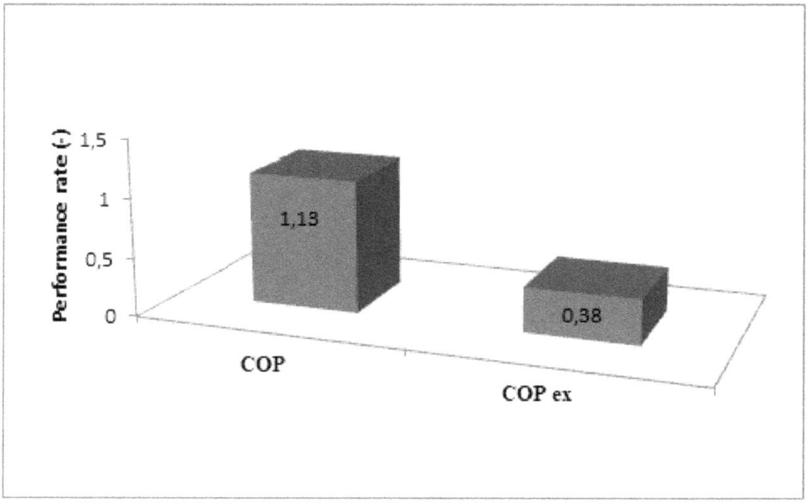

Figure 14. Sensibility of exergy destruction in desiccant wheel to dead state

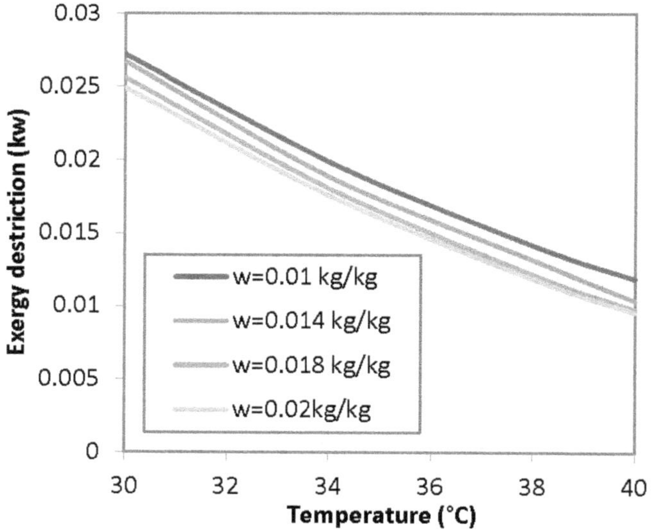

Figure 15. Sensibility of exergy destruction in heat source to dead state

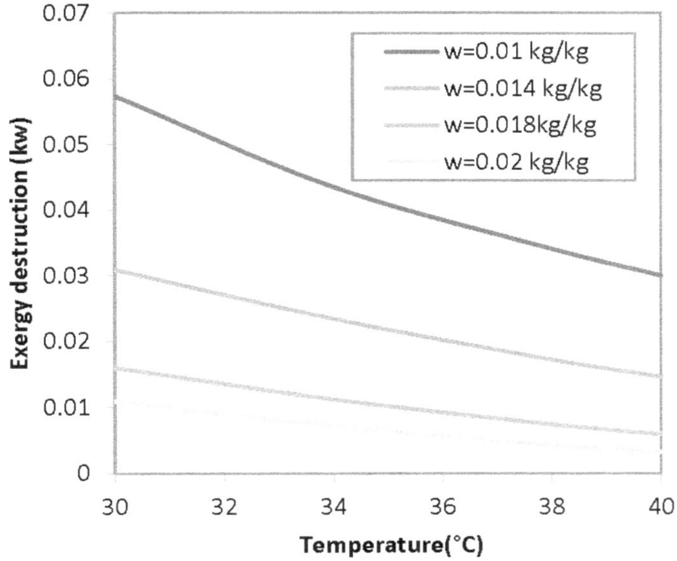

CONCLUSION

In this paper, A theoretical model for desiccant cooling system on ventilation mode is developed and experimentally validated. The system performance is evaluated based on the first and second law of thermodynamics. The influences of the individual irreversible processes of each component on the performance of desiccant cooling system under ventilation and recirculation modes are studied.

The major conclusions are as follows:

- The coefficient of performance of desiccant cooling system on ventilation mode which equal to 1.89 is higher than the corresponding exergy coefficient of performance 0.7.
- The coefficient of performance of desiccant cooling system on recirculation mode which equal to 1.13 is greater than the corresponding exergy coefficient of performance 0.38.
- The desiccant wheel and heat source responsible of major exergy destruction in solid air conditioning system.
- The exergy destruction decreases when the reference temperature increases which indicating that the actual exergy destruction in desiccant wheel and heat source is less than the theoretical exergy destruction.

NOMENCLATURE

COP ...... Coefficient of performance
COPex.th ...... Exergetic performance
Cp ...... Specific heat (kJ/kgK)
$\dot{Ex}$ ...... Exergy rate (kw)
F1 ...... First potential function characteristic
F2 ...... Second potential function characteristic
H ...... Specific enthalpy (kJ/kg)
Q ...... Heat rate (kw)
QCOOL ...... Specific cooling load (kJ/kg)
QRegeneration ...... Specific supplied heat (kJ/kg)
$\dot{m}$ ...... Mass flow rate (kg/s)
Ra ...... Gas constant (kJ/kgK)
Rv ...... Water vapour constant (kJ/kgK)
T ...... Air temperature (K, °C)
w ...... Absolute humidity of the air (kg water/kg dry air)

Greek Letters

ε ...... Effectiveness
φ ...... Relative humidity (%)

Subscripts and Superscripts

0 ...... Reference environment (dead state)
air ...... Air
ch ...... Chemical
cooling ...... Cooling
dest ...... Destruction
Dw ...... Desiccant wheel
exch ...... Heat exchanger
ex ...... exergy
heat ...... Heat
hum ...... Humidifier
Hs ...... Heat source
loss ...... Loss
th ...... Thermal
reg ...... Regeneration
sat ...... Saturation

water …… water

REFERENCES

Abbassi, Y., Ehsan, B., & Hossein, A. (2017). Comparative performance analysis of differ-ent solar desiccant dehumidification systems. *Energy and Building*, *150*, 37–51. doi:10.1016/j.enbuild.2017.05.075

Ali, M., Vukovic, V., Muhammad Ali, H., & Ahmed Sheikh, N. (2018). Performance analysis of solar-assisted desiccant cooling system cycles in world climate zones. *Journal of Solar Energy Engineering*, *140*(4), 041009. doi:10.1115/1.4039426

Ali, M., Vukovic, V., Sheikh, N. A., & Ali, H. M. (2015). Performance investigation of solid desiccant evaporative cooling system configurations in different climatic zones. *Energy Conversion and Management*, *97*, 323–339. doi:10.1016/j.enconman.2015.03.025

Belguith, Meddeb, & Ben Slama. (2020). Performance analysis of desiccant cooling systems in a hot and dry climate. *Euro-Mediterranean Journal for Environmental Integration*, *6*(2).

Belguith, Slama, Chaouachi, & Meddeb. (2020). Performance analysis of hybrid solid desiccant - vapor compression air conditioning system: Application and comparative study. *2020 11th International Renewable Energy Congress (IREC) Hammamet, Tunisia, 2020*, 1-6.

Bourdoukan, P., Wurtz, E., & Joubert, P. (2010). Comparison between the conventional and recirculation modes in desiccant cooling cycles and deriving critical efficiencies of components. *Energy*, *35*(2), 1057–1067. doi:10.1016/j.energy.2009.06.021

Caliskan, H., Lee, D. Y., & Hong, H. (2019). Enhanced thermodynamic assessments of the novel desiccant air cooling system for sustainable energy future. *Journal of Cleaner Production*, *211*, 213–221. doi:10.1016/j.jclepro.2018.11.174

Carpinlioglu, M. Ö. (2015). A comment on the concept of entransy (versus exergy) for The performance assessment of a desiccant cooling system. *Energy and Building*, *101*, 163–167. doi:10.1016/j.enbuild.2015.05.003

Chen, H., Yang, H., & Luo, Y. (2018). Investigation on solar assisted liquid desiccant dehumidifier and evaporative cooling system for fresh air treatment. *Energy*, *143*, 114–127. doi:10.1016/j.energy.2017.10.124

Daou, K., Wang, R. Z., & Xia, Z. Z. (2006). Desiccant cooling air conditioning: A review. *Renewable & Sustainable Energy Reviews*, *10*(2), 55–77. doi:10.1016/j.rser.2004.09.010

Dezfouli, M. M. S., Mat, S., Pirasteh, G., Sahari, K. S. M., Sopian, K., & Ruslan, M. H. (2014). Simulation analysis of the four confgurations of solar desiccant cooling system using evaporative cooling in tropical weather in Malaysia. *International Journal of Photoenergy*, *2014*, 1–14. doi:10.1155/2014/843617

Enteria, N., Yoshino, H., Mochida, A., Satake, A., & Takaki, R. (2015). Exergo-economic performances of the desiccant evaporative air-conditioning system at different regeneration and reference temperatures. *International Journal of Refrigeration*, *56*, 81–98. doi:10.1016/j.ijrefrig.2014.11.007

Goncalves, P., Angrisani, G., Sasso, M., Rodrigues Gaspar, A., & Gamerio da Silva, M. (2014). Exergetic analysis of a desiccant cooling system: Searching for performance improvement opportunities. *International Journal of Energy Research*, *38*(6), 714–727. doi:10.1002/er.3076

Hürdoğan, E., Büyükalaca, O., Yılmaz, T., & Hepbaşlı, A. (2010). Experimental investigation of a novel desiccant cooling system. *Energy and Building*, *42*(11), 2049–2060. doi:10.1016/j.enbuild.2010.06.014

Hwang, W. B., Choi, S., & Lee, D. Y. (2017). In-depth analysis of the performance of hybrid desiccant cooling system incorporated with an electric heat pump. *Energy*, *118*, 324–332. doi:10.1016/j.energy.2016.12.007

Joudi, K. A., & Dhaidan, N. (2001). Application of solar assisted heating and desiccant cool-ing systems for a domestic building. *Energy Conversion and Management*, *42*(8), 995–1022. doi:10.1016/S0196-8904(00)00111-4

Jurinak, J. J. (1982). *Open cycle solid desiccant cooling component models and system simulations* [PhD thesis]. University of Wisconsin-Madison, Madison, WI.

Kanoglu, M., Çarpinlioglu, M. Ö., & Yildirim, M. (2004). Energy and exergy analyses of an experimental open-cycle desiccant cooling system. *Applied Thermal Engineering*, *24*(5-6), 919–932. doi:10.1016/j.applthermaleng.2003.10.003

Maclaine-Cross, I. L., & Banks, P. J. (1972). Coupled heat and mass transfer in regenerators— Prediction using an analogy with heat transfer. *International Journal of Heat and Mass Transfer*, *15*(6), 1225–1242. doi:10.1016/0017-9310(72)90187-1

Panaras, G., Mathioulakis, E., Belessiotis, V., & Kyriakis, N. (2010). Experimental validation of a simplifed approach for a desiccant wheel model. *Energy and Building*, *42*(10), 1719–2172. doi:10.1016/j.enbuild.2010.05.006

Pandelidis, D., Pacak, A., Cichon, A., Anisimov, S., Drąg, P., Vager, B., & Vasilijevc, V. (2018). Multi-Stage desiccant cooling system for moderate climate. *Energy Conversion and Management*, *177*, 77–90. doi:10.1016/j.enconman.2018.09.061

Rafique, M., Gandhidasan, P., Luai, M., Hadhrami, A., & Rehman, S. (2016). Energy, exergy and anergy analysis of a solar desiccant cooling system. *Journal of Clean Energy Technologies*, *4*(1), 78–83. doi:10.7763/JOCET.2016.V4.257

Scheridan, J. C., & Mitchell, J. W. (1985). A hybrid solar desiccant cooling system. *Solar Energy*, *34*(2), 187–193. doi:10.1016/0038-092X(85)90179-3

Shahzad, M. K., Chaudhary, G. Q., Ali, M., Sheikh, N. A., Khalil, M. S., & Rashid, T. U. (2018). Experimental evaluation of a solid desiccant system integrated with cross flow Maisotsenko cycle evaporative cooler. *Applied Thermal Engineering*, *128*, 1476–1487. doi:10.1016/j.applthermaleng.2017.09.105

Sphaier, L. A., & Nóbrega, C. E. L. (2012). Parametric analysis of components effectiveness on desiccant cooling system performance'. *Energy*, *38*(1), 157–166. doi:10.1016/j.energy.2011.12.019

Tavakol, P., & Behbahaninia, A. (2018). Presentation of two new two-stage desiccant cooling cycles based on heat recovery and evaluation of performance based on energy and exergy analysis. *Journal of Building Engineering*, *20*, 455–466. doi:10.1016/j.jobe.2018.08.002

Threlkeld, J. L., Ramsey, J. W., & Kuehn, T. H. (1998). Thermal environmental engineering (3rd ed.). Prentice-Hall.

Xiong, Z. Q., Dai, Y. J., & Wang, R. Z. (2010). Development of a novel two-stage liquid desiccant dehumidification system assisted by CaCl2 solution using exergy analysis method. *Applied Energy*, *87*(5), 1495–1504. doi:10.1016/j.apenergy.2009.08.048

Zhu, J., & Chen, W. (2014). Energy and exergy performance analysis of a marine rotary desiccant air-conditioning system based on orthogonal experiment. *Energy*, *77*, 953–962. doi:10.1016/j.energy.2014.10.014

Chapter 6
Entropy Generation Minimization in a Vertical Porous Channel

Amel Tayari
Institut Supérieur des Sciences Appliquées et Technologies de Gabès, Tunisia

Mourad Magherbi
Institut Supérieur des Sciences Appliquées et Technologies de Gabès, Tunisia

ABSTRACT

This chapter is mainly focused on the minimization of the total entropy generation in a thermodynamic system, which concerns the heating of water when it passes through a saturated porous media. The heating process is assumed by an array of heating tubes immerged on the porous media and perpendicular to the water flow direction. This irreversibility calculation and minimization is carried out in dimensional form in order to have a real idea about the entropy production in the system. Also, the authors focus on several points regarding the rigor of entropy generation calculation, when they pass from dimensional form to dimensionless form.

INTRODUCTION

Entropy generation analyses have become a very useful tool for the thermodynamic systems design in order to decrease energy degradation. The sources of irreversibility are numerous and depend on constraints which act on the flow. These sources include heat gradient, concentration gradient, magnetic field effect, fluid viscous dissipation

DOI: 10.4018/978-1-7998-8801-7.ch006

effect and Darcy viscous dissipation effect. This latter cause of irreversibility appears in the fluid flow through a porous media.

Many studies have been published on entropy generation. These studies are covered different flow from laminar to turbulent and different geometric configuration and different causes of irreversibility and different media. (Bejan 1982, 1996) illustrated that the flow parameter could be selected in order to minimize the irreversibility associated with a specific convective heat transfer processes. For a mixed convective flow, Mahmud et al. (2003) gave a detailed analysis of the entropy generation nature and sources of irreversibility in a vertical non-porous channel with transverse hydromagnetic effect. Mahmud and Fraser (2002) studied the entropy generation minimization concept for different geometric configurations, flow situations, and thermal boundary conditions.

MATHEMATICAL MODELING

The schematic of the system under consideration is shown in Figure1. It consists of a laminar two-dimensional mixed convective flow inside a vertical saturated porous channel. The fluid is assumed to be incompressible.

Figure1. Physical problem

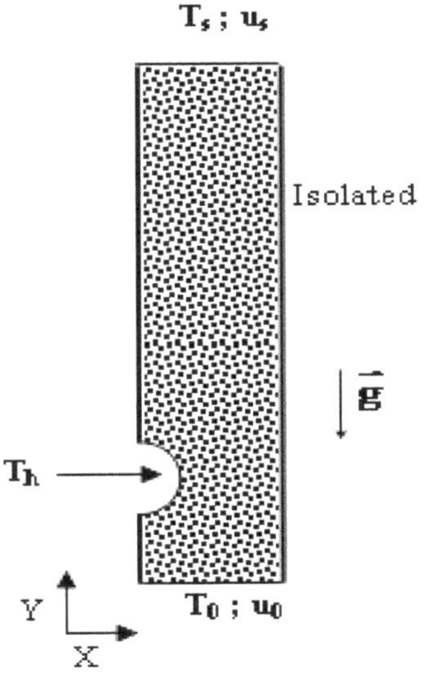

The local thermal equilibrium model is used, therefore the considered effective thermal conductivity can be written as the weighted arithmetic mean of the solid phase and the fluid phase conductivities.

$$k_{eff} = (1 - \varepsilon)k_s + \varepsilon k_f \tag{1}$$

Where, k_f and k_s are the thermal conductivities of the fluid and of the solid respectively.

Using the Darcy-Brinkman model, the Navier-Stokes and the energy balance equations are:

$$\nabla.\mathbf{u} = 0 \tag{2}$$

$$\frac{\rho}{\varepsilon^2}(\mathbf{u}\nabla)\mathbf{u} = -\nabla.\mathbf{p} + \frac{\mathbf{u}}{\varepsilon}\nabla.(\nabla\mathbf{u}) - \frac{\mu}{K}\mathbf{u} + \rho\mathbf{g} \tag{3}$$

$$\rho.c_p\mathbf{u}.\nabla T = k_{eff}.\Delta T \tag{4}$$

Where, μ is the effective viscosity, K is the medium permeability and ε is the medium porosity.

The initial conditions are:

- P=u=v=0 and T=298.15k in the whole domain of the porous medium

The boundary conditions are:

-At the isothermal semicircle walls: $T=T_h$.
-The other walls are thermal insulators
-At the inlet channel: u=5mm/s, v=0

ENTROPY GENERATION

The dimensional entropy generation equation in a porous medium is given by:

$$S_t = \frac{k_{eff}}{T^2}\left[\left(\frac{\partial T}{\partial x}\right)^2 + \left(\frac{\partial T}{\partial y}\right)^2\right] + \frac{\mu}{T.K}\left(v_x^2 + v_y^2\right) + \frac{\mu}{T}\left[2\left(\frac{\partial v_x}{\partial x}\right)^2 + 2\left(\frac{\partial v_y}{\partial y}\right)^2 + \left(\frac{\partial v_x}{\partial y} + \frac{\partial v_y}{\partial x}\right)^2\right] \tag{5}$$

On the right-hand side of equation (5), the first term represents the heat transfer entropy generation (Si_Ther), the second is the Darcy fluid entropy generation (Si_Darcy) and the third represents the clear fluid viscous entropy generation (Si_Visq).

The total entropy generation for the entire channel is obtained by integrating equation (5):

$$S_t = \int_V S_l dV \tag{6}$$

The dimensionless variables are:

$$X = \frac{x}{H};\quad Y = \frac{y}{H};\quad V_x = \frac{v_x}{u_0};\quad V_y = \frac{v_y}{u_0};\quad \theta = \frac{T - T_0}{T_1 - T_0} \tag{7}$$

Using these dimensionless variables, dimensionless entropy generation equation can be written as:

$$S_{lA} = \left(\frac{\partial\theta}{\partial X}\right)^2 + \left(\frac{\partial\theta}{\partial Y}\right)^2 + \frac{Br^*}{Da}\left(V_x^2 + V_y^2\right) + Br^*\left[2\left(\frac{\partial V_x}{\partial X}\right)^2 + 2\left(\frac{\partial V_y}{\partial Y}\right)^2 + \left(\frac{\partial V_x}{\partial Y} + \frac{\partial V_y}{\partial X}\right)^2\right] \tag{8}$$

NUMERICAL RESOLUTION PROCEDURE

In this work COMSOL multiphysics is introduced to solve momentum and energy equations. It is also noted that COSMOL multiphysics can be intended on the idea of discretization of equations using the finite element method (Petrov–Galerkin method). From the regarded, entropy generation due to thermal gradient, viscous fluid effect and the Darcy viscous fluid dissipation are evaluated. Many other techniques for solving the above equations are available in the literature (Farhan et al. (2020), Alkasassbeh et al. 2019).

RESULTS AND DISCUSSIONS

We start with the investigation of the irreversibility in the porous medium described in the figure1. The flow enters in the channel with a temperature of 298.15k and passes through two vertical insulators walls. It is heated by an isothermal semicircle wall at fixed temperature equal to 428.15k. The center of the semicircle, with radius equal to 5cm, is placed at 15cm from the bottom of the channel. Results show that the total irreversibility is confined around the semicircle wall, as seen in figure 2. This is due to the fact that the temperature gradient and the fluid velocity are important in this region. In these conditions the flow develops a total entropy generation equal to171.963W/k. This total entropy generation is the sum of the heat transfer, Darcy viscous fluid and clear viscous fluid entropies generations which are equals to 112.548W/k, 59.415W/k and 2.274 10^{-7}W/k respectively. Remark that the largest contribution is due to the heat transfer, following by the Darcy fluid friction. The clear fluid viscous irreversibility is insignificant and remains negligible, so it will be not considered in this work. Under the considered operatory conditions, numerical results indicate that the outlet flow temperature and the average temperature of the whole porous medium are 344.175k and 333.35k respectively.

Figure 3 shows the variation of the thermal irreversibility versus the y- coordinate at x-coordinate equal to 5.2cm. As can be seen from this figure the thermal local entropy generation is zero from the inlet of the channel until reaching the intersection

Figure 2. Total entropy generation map

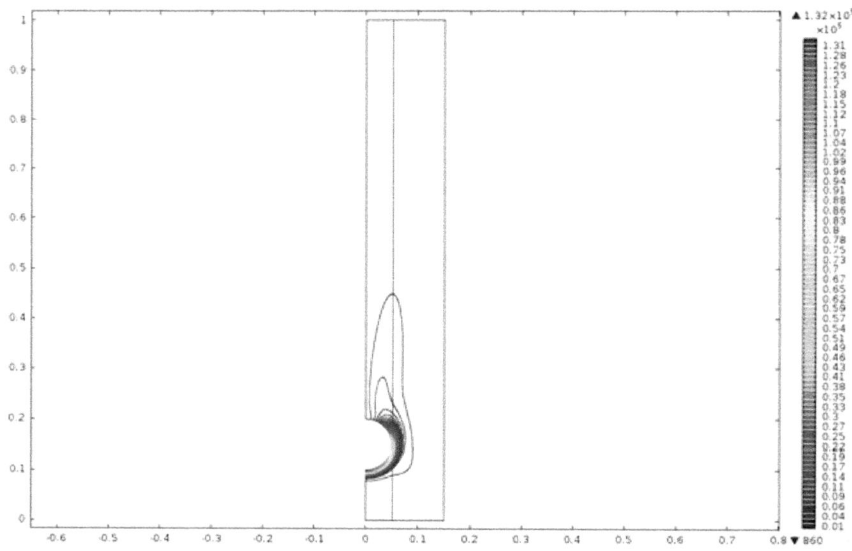

point, between the first half of the semicircle and the insulator vertical wall, where it considerably increases and reaches the maximum value of 60000W$m^{-3}k^{-1}$. Thermal local entropy generation decreases rapidly along the second half of the isothermal semicircle, after wards it continues to slightly decreasing until the outlet of the channel. Similar observation can be seen from the Figure 4 related to the Darcy fluid entropy generation, with the exception that the Darcy fluid irreversibility is not zero at the inlet of the channel. Also, the Darcy fluid irreversibility appears as a peak, which is in this case, centered on the isothermal semicircle wall (its maximum coincides with the center of the semicircle). The Main goal of this work is to minimize the entropy generation in the considered system.

After several tests, the adequate geometric configuration, with the three criteria cited above, is obtained by adding three other arrays of heating tubes (figure 5) and simultaneously reducing their heating temperature from 428.15k to 380.15k. In this

Figure 3. Thermal entropy production variation versus y-coordinate at x=5.2 cm

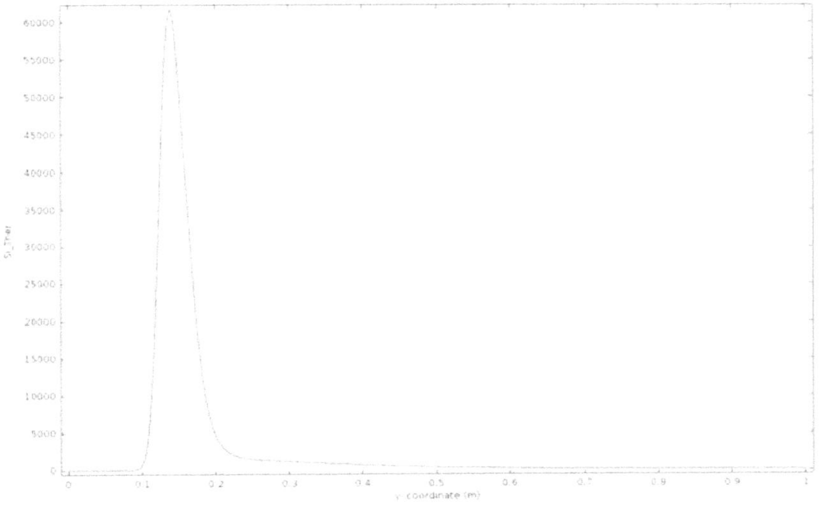

Figure 4. Darcy fluid entropy production variation versus y-coordinate at x=5.2 cm

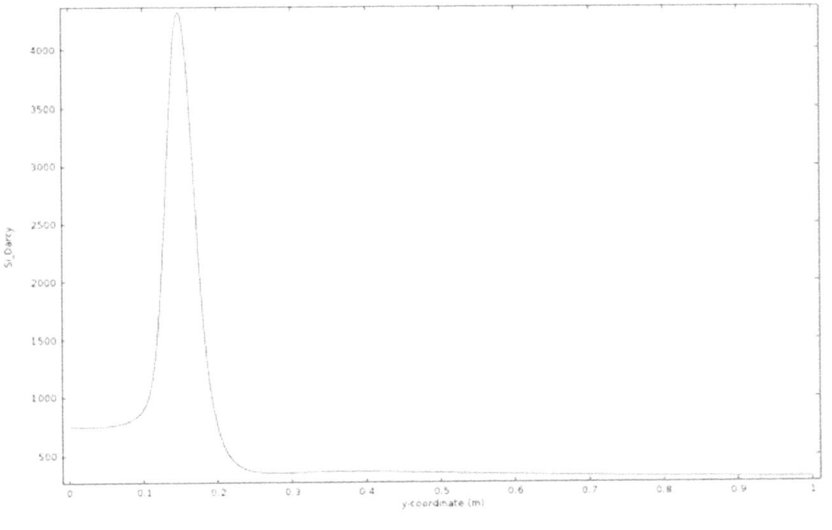

Figure 5. Geometric configuration of the second system

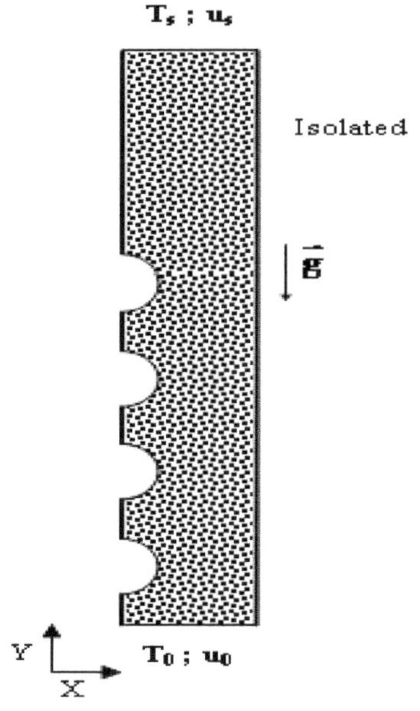

new design, the four semicircles have the same radius equal to 5cm, but their centers are placed at 10cm, 25cm, 40cm and 55cm from the bottom of the channel respectively. Results show that we have managed to diminish the total entropy generation in the porous medium from 171.963W/k to 139.675 W/k, which corresponds to a decrease of 18.77%. The outlet flow temperature and the average temperature of the whole porous medium are kept practically constant and equal to 344.044k and 333.125k respectively (insignificant variation). The inlet flow velocity remains identical and equal 5mms^{-1}. As seen in Figure 6, the total entropy generation is now confined around the four isothermal semicircle walls. This entropy generation is also due to the heat transfer and the Darcy fluid irreversibilities. This is well seen by the plot of the local heat transfer and the Darcy fluid irreversibilities in Figures 7 and 8 respectively. Figure 7 reveals that the pronounced peak of irreversibility related to the first geometric configuration is now shared into four peaks of irreversibility centered around the four isothermal semicircle walls with magnitudes equal to 18000 Wm^{-3}k^{-1}, 5000 Wm^{-3}k^{-1}, 2500 Wm^{-3}k^{-1} and 1700 Wm^{-3}k^{-1} from the bottom to the top of the saturated porous channel. The irreversibility reduction, between the two geometric configurations, close to the middle point of the first semicircle, where the local entropy generation is the highest, is about 70%. It's important to notice that the partition of the irreversibility, related to the new geometric configuration is not uniform, but accompanied by a diminution of 72%, 50% and 32% from the first to the second, from the second to the third and from the third to the fourth isothermal semicircle walls respectively. Similar observations of the Darcy fluid entropy generation can be seen from the Figure 8. As can be seen from this figure, there are four peaks of irreversibility centered on the active semicircle walls, with magnitudes equal to 3300 $Wm^{-3}k^{-1}$,2580 $Wm^{-3}k^{-1}$,2000 $Wm^{-3}k^{-1}$ and 1750 $Wm^{-3}k^{-1}$ from the bottom to the top of the porous channel.

The first peak of irreversibility exhibits a diminution of 23.25% compared with the only peak related to the first geometric configuration. Remark that, the partition of the Darcy fluid irreversibility is relatively more equitable then that associated to the heat transfer irreversibility. In fact, the reduction of irreversibility is equal to 21.8%, 22.5% and 12.5% when we pass from the first to the second, from the second the third and from the third to the fourth isothermal semicircle walls respectively.

Figure 6. Total entropy generation map

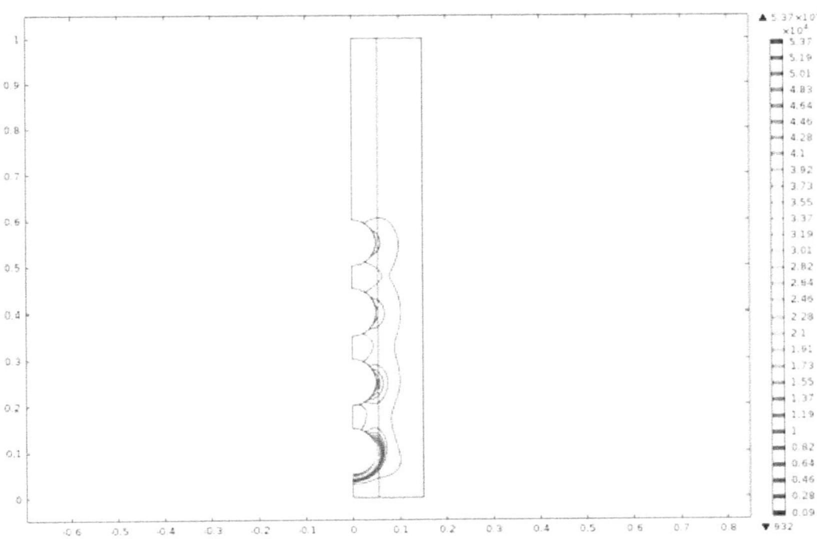

Figure 7. Thermal entropy production variation versus y-coordinate at x=5.2 cm

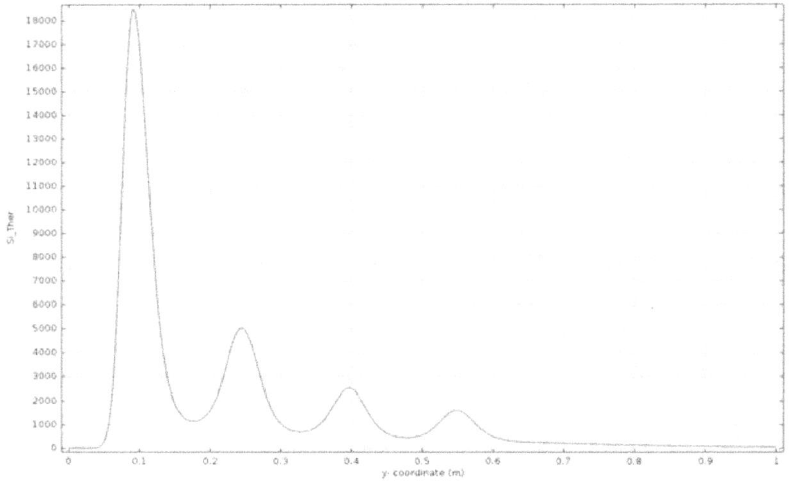

Figure 8. Darcy fluid entropy production variation versus y-coordinate at x=5.2 cm

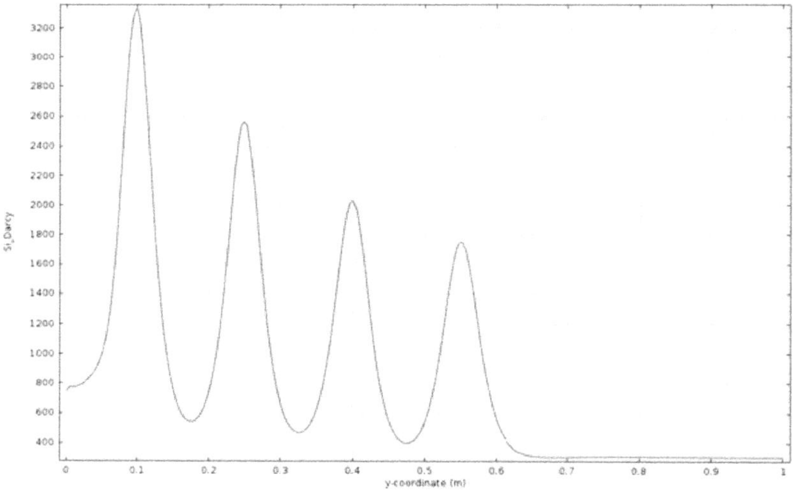

CONCLUSION

This work concerns a dimensional investigation and minimization of entropy generation in a saturated porous media. At first and for a given geometric configuration, the calculation and the contribution of each cause of the irreversibility on the total entropy generation was made. Also, the localization of the places where irreversibility is accentuated was deduced. In the same time, the average whole temperature of the porous medium and the outlet flow temperature are calculated. Secondly, a minimization operation was conducted, without disturb the operatory criteria of the system, which are chosen to be the average whole temperature of the porous medium, the outlet flow temperature and the inlet flow velocity. This operation, which affects the process design in its active walls, may leads to a partition of irreversibility accompanied by a reduction of its magnitude. In this context and under the fixed process criteria, the minimization of the entropy generation was successfully released by adding three others arrays of heating tubes and reducing their heat temperature. As a result, the new geometric configuration allows the partition of the irreversibility and simultaneously offers a decrease of 18.77% on the total entropy generation.

NOMENCLATURE

c_p ...... specific heat capacity (kJ/kg.K)
g ...... gravitational acceleration (m/s^2)
H ...... channel height (m)
K ...... permeability of the porous media (m^2)
k ...... thermal conductivity (J/m.s.K)
L .... channel length (m)
p ...... Pressure ($N.m^{-2}$)
T ...... temperature (K)
T_0 ...... inlet temperature (K)
T_s ...... outlet temperature (K)
T_h ...... temperature of the isothermal semicircle wall (K)
u_0 ...... inlet flow velocity($m.s^{-1}$)
u_s ...... outlet flow velocity($m.s^{-1}$)
S_l ...... entropy production per unit volume ($W.m^{-3}.K^{-1}$)
S_t ...... total entropy production ($W.K^{-1}$)
Si_Ther ...... thermal entropy production ($W.m^{-3}.K^{-1}$)
Si_Darcy ...... Darcy fluid entropy production ($W.m^{-3}.K^{-1}$)
Si_Visq ...... Clear fluid viscous entropy generation ($W.m^{-3}.K^{-1}$)
ΔT ...... temperature difference (T_h-T_0)
u ...... dimensional velocity vector ($m.s^{-1}$)
x, y ...... Cartesian coordinates (m)
v_x,v_y ...... velocity components in x and y directions respectively ($m.s^{-1}$)

Greek Symbols

ε ...... medium porosity
ρ ...... mass density ($kg.m^{-3}$)
μ ...... dynamic viscosity ($kg.m^{-1}s^{-1}$)

Subscripts

a ...... dimensionless
eff ...... effective
l ...... local
f ...... fluid
s ...... solid

REFERENCES

Alkasassbeh, M., Omar, Z., Mebarek-Oudina, F., Raza, J., & Chamkha, A. J. (2019). Heat transfer study of convective fin with temperature-dependent internal heat generation by hybrid block method. *Heat Transfer, Asian Research*, *48*(4), 1225–1244. doi:10.1002/htj.21428

Bejan, A. (1982). *Entropy Generation through Heat and Fluid Flow*. Wiley.

Bejan, A. (1996). *Entropy Generation Minimization*. CRC Press.

Farhan, M., Omar, Z., Mebarek-Oudina, F., Raza, J., Shah, Z., Choudhari, R. V., & Makinde, O. D. (2020). Implementation of one step one hybrid block method on nonlinear equation of the circular sector oscillator. *Computational Mathematics and Modeling*, *31*(1), 116–132. doi:10.100710598-020-09480-0

Mahmud, S., & Fraser, R. A. (2002). Second law analysis of heat transfer and fluid flow inside a cylindrical annular space. *Exergy*, *2*(4), 322–329. doi:10.1016/S1164-0235(02)00078-X

Mahmud, S., & Fraser, R. A. (2003). Mixed convection–radiation interaction in a vertical porous channel: Entropy generation. *Energy*, *28*(15), 1557–1577. doi:10.1016/S0360-5442(03)00154-3

Chapter 7
Entropy Generation Rate for Performance of Heat Transfer in Heat Exchangers:
A Comprehensive Review

Soraya Trabelsi
Independent Researcher, Tunisia

ABSTRACT

This chapter provides a specific study of the performance of thermal systems, principally heat exchangers, which are applied in several industrial applications such as chemical industry, energetic industry, industrial lasers, and so on. These thermodynamics systems were critical in transferring heat from a higher to a lower temperature fluid. They have been used for several years and are available currently for various designs. Thermodynamic properties influence the heat transfer and the performance of heat exchangers. Therefore, it is important during the design of heat exchangers to select primary the accurate operating conditions in terms of thermodynamics to provide a minimum amount of entropy generation in the system. In this study, the concept of entropy is used to analyze heat transfer processes from the thermodynamic viewpoint through the second law of thermodynamics. To assess heat exchanger performance, investigations are given for entropy generation, entropy generation number, and efficiency. These studies offer a new way to obtain well-designed heat exchangers.

DOI: 10.4018/978-1-7998-8801-7.ch007

1. INTRODUCTION

In the major industry applications, heat exchangers are extensively used such as chemical, mechanical and gas industries. They are planned to transfer energy between two or more fluids. Among the various devices, selecting the accurate heat exchanger is a complex task. In fact, the design and the choice of a heat exchanger is very important, since many elements have to be considered, such as the pressure drop, the rate of heat transfer, the efficiency...etc. In any thermal system, the enhancement of system efficiency is related to reducing losses during the process by analyzing irreversibility. As a result, the system performance is based on an investigation of the concept of the second law of thermodynamics. This concept had ample considerations in the previous investigations and also until today. It is applied in several research based on irreversibility analysis to give optimum conditions to design thermal systems and particularly balanced counter-flow heat exchangers.

(Bejan,1977a, 1977b) presented an extended study of the irreversibility process due to heat transfer and viscous effects for various flow configurations. In another researcher (Bejan,1980) he analyzed in details the irreversibility process, throw entropy generation concept, using the second law of thermodynamics and accounting for only heat transfer process. Particularly, he evaluated the entropy generation rate in a balanced counter-flow heat exchanger with zero pressure drop. More investigations of entropy generation in a counter-flow heat exchanger are found in the work of Bejan (1977c).

Moreover,(Sekulic,1986a) used irreversibility concept based on the second law of thermodynamic in a co-current and counter-current heat exchangers to analyze the optimization condition through the choice of minimum entropy generation. He also presented (Sekulic,1986b)the concept of entropy generation to evaluate the quality of heat transfer process in heat exchanger analysis. He used the quality called "Heat Exchange Reversibility Norm" (HERN), which measure the value of energy transformation of heat exchangers.

Many others researchers investigated the entropy generation on various type of heat exchangers and are found in the literature (Ordóñez (2000), Kolenda (2004), Guo (2010), Basak (2012), and Li (2013).

However, the physical performance of heat exchangers are optimized to reach maximum efficiency by optimizing many amounts of physical parameters for every application.

The use of thermodynamic analysis, for systems that involves the first and second laws of thermodynamics, allows to measure and to specify the degree of the performed processes.

On the foundation of the two laws of thermodynamics, the concept of entropy is applied to analyze the performance of heat exchangers. This study mainly aims to expose the different investigations on entropy generation to improve the efficiency of heat transfer in such devices.

Depending on the direction of the flows, two different operating modes are distinguished. Parallel and counter-flow heat exchangers. In parallel-flow (or co-current) the heating and the heated fluid flow in the same direction. In counter-flow (or counter-current) they flow in opposite directions. Therefore, the counter-flow mode gives more efficiency because the heat is distributed more uniformly across the heat exchanger and lets to extract the maximum amount of heat. This particularity allows the type of arrangement to be used more often. The degree of efficiency gained by using a counter-flow system depends on several parameters especially the flow rates and temperatures.

In this chapter, new researches based on thermodynamic optimization in terms of overall exchange coefficient as well as effectiveness and entropy generation in heat exchangers, are provided to give a comprehensive issue for better performance systems.

The current review is organized as follows: mechanisms of entropy transfer are given in section 2, where thermodynamic interaction between the system and its surrounding is treated with more details. In this section, entropy balance and entropy generation equations for various cases are presented. Application to heat exchanger counter-flow is given in section 3. In subsection 3.1 system of heat exchanger is described. Practical definitions for calculations of effectiveness and entropy generation rate are given in subsection 3.2 and 3.3 respectively. In the last part of section 3, we proposed the entropy generation number to evaluate the irreversibility process in heat exchanger. This chapter ended with diverse results of the very recent investigates that are summarized in section 4. In subsection 4.1, new results for heat exchangers using conventional fluids are exposed while results with nanofluids are presented in subsection 4.2.A conclusion is given in Section 5.

2. ENTROPY RATE BALANCE AND ENTROPY RATE GENERATION EQUATIONS

The fundamental property of entropy is a measure of molecular disorder of a system. It can be created ant not destroyed. So, in thermodynamics the concept of entropy is used to evaluate the irreversible and reversible processes.

The entropy is transferred to or from the system either by heat transfer or by mass flow. There is no entropy transferred by work.

The entropy transferred by heat transfer mechanism is written as:

$$S_{heat} = \int \frac{\delta Q}{T} \text{ (If absolute temperature is constant)} \tag{1}$$

$S_{heat} = \sum_j \frac{Q_j}{T_j}$ (If temperature is non uniform or different throughout the boundaries of the system). (2)

The entropy transferred by mass is:

$$S_{mass} = ms \tag{3}$$

However, when the properties of the mass vary during the process, Eq. (3) can be rewritten as:

$$S_{mass} = \int s\delta m = \int \dot{S}dt \tag{4}$$

Heat transfer occurs if two objects at different temperature are taken in contact. The heat is transferred from the hotter object to the colder one until the system reaches a state of thermodynamic equilibrium and the irreversible processes ended.

As soon as, the system starts to transfer entropy with its environments, the state of the system change (non equilibrium state)and irreversible processes occur creating entropy generation.

The entropy change of a system during a process between initial and final states is:

$$\Delta S_{syst} = S_{final} - S_{initial} \tag{5}$$

The entropy transferred outside the system boundary during a process 1-2 is given by:

$$dS_{1-2} = \int \frac{dQ}{T} \tag{6}$$

The mechanism of entropy transfer for a general system is illustrated in Fig.1.

It is clear that the entropy variation of the system is equal to the entropy transfer by heat and by mass. Therefore, the entropy balance is expressed as:

$$S_{out} - S_{in} = S_g = \Delta S_{syst} \tag{7}$$

Figure 1. Mechanisms of entropy transfer for a general system and its surroundings.

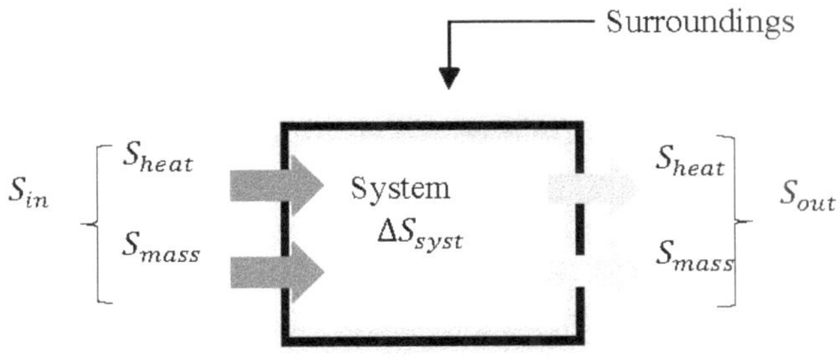

The term on the left-hand side of Eq.(7) shows the net entropy transfer by heat and by mass (Figure 1).

S_g is the entropy generation. $S_g>0$ for irreversible process and for process which is completely reversible $S_g=0$.

Therefore, when $S_g=0$ the entropy balance is reduced to $\Delta S_{syst} = S_{out} - S_{in}$ and the entropy variation of a system is equal to the net entropy transfer.

In engineering systems, the dissipation of energy depends on the range of irreversibility surviving during a process, which causes degradation of the performance of such systems. Consequently, entropy generation measure the dissipated energy.

The entropy balance in the rate form is given by:

$$\Delta \dot{S}_{syst} = \dot{S}_{in} - \dot{S}_{out} + \dot{S}_g \tag{8}$$

Where the rate of entropy transfer by mass is $\dot{S}_m = \dot{m}s$ and the rate of entropy transfer by heat is $\dot{S}_h = \dfrac{\dot{Q}}{T}$.

The entropy rate balance can be written on a unit-mass as follows:

$$\Delta \dot{s}_{syst} = \dot{s}_{in} - \dot{s}_{out} + \dot{s}_g \tag{9}$$

For closed systems, the mass is uniform and there is no mass transfer. The entropy transfer is only due to heat. The entropy balance equation is written as:

$$\Delta S_{syst} = S_{final} - S_{initial} = \sum_j \frac{Q_j}{T_j} + S_g \tag{10}$$

$\sum_j \frac{Q_j}{T_j}$ represents the sum of entropy transfer through the system boundary by heat transfer.

On the other hand, in rate form, Eq.(10) becomes:

$$\Delta \dot{S}_{syst} = \sum_j \frac{\dot{Q}_j}{T_j} + \dot{S}_g \tag{11}$$

Where $\dot{S}_g$ represents the time rate of entropy generation.

In the case of an adiabatic system (Q_j=0); the term $\frac{\dot{Q}_j}{T_j}$ is dropped out. Consequently, the entropy rate balance equation is reduced to:

$$\Delta \dot{S}_{syst} = \dot{S}_g \tag{12}$$

Eq.(12) indicates that neither mass nor heat exchange occurs. There is no entropy transfer.

Using open systems (Figure 2), entropy can be transferred into or out a control volume by streams of fluid.

Figure 2. Entropy balance for an open system

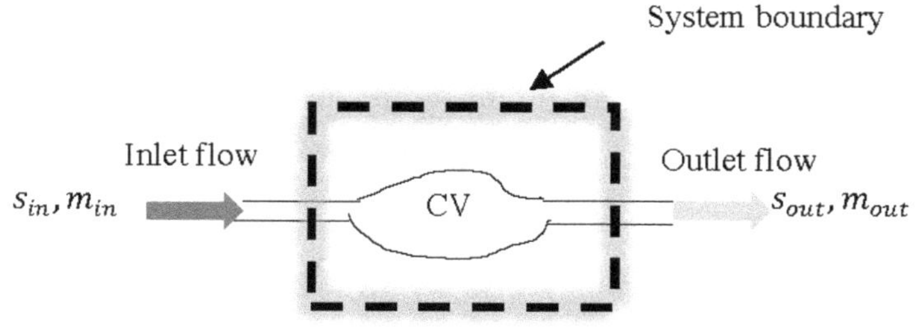

Entropy balance is obtained with considering the one of closed system and by taking into account for such entropy transfer at the boundaries into and out the control volume.

$$\frac{dS_{cv}}{dt} = \sum_j \frac{\dot{Q}_j}{T_j} + \sum_i \dot{S}_{in} - \sum_i \dot{S}_{out} + \dot{S}_g \tag{13}$$

Otherwise

$$\frac{dS_{cv}}{dt} = \sum_j \frac{\dot{Q}_j}{T_j} + \sum_i \dot{m}_{in} s_{in} - \sum_i \dot{m}_{out} s_{out} + \dot{S}_g \tag{14}$$

$\dot{m}_{in} s_{in}$ and $\dot{m}_{out} s_{out}$ are respectively rates of entropy transfer by mass flow at the boundaries into and out the control volume.

$\frac{dS_{cv}}{dt}$ characterizes the time rate of change of entropy inside the control volume.

At steady state flow process $\frac{dS_{cv}}{dt} = 0$, the entropy rate balance is reduced to:

$$\sum_j \frac{\dot{Q}_j}{T_j} + \sum_i \dot{m}_{in} s_{in} - \sum_i \dot{m}_{out} s_{out} + \dot{S}_g = 0 \tag{15}$$

In the situation where there is no heat transfer $\dot{Q}_j = 0$, Eq.(15) gives the relation $s_{out} - s_{in} \geq 0$ that can be explained by:

$s_{out} - s_{in} > 0$ the specific entropy rises as mass flows from inlet to outlet.
$s_{out} - s_{in} = 0$ and then $s_{out} = s_{in}$ this result means that mass streams from side to side of the control volume without modifying its entropy. It is a specific case, which correspond to no irreversibility inside the control volume.

In addition, using mass balance equation:

$$\dot{m} = \dot{m}_{in} = \dot{m}_{out} \tag{16}$$

In an adiabatic stream with one-inlet, one outlet control volume, Eq.(15) becomes:

$$\dot{m}\left(s_1 - s_2\right) + \dot{S}_g = 0 \tag{17}$$

Otherwise

$$\dot{S}_g = \dot{m}\left(s_1 - s_2\right) \geq 0 \tag{18}$$

As a result, we obtain:

$$s_1 \geq s_2 \tag{19}$$

The entropy balance, with its various form, is proved to be very commanding since it is used to evaluate the entropy generation term that plays an important role in the optimization of thermodynamics systems. The effects of irreversible process taking place in these systems can be analyzed.

3. APPLICATION OF THE PHYSICAL LAWS – HEAT EXCHANGER COUNTER-FLOW

Heat exchangers are used in the most energy systems (solar thermal collectors, steam turbine, air compressor…etc). They are applied to exchange thermal energy between two or more fluids streams at different temperatures without allowing their mixture. They are integrally operated irreversibly so the heat transfer, throw a temperature difference, is often accompanied by entropy generation. In this section, some of the elementary concepts of thermal design of heat exchangers are presented referring to the principal laws of thermodynamics.

3.1 System of Heat Exchanger

The system of heat exchanger is shown in figure3.The working fluids are separated by a metal plate. They flow in the opposite directions. Heat is transferred from hot fluid to cold fluid throw the metal plate in the z-direction.

Figure 3. (a) Schematic description of control volume for a counter-flow heat exchanger. (b) Temperature distributions.

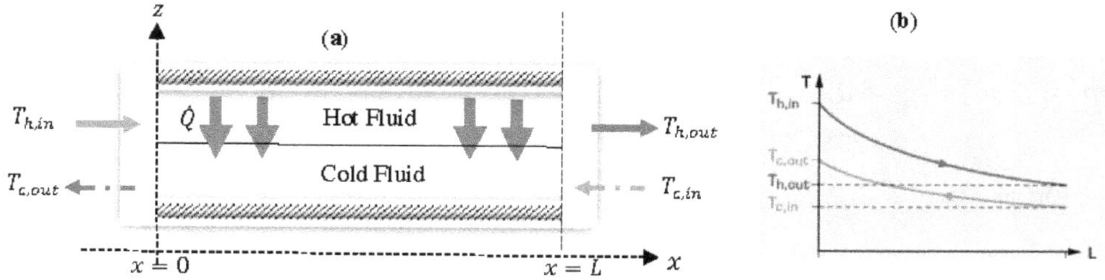

In this section, some assumptions are used in the developing of thermal relation formulas characteristic of the heat exchanger. They are summarized below:

1. the global heat transfer coefficient (combined heat transfer convection and conduction) is constant through the heat exchanger,
2. absence of phase change,
3. steady flow rates for the two streams,
4. losses due to heat exchange between the system and its surroundings are negligible,
5. flow for the two fluids is uniformly distributed along the transfer area.

3.2 Heat Transfer Effectiveness

The heat exchanger effectiveness is expressed as:

$$\varepsilon = \frac{\dot{Q}}{\dot{Q}_{max}} \tag{20}$$

Where $\dot{Q}$ is the actual heat transfer and $\dot{Q}_{max}$ correspond to the maximum heat transfer.

Actual heat transfer is given by equation balance of heat:

Energy lost by hot fluid = Energy gained by cold fluid

$$\dot{Q} = \dot{m}_h c_{p,h}\left(T_{h,in} - T_{h,out}\right) = \dot{m}_c c_{p,c}\left(T_{c,out} - T_{c,in}\right) \tag{21}$$

The maximum heat transfer is:

$$\dot{Q}_{max} = C_{min}\left(T_{h,in} - T_{c,in}\right) \tag{22}$$

Where $C_{min} = \left(\dot{m}c_p\right)_{min}$ is the minimum heat capacity.

The minimum fluid may be either the cold or the hot fluid. Accordingly:

$$\dot{Q}_{max} = \dot{m}_h c_{p,h}\left(T_{h,in} - T_{c,in}\right) \text{ if } C_c < C_h \tag{23}$$

$$\dot{Q}_{max} = \dot{m}_c c_{p,c}\left(T_{h,in} - T_{c,in}\right) \text{ if } C_h < C_c \tag{24}$$

Therefore, Eq.(20) is written as:

$$\varepsilon = \frac{\dot{m}_h c_{p,h}\left(T_{h,in} - T_{h,out}\right)}{C_{min}\left(T_{h,in} - T_{c,in}\right)} = \frac{\dot{m}_c c_{p,c}\left(T_{c,out} - T_{c,in}\right)}{C_{min}\left(T_{h,in} - T_{c,in}\right)} \tag{25}$$

The two more useful method analyzing the performance of heat exchangers are the effectiveness-number of transfer unit method (ε–NTU) and the logarithmic mean temperature difference method (LMTD). The first one does not require the knowledge of outlet temperature of the streams. Only inlet temperature of streams is sufficient to describe heat transfer rate of heat exchanger.

The dimensionless number of heat transfer (NTU) is defined by:

$$NTU = \frac{UA}{C_{min}} \tag{26}$$

U and A are respectively the global heat transfer coefficient and heat transfer surface area.

We defined the heat capacity ratio by:

$$C_r = \frac{C_{min}}{C_{max}} \leq 1 \tag{27}$$

The relation between effectiveness and number of heat transfer units for counter-flow heat exchanger is:

$$\varepsilon = \frac{1 - e^{-NTU(1-C_r)}}{1 - C_r e^{-NTU(1-C_r)}} \tag{28}$$

C_r=0 is a particular case. The effectiveness reduces to:

$$\varepsilon = 1 - e^{-NTU} \tag{29}$$

The last relation is available for counter-flow as well as for parallel-flow heat exchangers.

3.3 Entropy Generation of Heat Exchanger

Generally, there are different kinds of losses that occur in heat exchangers:

1. losses characterized by friction pressure drop in the ducts,
2. losses caused by finite temperature difference,
3. losses associated to heat exchange between the system and its surroundings, and
4. losses due to manufacture of heat exchanger.

The last losses are generally negligible.

Diverse methods are given in the literature that attempt to minimize these losses. Previously, the most famous method was carried out by Bejan (1980c). The entropy generation balance was applied to gas-gas heat exchanger. He formulated the equation of entropy generation rate balance for an ideal gas with constant specific heat as follows:

$$\dot{S}_g = \dot{S}_{g,\Delta T} + \dot{S}_{g,\Delta P} \tag{30}$$

The first term on the right side of Eq.(30) represents entropy generation rate due to heat transfer irreversibility and the second term is for the fluid friction irreversibility.

Bejan (1980c) explained that $\dot{S}_g = 0$ correspond to the highest quality whereas $\dot{S}_g > 0$ correspond to poor quality.

According to the second law of thermodynamics, the expression of entropy generation rate changes with the nature of the working fluid.

If the fluid is an ideal gas, the entropy generation rate changes in terms of the end temperatures and pressures. The total entropy generation rate, $\dot{S}_g$, has the form:

$$\dot{S}_g = C_{p,h} ln\left(\frac{T_{h,out}}{T_{h,in}}\right) + C_{p,c} ln\left(\frac{T_{c,out}}{T_{c,in}}\right) - \left(\dot{m}_h R\right) ln\left(\frac{P_{h,out}}{P_{h,in}}\right) - \left(\dot{m}_c R\right) ln\left(\frac{P_{c,out}}{P_{c,in}}\right) \tag{31}$$

Where $C_{p,h} = \dot{m}_h c_p$; $C_{p,c} = \dot{m}_c c_p$ and *R* is the ideal gas constant.

In the right hand of Eq.(31), the first two terms correspond to the heat transfer entropy generation rate, $\dot{S}_{g,\Delta T}$, and the last two terms represents the pressure drop entropy generation rate, $\dot{S}_{g,\Delta P}$.

If the fluid is a liquid (incompressible fluid), the entropy generation becomes:

$$\dot{S}_g = C_{p,h} ln\left(\frac{T_{h,out}}{T_{h,in}}\right) + C_{p,c} ln\left(\frac{T_{c,out}}{T_{c,in}}\right) + \left(\frac{\dot{m}}{\rho}\right)_h \Delta P_h - \left(\frac{\dot{m}}{\rho}\right)_c \Delta P_c \tag{32}$$

The pressure drop is defined by:

$$\Delta P = P_{in} - P_{out} \tag{33}$$

Eqs.(31) and (32) measure the irreversible processes inside the heat exchanger.

In heat exchangers under the consideration of negligible pressure drop and taking into account of Eq.(21), the expression of entropy generation rate reduces to:

$$\dot{S}_g = \dot{m}_h c_{p,h} ln\left(\frac{-\dot{Q}}{\dot{m}_h c_{p,h} T_{h,in}} + 1\right) + \dot{m}_c c_{p,c} ln\left(\frac{\dot{Q}}{\dot{m}_c c_{p,c} T_{c,in}} + 1\right) \geq 0 \tag{34}$$

In order to analyze the prominence of the two irreversibility mechanisms, Bejan (1980c) defined the irreversibility distribution ratio, ϕ, by the following relation:

$$\phi = \frac{\dot{S}_{g,\Delta P}}{\dot{S}_{g,\Delta T}} \tag{35}$$

Combining Eq.(30) and Eq.(35) yields:

$$\dot{S}_g = (1+\phi)\dot{S}_{g,\Delta T} \tag{36}$$

For $0 \leq \phi < 1$ the irreversibility is dominated by the heat transfer effects while for $\phi > 1$ the irreversibility is dominated by fluid friction effects.

Note that $\phi = 1$ is a specific case. Irreversibility by heat transfer and irreversibility by fluid friction have the same contribution on entropy generation rate.

In addition, further investigators proposed an alternative irreversibility distribution parameter and called it Bejan number (B*e)* defined by:

$$Be = \frac{\dot{S}_{g,\Delta T}}{\dot{S}_{g,\Delta T} + \dot{S}_{g,\Delta P}} \tag{37}$$

Or using Eq.(36), the last relation takes the form below:

$$Be = \frac{1}{1+\phi} \tag{38}$$

Consequently, the irreversibility process is dominated by the heat transfer effects for Be=1 while irreversibility is dominated by the fluid friction effects for Be=0. Noticeably, the heat transfer irreversibility and the fluid friction irreversibility have the same magnitude for a value of Bejan number equal to 0.5.

3.4 Entropy Generation Number

The entropy generation number is defined as:

$$N_s = \frac{\dot{S}_g}{C_{min}} \tag{39}$$

The entropy generation number N_s is a dimensionless number. A very small value of N_s correspond to very small losses in the heat exchanger. Whereas, an increase in N_s implies an increase in the losses.

Bejan (1980c) has revealed the importance of the number of entropy generation N_s for assessing the performance of a heat exchanger surface directly in terms of the quantity of usable energy missed by the heat exchanger. He gave an expression for the entropy generation number in balanced counter-flow heat exchanger with zero pressure drop as:

$$N_s = ln\frac{\left(1+\frac{T_{c,in}}{T_{h,out}}NTU\right)\left(1+\frac{T_{h,out}}{T_{c,in}}NTU\right)}{\left(1+NTU\right)^2} \tag{40}$$

He explained that the entropy generation number is a more suitable measure of thermodynamic deficiency and offers a more comprehensive representation about how various design parameters influence the thermal performance.

Consequently, N_s=0 for NTU=0 and $NTU\rightarrow\infty$.
N_s reaches its maximum at NTU=1; in this case the maximum N_s is written as follows:

$$N_s = ln\left(\frac{1}{2} + \frac{1}{4}\left[\frac{T_{c,in}}{T_{h,out}} + \frac{T_{h,out}}{T_{c,in}}\right]\right) \tag{41}$$

In addition, augmentation entropy generation dimensionless number, $N_{s,a}$, is expressed by:

$$N_{s,a} = \frac{\dot{S}_{g,a}}{\dot{S}_{g,0}} \tag{42}$$

For given mass flow and heat flux, $N_{s,a}$ turn into:

$$N_{s,a} = \frac{1}{1+\phi_0} N_{s,\Delta T} + \frac{\phi_0}{1+\phi_0} N_{s,\Delta P} \tag{43}$$

Where $\phi 0$ designs the irreversibility distribution ratio of the reference design, $N_{s,\Delta T}$ and $N_{s,\Delta P}$ are the values of $N_{s,a}$ in the limits of respectively pure heat transfer irreversibility and pure fluid-flow one.

4. RESULTS

4.1 Performance Heat Exchangers Using Conventional Fluids

In several heat exchangers, the use of water, air and oil, considered as conventional fluids, played an important part for the improvement of the thermodynamic performance of these thermal systems. In this section, the newest investigations to enhance heat transfer of heat exchangers using conventional fluids are presented.

Kurnia et al.(2019) exposed a numerical investigation of heat transfer and entropy generation in a double pipe helical heat exchanger with various cross-sections. They used CFD model and adopted the second thermodynamic law view to capture entropy generation. The working fluid for this investigation is water. Effects of various parameters are considered. They found that for low flow rate applications the concentric configuration is advisable whereas eccentric outer configuration is suitable for high flow rate. Furthermore, square pipe offered reasonably high heat transfer for all configurations. It generated the smallest entropy generation and had the lowest pressure drop, which rigorously characterize the required pumping power.

They highlighted that square cross-section is the better to be adopted in double pipe heat exchanger.

Caoet al. (2020) used the second law of thermodynamics to evaluate the irreversibility in helical baffle heat exchangers. They have tested fluid flow and heat transfer performances for heat exchangers with eight helical baffle configurations, involving the continuous helical (CH) scheme, quadrant helical (QH) scheme with diverse axial overlapped ratios and novel sextant helical (SH) scheme. The working mediums are respectively water assigned as the tube-side and conductive oil as shell-side. They found that SH scheme possessed the more performance than the CH and QH schemes. Furthermore, the smallest entropy generation gave with the SH and QH schemes at low Reynolds number. They concluded that SH baffle heat exchanger could be considered as a suitable improved design since it offered excellent behaviors in terms of thermodynamic structures.

Nematiet al. (2020)looked for the optimization of a regular, four-row heat exchanger by varying locally the shape of the annular fins. The CFD model is used to predict thermodynamics parameters (heat transfer, pressure drop and entropy generation). The analyze is based on four distinct objective functions to evaluate the local fin-shape distributions through the bundle that: (i) maximize the heat flux, (ii) minimize the pressure drop while maximizing the total heat transfer rate, (iii) minimize the overall pressure drop, or (iv) minimize the entropy generation. Results delivered by the objective functions with reducing the overall pressure drop while maximizing the total heat transfer rate, are particularly interesting. Overall optimized cases, the entropy generation rate is reduced which confirmed the thermodynamic enhancement in tube bundle performance.

Arasteh et al. (2020)presented a numerical study of heat transfer and flow in a counter-flow sinusoidal parallel-plate heat exchanger using metal foam in the divergent sections of channels. The cold fluid is water and hot fluid is oil. The sinusoidal heat exchanger contains porous media. Darcy–Brinkman–Forchheimer flow model and local thermal non-equilibrium model are used. Calculations are done with two-wave amplitudes two wavelengths, three porous particle diameters and three porous thicknesses. A dimensionless number called performance evaluation criteria (PEC) is introduced to trade off the effects of the sinusoidal wall and porous media. They found that rising the heat exchanger wave amplitude and decreasing wavelength without porous medium lead to heat transfer enhancement. While using metal foam with various thicknesses and particle diameters displays diverse trends in heat transfer as wave amplitude and wavelength increase. They revealed that introducing the metal foam caused an augmentation in heat exchanger pumping power, while increasing the porous particle diameter lead to a subtle augmentation in pumping power. They also developed four correlations to predict the effectiveness of a porous inserted counter-flow sinusoidal parallel-plate heat exchanger based on the number of thermal units.

A numerical analysis of a plate- fin heat exchanger with offset strips in water-water heat transfer processes is presented by Nascimento et al. (2020). They used the CFD method to resolve the turbulent flow with the Shear Stress Transport SST k- ω turbulence model combining with random vector functional-link(RVFL) network and the many-objective optimization method called NSGA-III in the form of a surrogate approach in order to reduce computational time in comparison to the CFD simulations. The optimized results by NSGA-III are validated with results obtained by SSTk- ω turbulence model. They indicated that the volume and effectiveness values are in agreement with the literature while the pressure drop at the hot and cold side are reduced. This result discovered that the convective heat transfer are improved while the flow resistance are considerably reduced.

Goh et al. (2021) studied the effect of inter-turbulator distance of rotating turbulator inserts on the heat transfer performance and entropy generation of forced convection in the turbulent flow region of a heat exchanger using water as working fluid. Results indicated that from the second-law point of view, entropy generation rate of the non-rotating case augmented up to 202% in comparison with the rotating case. They also discussed the Bejan number. They found that the entropy generation is mainly attributed to the heat transfer irreversibility compared to the fluid friction irreversibility. They divulged that the rotating turbulator insert is the most thermodynamically beneficial configuration. It improved the thermal performance significantly and reduced the entropy generation considerably.

4.2 Heat Exchangers Using Nanofluids

Nanofluids are innovative class of heat transfer fluids and show higher potential to improve heat transfer in heat exchangers. They are characterized by high heat transfer coefficient and they offered an opening for developing improvement heat exchangers. In this section, we present the latest investigates based on the evaluation of heat transfer in heat exchangers using nanofluids as working fluids.

Several experimental and numerical researches revealed better enhancement when nanofluids are used in thermal systems (double pipe heat exchangers, plate heat exchangers, shell and tube heat exchangers…etc).

Khanlari et al. (2019) studied experimentally the effect of utilizing kaolin/deionized water nanofluid on the thermal efficiency of two different heat exchangers. A counter-flow concentric tube (CFCT) heat exchanger and parallel-flow concentric tube (PFCT) heat exchanger. They proved that working kaolin/deionized water nanofluid in the heat exchangers improved heat transfer performance considerably in comparison with deionized water. They approved the use of kaolin/deionized water in the concentric tube type heat exchanger.

Bendaraa et al. (2021) evaluated the thermal performance of double-pipe heat exchanger, with a nanofluid based on alumina suspension, using numerical simulations through an application developed on COMSOL Multiphysics environment. They indicated that the addition of nanofluid with an increase in its volume fraction lead to an increase in the overall exchange coefficient, the convective heat transfer coefficient, as well as the efficiency and the power of the exchanger. Moreover, the variation of the concentration of nanofluids lead to a regression of the pinch points of the inlet andoutlet temperatures, which signify that nanofluids are more efficient in cooling temperatures compared to conventional fluids.

Qi et al. (2019) investigated experimentally thermal performance and pressure drop of nanofluids ($TiO_2 - H_2O$) in the smooth and corrugated double-tube heat exchangers. They analyzed the influence of various parameters. They showed that nanofluids(TiO_2-H_2O) have better thermal performance than the deionized water. The number of transfer units (NTU) and effectiveness are also enhanced. The corrugated double-tube heat exchanger have stronger thermal performance than the smooth double-tube heat exchanger. On the other hand, the pressure drop of nanofluids in the corrugated double-tube heat exchanger is also considerably stronger. Other result is found, when using both nanofluids and corrugated tube, the overall thermal performance is enhanced, which reveals in the increase of the NTU and effectiveness. Nevertheless, for thermal fluid in the shell-side, the NTU and effectiveness decreased initially and then increased with the increase of Reynolds number.

5. CONCLUSION

The present chapter offer a comprehensive review on the researches improvement of heat transfer performance for different configurations of heat exchangers using conventional fluids or nanofluids as working fluids. Therefore, numerical and experimental investigations are carried out for different configurations of heat exchangers. This review summarizes the advanced approaches to assess the thermodynamics performance and to enhance heat transfer in heat exchangers. All the studies proposed are based on the second concept of thermodynamics. In order to evaluate the efficiency of heat transfer inside heat exchangers the investigators examined the thermodynamic characteristics of the thermal system and presented their results fundamentally in terms of entropy generation, entropy generation number, effectiveness and number of transfer units.

In fact, this literature review is considerably vast. We focused to present the studies from past three years to indicate recent trends and to highlight the *new* design of heat exchangers suitable for industrial applications as well as the efficient use to advance manufacturing processes.

REFERENCES

Arasteh, H., Mashayekhi, R., Ghaneifar, M., Toghraie, D., & Afrand, M. (2020). Heat transfer enhancement in a counter-flow sinusoidal parallel-plate heat exchanger partially filled with porous media using metal foam in the channels' divergent sections. *Journal of Thermal Analysis and Calorimetry*, *141*(5), 1669–1685. doi:10.100710973-019-08870-w

Basak, T., Kumar, P., Anandalakshmi, R., & Roy, S. (2012). Analysis of entropy generation minimization during natural convection in trapezoidal enclosures of various angles with linearly heated side wall (s). *Industrial & Engineering Chemistry Research*, *51*(10), 4069–4089. doi:10.1021/ie201107f

Bejan, A. (1977a). *The concept of irreversibility in heat exchanger design: counterflow heat exchangers for gas-to-gas applications*. Academic Press.

Bejan, A. (1978b). General criterion for rating heat-exchanger performance. *International Journal of Heat and Mass Transfer*, *21*(5), 655–658. doi:10.1016/0017-9310(78)90064-9

Bejan, A. (1980c). Second law analysis in heat transfer. *Energy*, *5*(8-9), 720–732. doi:10.1016/0360-5442(80)90091-2

Bejan, A. (2016d). *Advanced engineering thermodynamics*. John Wiley & Sons. doi:10.1002/9781119245964

Bendaraa, A., Charafi, M. M., & Hasnaoui, A. (2021). Numerical and experimental investigation of alumina-based nanofluid effects on double-pipe heat exchanger thermal performances. *SN Applied Sciences*, *3*(2), 1–10. doi:10.100742452-021-04195-2

Cao, X., Chen, D., Du, T., Liu, Z., & Ji, S. (2020). Numerical investigation and experimental validation of thermo-hydraulic and thermodynamic performances of helical baffle heat exchangers with different baffle configurations. *International Journal of Heat and Mass Transfer*, *160*, 120181. doi:10.1016/j.ijheatmasstransfer.2020.120181 do Nascimento, C. A. R., Mariani, V. C., & dos Santos Coelho, L. (2020). Integrative numerical modeling and thermodynamic optimal design of counter-flow plate-fin heat exchanger applying neural networks. *International Journal of Heat and Mass Transfer*, *159*, 120097. doi:10.1016/j.ijheatmasstransfer.2020.120097

Goh, L. H. K., Hung, Y. M., Chen, G. M., & Tso, C. P. (2021). Entropy generation analysis of turbulent convection in a heat exchanger with self-rotating turbulator inserts. *International Journal of Thermal Sciences*, *160*, 106652. doi:10.1016/j.ijthermalsci.2020.106652

Guo, J., Cheng, L., & Xu, M. (2010). Multi-objective optimization of heat exchanger design by entropy generation minimization. *Journal of Heat Transfer*, *132*(8), 081801. doi:10.1115/1.4001317

Khanlari, A., Sözen, A., Variyenli, H. I., & Gürü, M. (2019). Comparison between heat transfer characteristics of TiO 2/deionized water and kaolin/deionized water nanofluids in the plate heat exchanger. *Heat Transfer Research*, *50*(5), 435–450. doi:10.1615/HeatTransRes.2018026288

Kolenda, Z., Donizak, J., & Hubert, J. (2004). On the minimum entropy production in steady state heat conduction processes. *Energy*, *29*(12-15), 2441–2460. doi:10.1016/j.energy.2004.03.049

Kurnia, J. C., Ghoreishi-Madiseh, S. A., & Sasmito, A. P. (2019). Heat transfer and entropy generation in concentric/eccentric double-pipe helical heat exchangers. *Heat Transfer Engineering*, *41*(18), 1552–1575. doi:10.1080/01457632.2019.1661666

Li, M., & Lai, A. C. (2013). Thermodynamic optimization of ground heat exchangers with single U-tube by entropy generation minimization method. *Energy Conversion and Management*, *65*, 133–139. doi:10.1016/j.enconman.2012.07.013

Nemati, H., Moghimi, M. A., Sapin, P., & Markides, C. N. (2020). Shape optimisation of air-cooled finned-tube heat exchangers. *International Journal of Thermal Sciences*, *150*, 106233. doi:10.1016/j.ijthermalsci.2019.106233

Ordóñez, J. C., & Bejan, A. (2000). Entropy generation minimization in parallel-plates counter-flow heat exchangers. *International Journal of Energy Research*, *24*(10), 843–864. doi:10.1002/1099-114X(200008)24:10<843::AID-ER620>3.0.CO;2-M

Qi, C., Luo, T., Liu, M., Fan, F., & Yan, Y. (2019). Experimental study on the flow and heat transfer characteristics of nanofluids in double-tube heat exchangers based on thermal efficiency assessment. *Energy Conversion and Management*, *197*, 111877. doi:10.1016/j.enconman.2019.111877

Sekulic, D. P. (1986). Entropy generation in a heat exchanger. *Heat Transfer Engineering*, *7*(1-2), 83–88. doi:10.1080/01457638608939647

Sekulic, D. P. (1990). *The second law quality of energy transformation in a heat exchanger*. Academic Press.

APPENDIX

Nomenclature

A …… surface of heat transfer area (m^2)
Be …… Bejan number
C …… heat capacity rate ($W.K^{-1}$)
c_p …… specific heat at constant pressure $J.kg^{-1}.K^{-1}$
C_r …… ratio of heat capacity rates
L …… length of the heat exchanger (m)
N_s …… entropy generation number
$N_{s,\Delta P}$ …… entropy generation number due to friction
$N_{s,\Delta T}$ …… entropy generation number due to heat transfer
$\dot{m}$ …… mass flow rate ($kg.s^{-1}$)
P …… pressure (Pa)
$\dot{Q}$ …… heat transfer rate (W)
R …… ideal gas constant
s …… specific entropy ($J.kg^{-1}.K^{-1}$)
S …… entropy ($J.K^{-1}$)
$\dot{S}_g$ …… entropy generation rate ($W.K^{-1}$)
T …… temperature (K)
U …… overall heat transfer coefficient ($W.m^{-2}.K^{-1}$)
ΔP …… pressure drop (Pa)
NTU …… Number of Transfer Units

Greek Symbols

ρ …… Fluid density K$g.m$-^{3}
ε …… heat transfer effectiveness
ϕ …… irreversibility distribution ratio

Subscripts

c …… cold
h …… hot
in …… inlet
out …… outlet
min …… minimum value
max …… maximum value

Chapter 8
Irreversibility and Heat Transfer in Darcy-Forchheimer Magnetized Flow in a Porous Double Lid-Driven Cavity Filled With Copper-Water Nanofluid

Souad Marzougui
Institut Supérieur des Sciences Appliquées et Technologies de Gabès, Tunisia

Mourad Magherbi
Institut Supérieur des Sciences Appliquées et Technologies de Gabès, Tunisia

ABSTRACT

The present work reports a numerical simulation of entropy generation and heat transfer in a lid-driven porous cavity filled with a nanofluid using Darcy-Forchheimer model. Given the large number of dimensionless parameters related to this problem, some of them are kept constant and therefore the other governing dimensionless number such as number, the Hartman number, and the nanoparticles volume fraction, 0£Ha£50, 2%£φ£8%, respectively. The effects of the nanoparticles volume fraction and Hartman number on the different irreversibilities are studied. Results show that the entropy generation is strongly affected by the increase of Hartmann number and the volume fraction. Results reveal that the irreversibility in the nanofluid decrease with the nanoparticle volume fraction for different Hartmann numbers.

DOI: 10.4018/978-1-7998-8801-7.ch008

INTRODUCTION

The mixed magneto-convection heat transfer in square lid-driven cavities filled with a nanofluid saturated porous media has received extensive attention in the literature according to their wide variety of engineering applications such as the cooling of electronic devices, solar collectors and energy storage, crystal growth, food processing and nuclear reactors, etc. The literature concerning convective flow in porous media are books of Bejan et al. (Nield & Bejan, 2006), Ingham and Pop (Ingham & Pop, 2005) and Vafai (Vafai, 2005).

Rudraiah *et al.* (Rudraiah et al., 1995) and Chamkha (Chamkha, 2002) investigated numerically the natural convection of an electrically conducting fluid in the presence of a magnetic field.

M.A. Ismael et al (Ismael et al., 2014)studied numerically steady laminar mixed convection inside lid-driven with partial slip imposed in these two moving walls neglected viscous dissipation they found that in the absence of partial slip the feature of the isotherms is not modified for varying Richardson number, they noted that there are critical values of the partial slip parameter depend on Richardson number and the direction of moving wall and the mixed convection is the dominance over the natural convection in the absence of the partial slip effect. Mekroussi et al (Said et al., 2013) investigated numerically to analyse the mixed convection flow and heat transfer in a lid-driven cavity with a sinusoidal wavy bottom Surface. The results of this investigation illustrate that the average Nusselt number at the heated surface increases with an increase in the number of undulations as well as the angle of inclination. Habeeb (Habeeb, 2012) studied the free convective two-dimensional flow and heat transfer in an anisotropic fluid-filled porous rectangular enclosure have been investigated using the Lattice Boltzmann method for the non-Darcy flow model. Effects of the moving lid direction. The results obtained are discussed in terms of the Nusselt number, vectors, contours, and isotherms. The numerical study of hydromagnetic double-diffusive mixed convection in a square lid-driven cavity is performed by Dawood and Teamah (Dawood & Teamah, 2012). They found that the direction of the lid is more effective on heat and mass transfer and fluid flow with increasing of a magnetic field for all studied parameters. Hydrodynamic mixed convection heat transfer in a lid-driven cavity heated from the top with the wavy bottom surface is numerically studied by Saha et al. (Saha et al., 2015). Chamkha (Chamkha, 2002) investigated numerically the natural convection of an electrically conducting fluid in the presence of a magnetic field. Al-Salem et al (Al-Salem et al., 2012) investigated the effects of moving the top wall direction in a linearly heated square cavity. Both forced and mixed convection cases. They found Both moving lid direction and Hartmann number can be utilized to control energy passively. Thus, energy saving is possible with these passive methods. The Increase of Hartmann number decreases the flow strength and constricts the heat transfer rate. It is found that the moving lid direction affects heat and fluid flow when mixed convection is the dominant mechanism of heat transfer. This indicates that moving lid direction can be a control parameter of heat and mass transfer. The linearly varying temperature boundary condition at the bottom wall augmented the effect of moving lid direction on heat and mass transfer, especially at higher Hartmann numbers. Khanafer and Vafai (Khanafer & Vafai, 2002) presented a numerical study of mixed-convection heat and mass transport in a lid-driven square enclosure filled with a non-Darcian fluid-saturated porous medium by using the finite volumes technique. Mixed convection in a lid-driven porous cavity in the presence of a magnetic field is studied numerically by Muthtamilselvan et al. (Muthtamilselvan et al., 2009) and Rahman et al. (Rahman et al., 2010). It is found that the heat transfer is strongly dependent on the strength of the magnetic field and the Darcy number. The numerical study of hydromagnetic double-diffusive mixed convection in a square lid-driven cavity is performed by Dawood and Teamah (Dawood & Teamah, 2012). They found that the direction of the lid is more effective on heat and mass transfer and fluid flow with increasing of the magnetic field for all studied parameters. Hydrodynamic mixed convection heat transfer in a lid-driven cavity heated from the top with a wavy bottom surface is numerically studied by Saha et al. (Saha et al., 2015). It is observed that the wavy lid-driven cavity can be considered as an effective heat transfer mechanism in presence of a magnetic field at larger wavy surface amplitudes and low Richardson numbers. Mchirgui et al. (Chamkha, 2002) reported a numerical study of the entropy generation in double-diffusive convection through a square porous cavity saturated with a binary perfect gas mixture submitted to horizontal thermal and concentration gradients.

In our facts, the effects of nanoparticles volume fraction and magnetic field on the different causes of irreversibility are not investigated yet. This numerically study trades principally with discoveries the impacts of the nanoparticles volume fraction and magnetic field over magnetic, fluid viscous, Darcy viscous, thermal and total irreversibilities in the flow of nanofluid inside the double lid- driven using Darcy Forchheimer model.

MATHEMATICAL FORMULATION

Consider a two-dimensional laminar flow in a square lid-driven porous cavity filled with a Newtonian, incompressible nanofluid. The cavity has a height of H as shown in Fig. 1.

Figure 1. Schematic view of the physical model coordinate system.

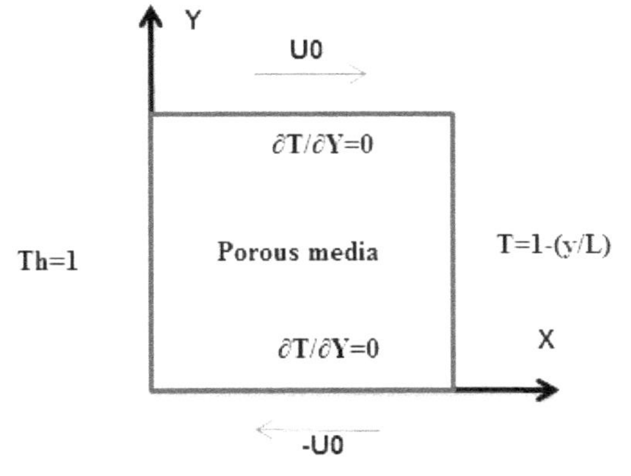

The top and bottom walls are adiabatic. The top wall is assumed to be moving from left to right at constant speed U_0 and the bottom is assumed to be moving in the reverse direction. The right and the left wall is maintained at constant hot temperature (Th) and the left wall is linearly heated. A uniform magnetic field of strength B_0, making an inclination angle (α) with the horizontal, is applied in the x-y plane of the cavity. The physical properties of the fluid are considered to be constant except the density, which satisfies the Boussinesq approximation as:

$$\rho = \rho 0[1 - \beta_\theta(\theta - \theta_0)] \tag{1}$$

Where ρ_0 is the fluid density at temperature θ_0 and β_θ is the thermal expansion coefficient such that:

$$\beta_\theta = -\frac{1}{\rho_0}\left(\frac{\partial \rho}{\partial \theta}\right)_P$$

The governing equations of continuity, momentum and energy balances can be written in the dimensionless form as:

Continuity equation:

$$\frac{\partial U}{\partial \mathrm{X}} + \frac{\partial V}{\partial \mathrm{Y}} = 0 \tag{2}$$

X-Momentum equation:

$$\frac{1}{\varepsilon}\frac{\partial U}{\partial Ä}+\frac{1}{\varepsilon^2}U\frac{\partial U}{\partial X}+\frac{1}{\varepsilon^2}V\frac{\partial U}{\partial Y}=\frac{\rho_f}{\rho_{nf}}\left[-\frac{\partial P}{\partial X}-\frac{\mu_{nf}}{\mu_f}\frac{1}{\varepsilon \mathrm{Re}}\nabla^2 U-\frac{\mu_{nf}}{\mu_f}\frac{U}{Da\,\mathrm{Re}}\right]$$
$$+\frac{\rho_f}{\rho_{nf}}\frac{\sigma_{nf}}{\sigma_f}\frac{\mathrm{Ha}^2}{\mathrm{Re}}\left(V\sin\gamma\cos\gamma-U\sin^2\gamma\right)-\frac{F_c}{\sqrt{Da}}U\sqrt{U^2+V^2} \tag{3}$$

Y-Momentum equation:

$$\frac{1}{\varepsilon}\frac{\partial V}{\partial Ä}+\frac{1}{\varepsilon^2}V\frac{\partial U}{\partial X}+\frac{1}{\varepsilon^2}V\frac{\partial V}{\partial Y}=\frac{\rho_f}{\rho_{nf}}\left[-\frac{\partial \mathbf{P}}{\partial Y}-\frac{\mu_{nf}}{\mu_f}\frac{1}{\varepsilon \mathrm{Re}}\nabla^2 V-\frac{\mu_{nf}}{\mu_f}\frac{V}{Da\,\mathrm{Re}}+\frac{(\rho\beta)_{nf}}{(\rho\beta)_f}Ri\theta\right]$$
$$+\frac{\rho_f}{\rho_{nf}}\frac{\sigma_{nf}}{\sigma_f}\frac{\mathrm{Ha}^2}{\mathrm{Re}}\left(U\sin\gamma\cos\gamma-V\cos^2\gamma\right)-\frac{F_c}{\sqrt{Da}}V\sqrt{U^2+V^2} \tag{4}$$

Energy equation:

$$\frac{\partial T}{\partial Ä}+U\frac{\partial T}{\partial X}+V\frac{\partial T}{\partial Y}=\frac{\alpha_{nf}}{\alpha_f}\frac{1}{\mathrm{Pr}\,\mathrm{Re}}\left(\frac{\partial^2 T}{\partial^2 X}+\frac{\partial^2 T}{\partial^2 Y}\right) \tag{5}$$

Where the dimensionless parameters are defined in the following forms,

$$X=\frac{x}{H};Y=\frac{y}{H};\mathrm{U}=\frac{u}{U_0};V=\frac{v}{U_0};P=\frac{p}{\rho_f U_0^2};\tau=\frac{U_0 t}{H};T=\frac{\theta-\theta_c}{\theta_h-\theta_c};\mathrm{Pr}=\frac{\nu_f}{\alpha_f};\mathrm{Re}=\frac{U_0 H}{\nu_f};$$
$$Da=\frac{K}{H^2};Ha=B_0 H\sqrt{\frac{\sigma_f}{\mu_f}};Ri=\frac{Ra}{Pe\,\mathrm{Re}};Ra=\frac{g\beta_\theta(\theta_h-\theta_c)L^3}{\nu_f\alpha_f};F_c=\left(1.75/\sqrt{150\varepsilon^2}\right)$$

Pe = Re Pr

The parameters F_C, Ha, Ri, Pe and Pr are respectively Forchheimer, Hartman, Richardson, Peclet and Prandtl numbers.

The expressions of density, thermal expansion, heat capacitance, dynamic viscosity, electrical conductivity, thermal conductivity and thermal diffusivity of the nanofluid are given as follows (Sharif, 2007):

$$\rho_{inf}=(1-\varphi)\rho_f+\varphi\rho_s \tag{6}$$

$$(\rho\beta)_{nf} = (1-\varphi)\,(\rho\beta)_f + \varphi(\rho\beta)_s \tag{7}$$

$$(\rho c_p)_{nf} = (1-\varphi)\,(\rho c_p)_f + \varphi(\rho c_p)_s \tag{8}$$

$$\frac{\mu_{nf}}{\mu_f} = \frac{1}{(1-\varphi)^{2,5}} \tag{9}$$

$$\frac{\sigma_{eff}}{\sigma_f} = 1 + \frac{3(\alpha-1)\varphi}{(2+\alpha)-(\alpha-1)\varphi};\alpha = \frac{\sigma_s}{\sigma_f} \tag{10}$$

$$\frac{k_{nf}}{k_f} = \frac{k_s + 2k_f - 2\varphi(k_f - k_s)}{k_s + 2k_f + \varphi(k_f - k_s)} \tag{11}$$

$$\alpha_{nf} = \frac{k_{nf}}{\left(\rho c_p\right)_{nf}} \tag{12}$$

The initial and boundary conditions, expressed in dimensionless form are:

Initial conditions

U (X,Y) = 0; V(X,Y)= 0; T (X,Y) =0 (13-a)

Vertical adiabatic walls

U (1,Y) = 0; V (1,Y)= 0; T (1,Y) = 1-Y/L (13-b)

U (0,Y) = 0; V(0,Y) = 0; T (0,Y) = 0 (13-c)

Horizontal active walls

U (1,y) = 1; V (1,y) = 0; $\frac{\partial T}{\partial X} = 0$ (13-d)

U (0,Y) = -1; V (0, y)= 0; $\frac{\partial T}{\partial X} = 0$ (13-e)

The Nusselt number is defined as follow:

$$\mathrm{Nu} = -\frac{k_{nf}}{k_f}\int_0^1 \left(\frac{\partial T}{\partial Y}\right) dX \tag{14}$$

SECOND LAW FORMULATION

The local entropy generation in porous media is given by Woods (1975). On the right-hand side of Eq. (14), the first term is the irreversibility due to heat transfer, the second represents the Darcy viscous dissipation term, the third is compatible with the clear fluid friction and the fourth is relative to the magnetic effect.

$$S_l = \left(\frac{k_{nf}}{k_f}\right)\left[\left(\frac{\partial T}{\partial X}\right)^2 + \left(\frac{\partial T}{\partial Y}\right)^2\right] + \left(\frac{\mu_{nf}}{\mu_f}\right)\frac{Br^*}{Da}\left(U^2+V^2\right)$$
$$+\left(\frac{\mu_{nf}}{\mu_f}\right)Br^*\left(2\left(\frac{\partial U}{\partial X}\right)^2 + 2\left(\frac{\partial V}{\partial Y}\right)^2 + \left(\frac{\partial U}{\partial Y}+\frac{\partial V}{\partial X}\right)^2\right) + Br^* Ha^2\left(\frac{\sigma_{nf}}{\sigma_f}\right)\left(Usin(\alpha)+Vcos(\alpha)\right)^2 \tag{15}$$

The modified Brinkman number Br* is expressed as:

$$Br^* = \frac{\mu_f \theta_0}{k_f}\left(\frac{U_0}{\theta_h - \theta_c}\right)^2 \tag{16}$$

The dimensionless total entropy generation is given by:

$$S = \int_W S_l \, dW \tag{17}$$

NUMERICAL METHOD AND VALIDATION

We solve the model of Navier Stokes and energy equation in non-dimensional form for an incompressible Newtonian fluid, consisting of the equations of conservation of momentum, with the appropriate boundary conditions. COMSOL Multiphysics bases the discretization of the equations on the finite element method. The non-slip boundary condition is used on the walls of the cavity. To validate our results, the solutions given by the present code are compared with those obtained by Basak et al. (Al-Salem et al., 2012) for Pr=0.015, Re=1, Da= 10-4, Gr=105, We use the same parameter of Basak et al (Al-Salem et al., 2012) with the same boundary conditions. The results of this comparison in terms of streamlines, isotherms and the problem of mixed convection in a lid-driven enclosure (Sharif, 2007; Sivanandam et al., 2020) is engaged to match the results of the current code (Table 2). From these results, it is evident that the present code generates results that are in good agreement with those of Basak et al (Al-Salem et al., 2012), Sivanandam et al (Sivanandam et al., 2020) and Sharif(Sharif, 2007). Hence the present code is considered completely reliable.

Figure 2. Validation of the numerical results

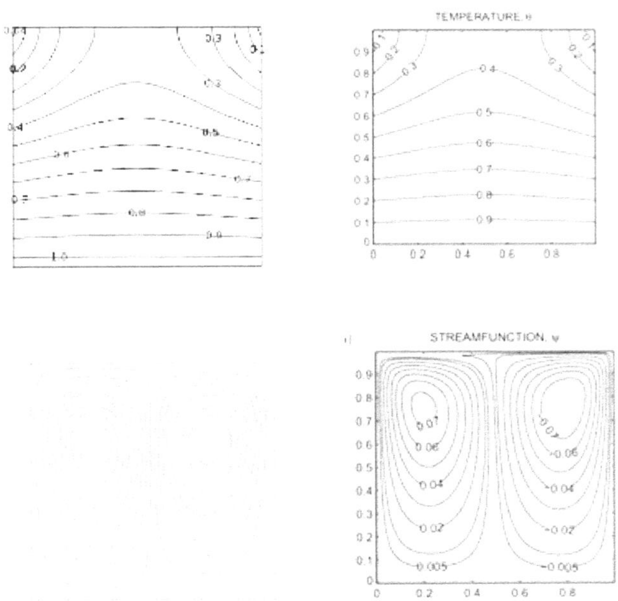

Present code

Baska et al [12]

RESULTS AND DISCUSSION

a-Effect of Volume Fraction

Figures 3-4 shows that the magnetic entropy generation and the viscous fluid entropy generation increase with increasing volume fraction for all values of the Hartman number. For fixed Volume fraction the magnetic and the viscous fluid entropy generation increase when the Ha increase. The Grashof number is taken constant equal to Gr= 10^5.

Table 2. comparison of average Nusselt number for mixed convection in a lid-driven cavity

Gr	Present Work	S. Sivanandam et al. (2020)	Sharif (2007)
Re=400			
10^2	4.0072	4.09	4.05
10^4	3.9006	3.85	3.82
Re=1000			
10^2	6.4262	6.48	6.55
10^4	6.4291	6.47	6.50

Figure 3. Effect of the volume fraction on the magnetic entropy generation

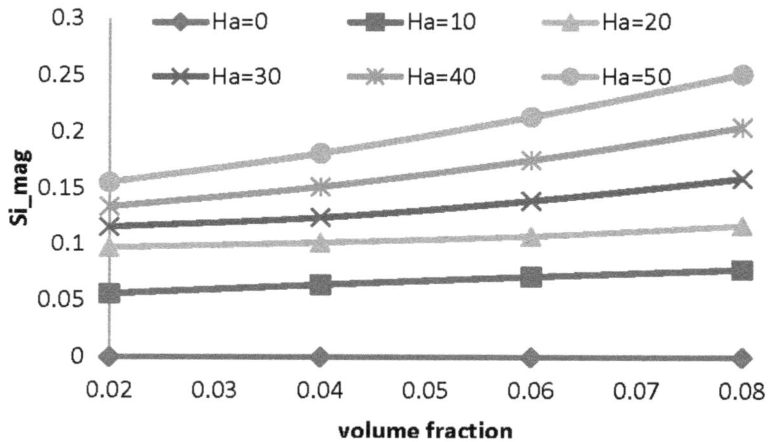

Indeed, for a fixed value of the Hartman number, the increase of the nanofluid velocity following the improvement of the convection via the increase of the volume fraction is in this case directly compensated by the extrinsic effect of the magnetic field through the Lorentz force on the flow which tends to slow down the fluid. It is obvious, as mentioned above, that when the Hartmann number becomes relatively large, the Lorentz force becomes predominant via its extrinsic effect on the flow, which results in a significant reduction of convection and consequently of the movement and the speed of the fluid.

In this respect, an increase in the volume fraction of the nanoparticles can only further reduce the convection by increasing the effective dynamic viscosity and consequently the viscous forces.

Figure 4. Effect of the volume fraction on the viscous fluid entropy generation

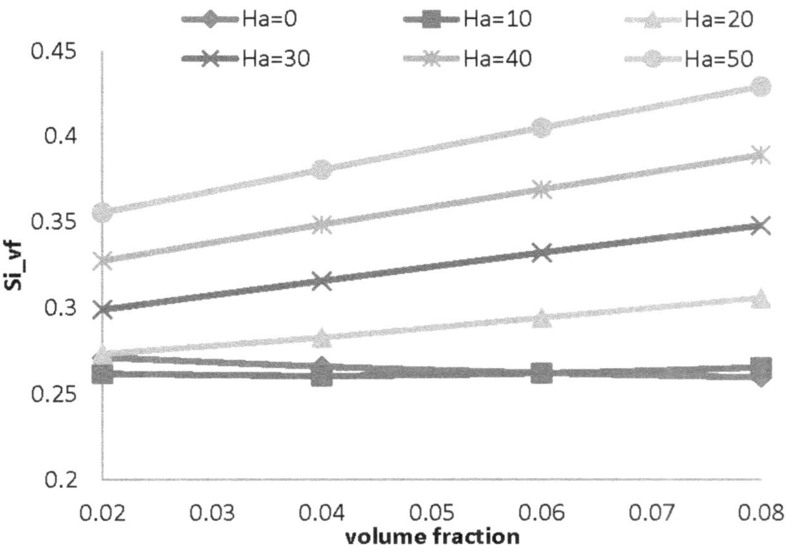

At this stage, and for a high but constant Hartmann number, the decrease in entropy generation when the volume fraction increases can be explained by the dominance of the extrinsic effects of the effective electrical conductivity and the effective dynamic viscosity at the force level of Lorentz and the fluid viscous force concerning the intrinsic (direct) effect of the electrical conductivity on magnetic entropy generation. Indeed, this reduction of the fluid velocity following the reduction of the convection by the addition of nanoparticles to a fixed but high Hartmann number is confirmed by the plots of the velocity vector components in Figure 4.

Figures 5-6 gives the variations of the components (U and V) velocity as a function of the abscissa X at Y = 0.5 for different volume fractions for a Hartmann number equal to 10 and 40. This figure shows a decrease of the vector components speed and therefore a reduction of convection in the cavity, following the increase in the volume fraction as mentioned above.

For a constant volume fraction, Figure .3 reveals that the magnetic irreversibility increases with the Hartmann number. Indeed, the increase of the magnetic entropy generation, with constant volume fraction, when the number of Hartmann increases is the result of the dominance of the intrinsic effect of the magnetic field, which tends to increase the magnetic irreversibility, on its extrinsic effect via the Lorentz force which tends to diminish it.

Figure 5. Variations of the velocity vector components: a) u; b) v c) velocity magnitude along the mediator plane (Y= 0.5) for different volume fraction Ha=10 and γ=15°

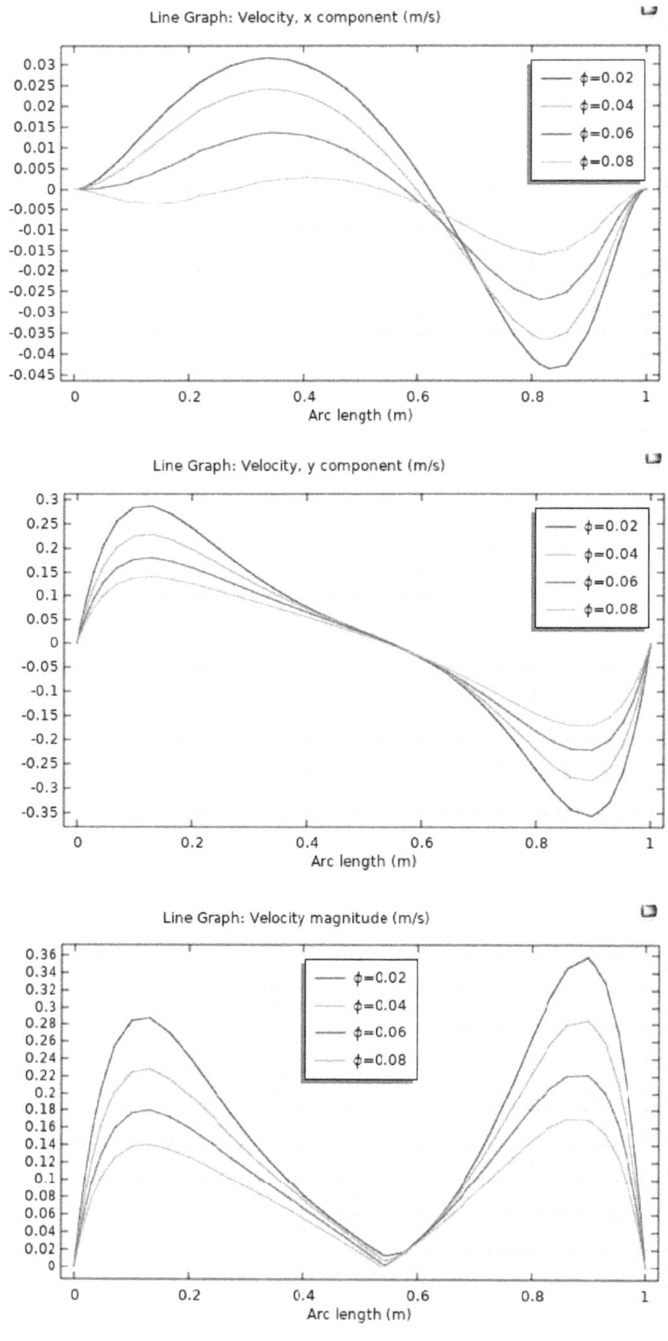

Figure 6. Variations of the velocity vector components: a) u; b) v c) velocity magnitude along the mediator plane (Y = 0.5) for different volume fraction Ha=40

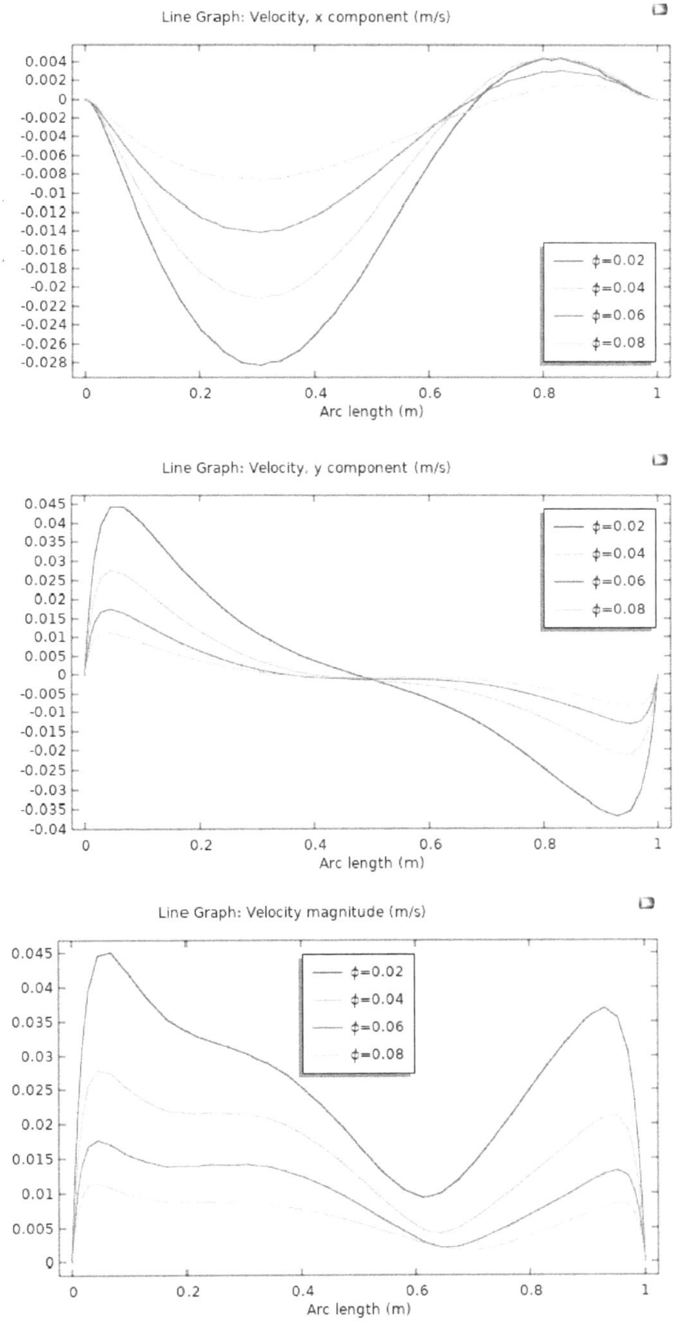

Figure 7. Effect of the volume fraction on the thermal entropy generation

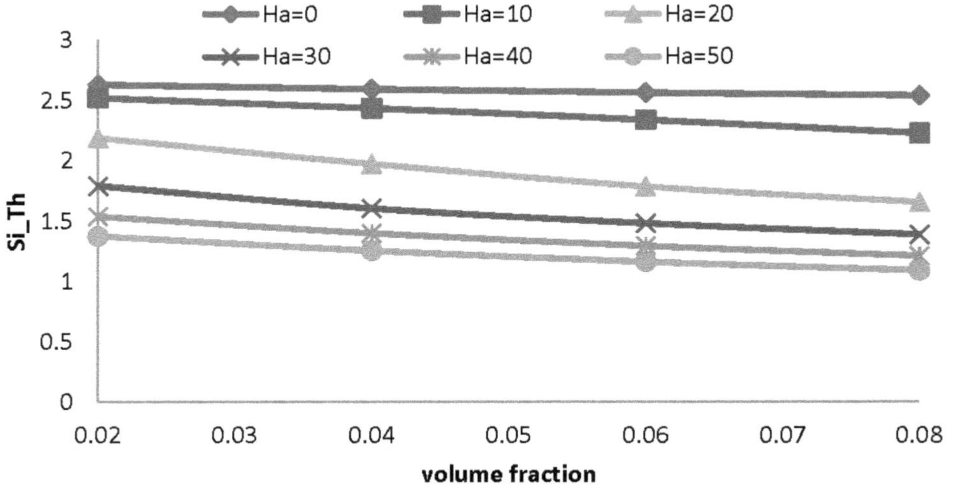

Figure 7 indicated that the thermal entropy generation for all Hartman number decrease when the volume fraction increase for a fixed value of Hartman number. The same for the entropy generation viscous Darcy and the total entropy generation illustrated in Figures 8 and 9.

Figure 8. Effect of the volume fraction on the viscous Darcy entropy generation

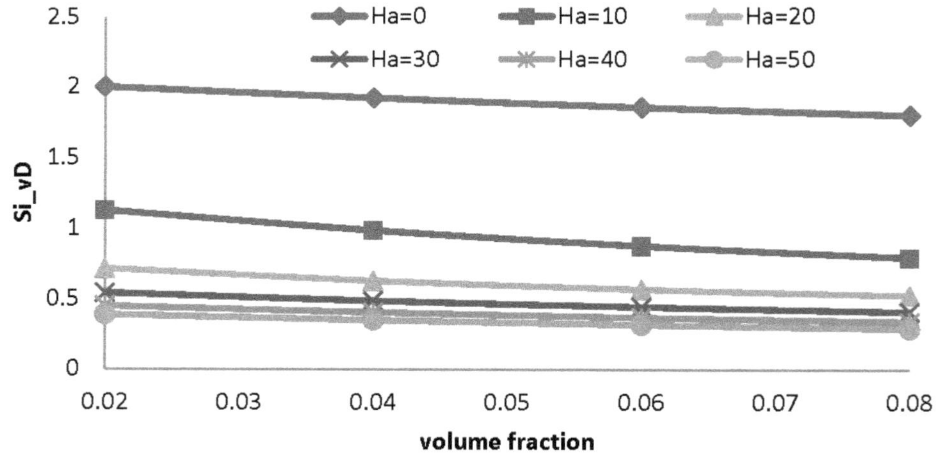

Figure 9. Effect of the Volume fraction on the total entropy generation

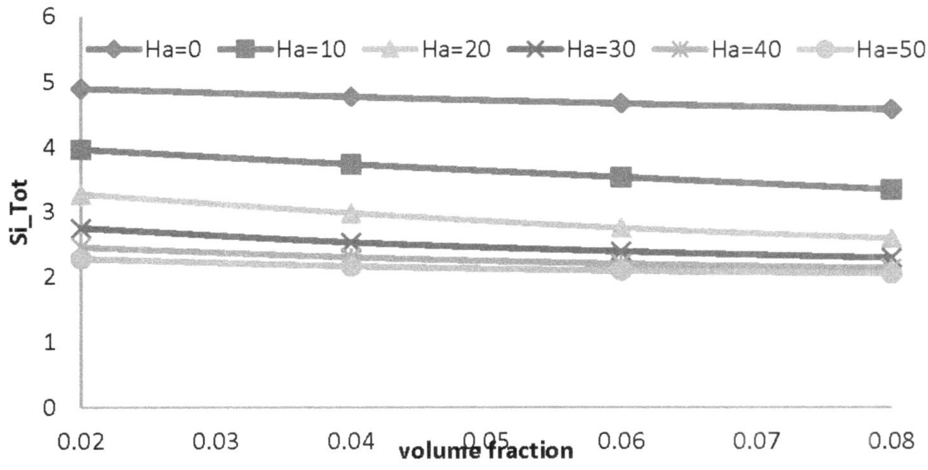

The average Nusselt number reduce when increasing the volume fraction. The effect of the Hartmann number in the average Nusselt number is more than the effect of volume fraction when the Ha increase the average Nusselt number decrease.

CONCLUSION

In the present article, the effects of volume fraction and magnetic field in mixed convection in porous media filled with nanofluid is studied. The variation of relevant parameters on the flow quantities is outlined and deliberated. The different irreversibilities, the average Nusselt Number, the velocity and temperature profiles are illustrated graphically. Some of the imperative outcomes are listed below:

1. The magnetic entropy generation and the viscous fluid entropy generation increase with increasing volume fraction and Hartman number
2. The viscous Darcy entropy generation, the thermal en entropy generation, and the total entropy generation decrease with increasing volume fraction and Hartman number.
3. The magnitude velocity decrease with increasing of the volume fraction and
4. The average Nusselt number reduce when increasing the volume fraction and the Hartmann number
5. The profile of temperature varied in a linear way when Ha is great.

Figure 10. Variations of the temperature along the mediator plane (X = 0.5) for different volume fraction. a) Ha=10 b) Ha=40

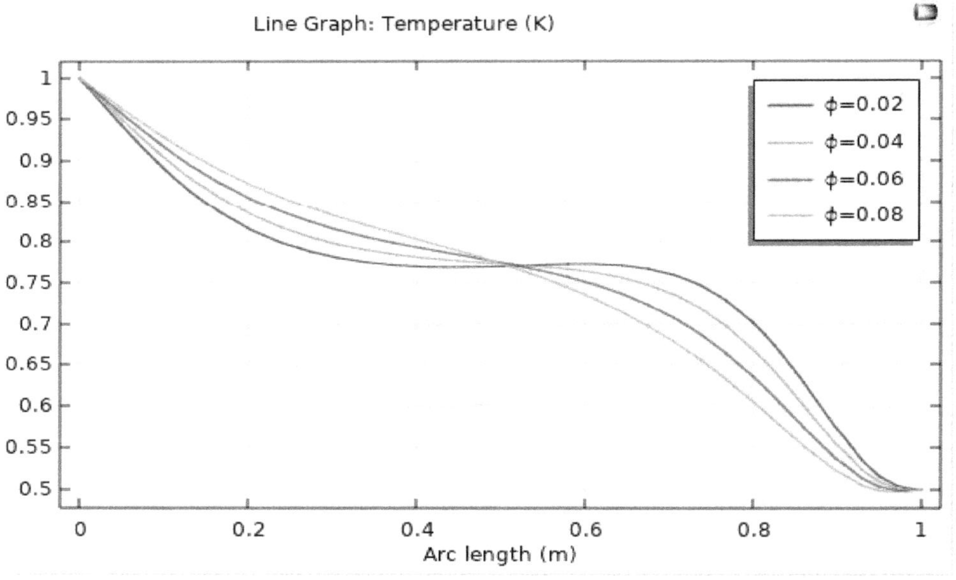

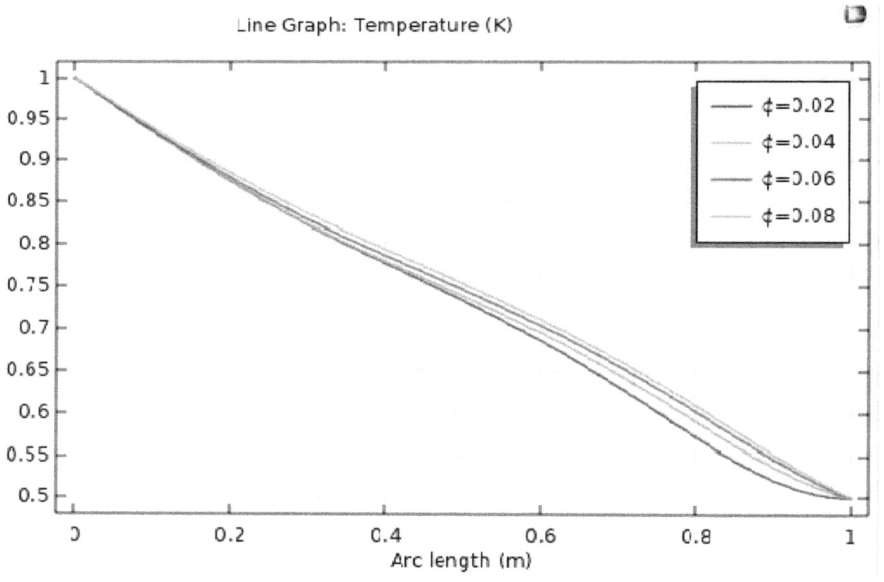

Figure 11. The average Nusselt number versus volume fraction for different Hartmann number

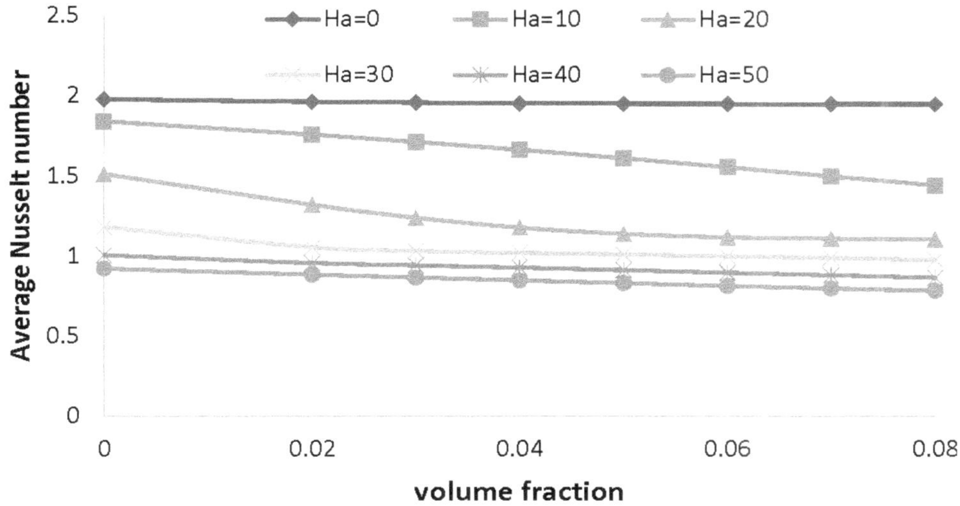

Figure 12. Variations of the velocity vector components and temperature: a) u; b) v c) velocity magnitude along the mediator plane; d) Temperature (X = 0.5) for different Hartmann number

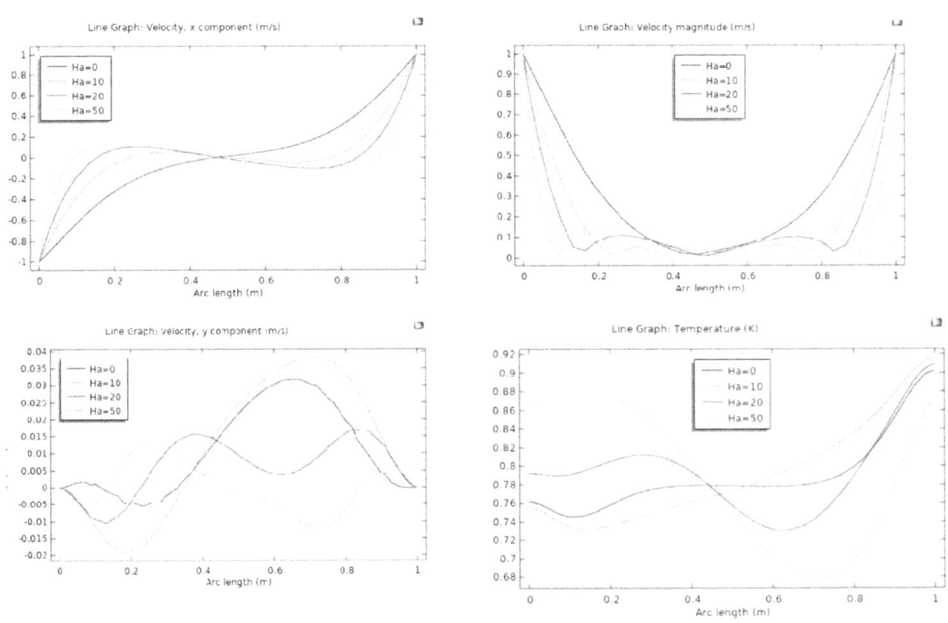

Figure 13. Variations of the velocity vector components and temperature: a) u; b) v c) velocity magnitude along the mediator plane; d) Temperature (Y = 0.5) for different Hartmann number

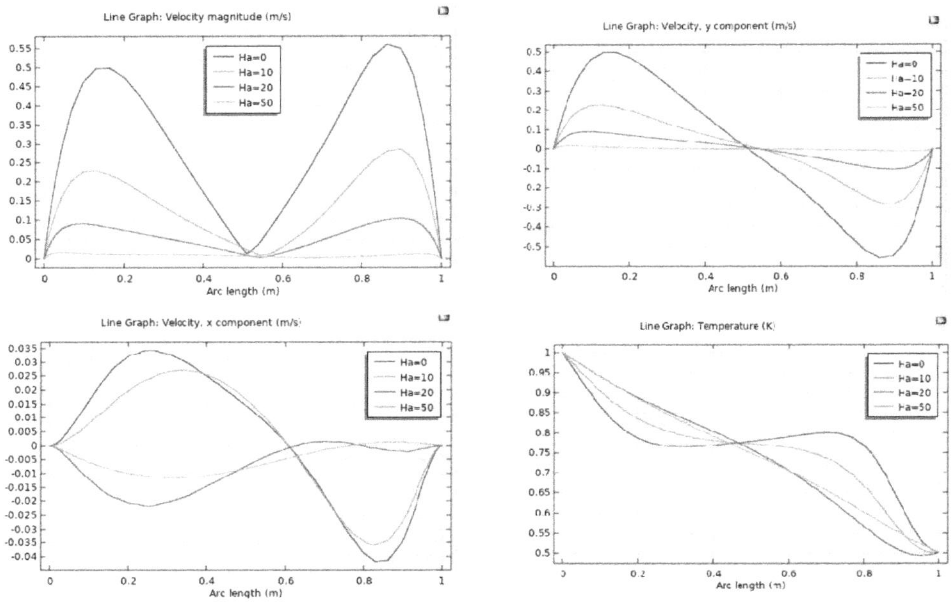

Figure 14. Streamlines and isotherms for different Hartmann number $\gamma=15°$

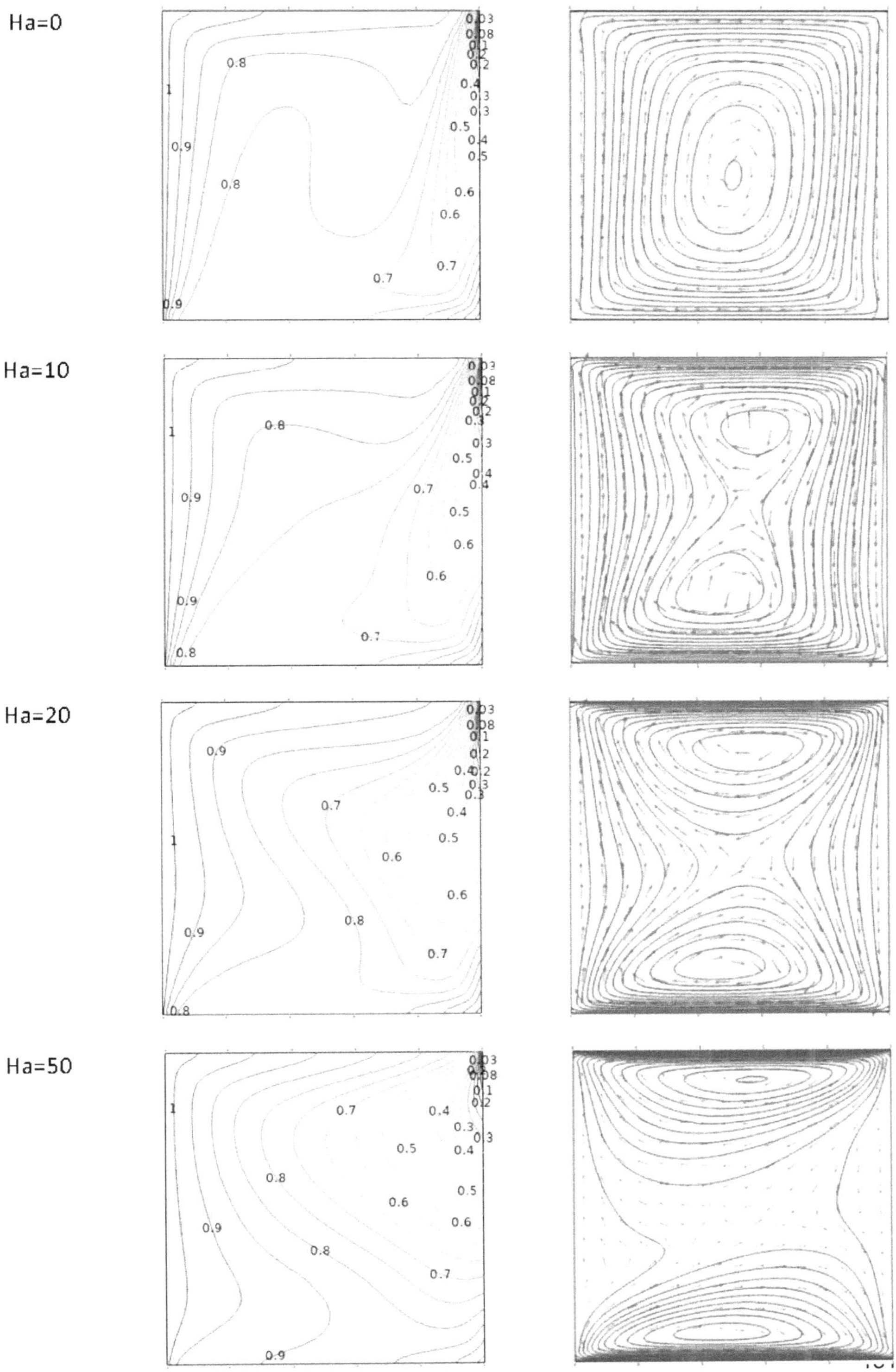

REFERENCES

Al-Salem, K., Oztop, H. F., Pop, I., & Varol, Y. (2012). Effects of moving lid direction on MHD mixed convection in a linearly heated cavity. *International Journal of Heat and Mass Transfer*, *55*(4), 1103–1112. doi:10.1016/j.ijheatmasstransfer.2011.09.062

Chamkha, A. J. (2002). Hydromagnetic combined convection flow in a vertical lid-driven cavity with internal heat generation or absorption. *Numerical Heat Transfer Part A-Application*, *41*(5), 529–546. doi:10.1080/104077802753570356

Dawood, M. M. K., & Teamah, M. A. (2012). Hydro-Magnetic Mixed convection double diffusive in a lid driven square cavity. *European Journal of Scientific Research*, *85*(3), 336–355.

Habeeb. (2012). Numerical Simulation of Convective Heat Transfer and Fluid Flow through Porous Media with Different Moving and Heated Walls. *International Journal of Mathematical, Computational, Physical, Electrical and Computer Engineering*.

Ingham, D. B., & Pop, I. (Eds.). (2005). *Transport Phenomena in Porous Media*. Elsevier.

Ismael, M. A., Pop, I., & Chamkha, A. J. (2014). Mixed convection in a lid-driven square cavity with partial slip. *International Journal of Thermal Sciences*, *82*, 47–61. doi:10.1016/j.ijthermalsci.2014.03.007

Khanafer, K., & Vafai, K. (2002). Double-diffusive mixed convection in a lid-driven enclosure filled with a fluid saturated porous medium. *Numerical Heat Transfer Part A*, *42*(5), 465–486. doi:10.1080/10407780290059657

Muthtamilselvan, M., Kandaswamy, P., & Lee, J. (2009). Hydromagnetic mixed convection in a lid-driven cavity filled with a fluid-saturated porous medium. *International Journal of Applied Mathematics and Mechanics*, *5*(7), 28–44.

Nield, D. A., & Bejan, A. (2006). *Convection in Porous Media*. Springer.

Rahman, M. M., Billah, M. M., Mamun, M. A. H., Saidur, R., & Hasanuzzaman, M. (2010). Reynolds and Prandtl numbers effects on MHD mixed convection in a lid-driven cavity along with joule heating and a centred heat conducting circular block. *Int. J. of Mech. and Mat. Eng.*, (5), 163–170.

Rudraiah, N., Barron, R. M., Venkatachalappa, M., & Subbaraya, C. K. (1995). Effect of a magnetic field on free convection in a rectangular enclosure. *International Journal of Engineering Science*, *33*(8), 1075–1084. doi:10.1016/0020-7225(94)00120-9

Saha, L.K., Somadder, M.C., & Roy, N.C. (2015). Hydro-magnetic mixed convection flow in a lid-driven cavity with wavy bottom surface. *American Journal of Applied Mathematics, 3*(1), 8-19.

Said, M., Nehari, D., Bouzit, M., & Chemlou, N.-E. S. (2013). Analysis of mixed convection in an inclined lid-driven cavity with a wavy wall. *Journal of Mechanical Science and Technology*, *27*(7), 2181–2190. doi:10.100712206-013-0533-9

Sharif, M. A. R. (2007). Laminar mixed convection in shallow inclined driven cavities with hot moving lid on top and cooled from bottom. *Applied Thermal Engineering*, *27*(5-6), 1036–1042. doi:10.1016/j.applthermaleng.2006.07.035

Sivanandam, S., Chamkha, A. J., Mallawi, F. O. M., Alghamdi, M. S., & Alqahtani, A. M. (2020). Effects of entropy generation, thermal Radiation and moving-wall direction on mixed convective flow of nanofluid in an Enclosure. *Mathematics*, *8*, 1471. doi:10.3390/math8091471

Vafai, K. (Ed.). (2005). *Handbook of Porous Media*. Taylor & Francis. doi:10.1201/9780415876384

Chapter 9
Microscopic and Macroscopic Interpretations of the Entropy Within the Framework of Quantum Mechanics:
Quantum Computer, Coulomb Crystal, Chaos, and Cosmology

ibtissem Jendoubi
Faculty of Sciences of Bizerte, Tunisia

Elhoucine Essefi
University of Gabes, Tunisia

ABSTRACT

The objective of this work was to microscopically and macroscopically interpret entropy within the framework of quantum mechanics: quantum computer, Coulomb crystal, chaos, and cosmology. Indeed, in quantum physics, the concept of information is the very basis of the minimal interpretation of the concept of state vector as a contextual prediction tool. The Coulomb crystal is the basic element for the development of a quantum computer. For example, the Coulomb crystal represents the basic element of high precision clocks, provides a favorable environment for the detailed study of chemical reactions, and constitutes an original technology for the development of a quantum computer. In addition, the combination of chaos with the recent definition of entropy allows us to understand very small systems at the atomic and quantum microscopic level, as well as very large systems at the macroscopic level of galaxies and black holes.

DOI: 10.4018/978-1-7998-8801-7.ch009

INTRODUCTION

Historically, the concept of entropy has been discussed from different viewpoints since 1870 (Ribeiro et al., 2021). It gives an idea about the evolution and reversibility of microscopic and macroscopic systems (Arias-Gonzalez, 2021). In the literature, the second law is controversially expressed in different ways (Jarzynski, 2011; Shahsavar et al., 2021). According to Clausius hypothesis 'heat does not spontaneously transfer from a cold body to a hot body'; while the quantum correlations may be used to prove this reversed heat flow (Micadei et al., 2019). On the other hand, Thomson's statement argues that 'A system in contact with a single source of heat can, in the course of a cycle, only receive work and supply heat'. These rules essentially define a kind of irreversibility; i.e., there is a state function generally called entropy S, which can only be increased in a closed system. At the macroscopic level, the second law allows us to calculate the equation of state based on the requirement that entropy must be maximized under a given wide range of variables in order to obtain the thermodynamic potential as a function of them (Tovbin, 2021; Valente, 2021; te Vrugt, 2021). At the microscopic level, irreversibility conflicts with the well-known reversibility of all the fundamental laws of physics (Strasberg & Winter, 2021; Melkikh, 2021).

Recently, the meaning of information is a pivotal concept in the modern physics (Tributsch et al., 2021). Certain physicists even interpreted the thermodynamic term of entropy of information loss in the framework of quantum mechanics like a measurement of the microscopic state of the system (Seshadr et al., 2021). This novel interpretation has led to new finding in the field of computer sciences (Ulyanov, 2021; Frank & Shukla, 2021), biology (Sarkar et al., 2021), cosmology (Tu et al., 2021; Weinstein et al., 2021; Jalalzadeh et al., 2021) and environmental sciences (Rapf & Kranert, 2021). Additionally, the second law of thermodynamics would only model « *the process of information loss* » correlating with the evolution of a system towards its equilibrium state. Recently, a sort of constructive skepticism regarding the explanatory value of the main informationalist trends in statistical thermophysics has been adopted (Anta, 2021a). More generally, informational interpretations of physics, or even attempts to reconstruct physics from an informational point of view, are in a constant development (Javier, 2021; Anta, 2021b). As it has been shown by some recent studies dealing with quantum information, they were proven to be promising (Wang et al., 2021). The concept of statistical information allowed the development of a coherent interpretation of thermodynamics by providing justification for the definition of thermodynamic entropy introduced by Boltzmann and Gibbs in statistical mechanics (Xu, 2021; Rajan, 2021).

Giving insights on recent definition of entropies, this chapter explains a widest interpretation of the second law of thermodynamics in the framework of quantum mechanics. In doing so, we present applications at the microscopic (quantum computer and coulomb crystal) and macroscopic (cosmology) levels. Further, combined application of Chaos Theory and entropies definitions were investigated to show their application in microscopic, macroscopic, natural and artificial phenomena.

THE SECOND LAW OF THERMODYNAMICS AND ITS INTERPRETATION IN THE FRAMEWORK OF STATISTICAL MECHANICS

The second law of thermodynamics, as formulated in the middle of the 19th century, is based on the fact that it is impossible to obtain mechanical energy (work) from heat flowing from a cold to a hot body. It is first statements, relating to the specifically thermal domain, are due to Clausius 'It is impossible to construct a device which operates on a cycle and whose sole effect is the transfer of heat from a cooler body to a hotter body' (Júnior, 2021), Kelvin-Planck 'It is impossible to construct a device which operates on a cycle and produces no other effect than the production of work and the transfer of heat from a single body' (Júnior, 2021), Thomson 'It is impossible to obtain useful work by cooling of a body with lowest temperature. These formulations depend on the terms "hot," "cold," and "a body with lowest temperature" or Carnot 'No heat engine operating between two heat reservoirs can be more efficient than a reversible heat engine operating between the same two reservoirs' (Bychkov, 2021).

For any thermodynamic system, open or closed, it is possible to define a state function S, called "entropy", which verifies the following properties: S is an extensive quantity, and its variation during any transformation is expressed as a sum of two terms:

$$\Delta S = \Delta S_e + \Delta S_i \tag{1}$$

$$\Delta S_e = \frac{\delta Q}{T} \tag{2}$$

whereas $\Delta S_i > 0$ results from internal modifications of the system associated with irreversible phenomena (Wang, 2021).

Molecular Chaos Hypothesis [Boltzmann 1872]

The molecular chaos hypothesis is the assumption that the velocities of colliding particles are uncorrelated, and independent of position.

Boltzmann invented a physical quantity noted:

$$S = -k_B \sum_{(l)} P_l ln P_l \tag{3}$$

with k_B is constant, called Boltzmann constant and $k_B = \frac{R}{\mathcal{N}_A} \simeq 1.38 \times 10^{-23} \times JK^{-1}$

S has the following properties:

- *S* is a positive quantity.
- when $S = 1$, i.e. when the system is frozen in a single micro-state.
- This quantity is -like any thermodynamic quantity- an average quantity.
- This quantity depends only on the distribution law of the micro-states P_l. In the sense of information theory, *S* is a measure of the lack of information.
- When an isolated system is allowed to evolve, Boltzmann showed that *S* grows (Theorem H).

For an isolated system, given that $P_l = \frac{1}{©}$, the relation (3) takes the famous form (engraved on Boltzmann's tombstone)

$$S = k_B ln\Omega \tag{4}$$

Where Ω is the number of possible configurations or "accessible" physical states.

Coarse-graining Hypothesis [Gibbs 1902]

The entropy determined by Boltzmann does not consider the interaction between the components of the liquid (except at the moment of shocks); it is therefore only suitable for perfect gases and not for real fluids. This is what led Gibbs to consider a more general definition of the thermodynamic entropy (te Vrugt, 2021). Gibbs defined statistical entropy for an arbitrary phase space probability density function (Gibbs, 1902) (*pdf*) $\rho(x)$ as:

$$S_G[\rho(x)] = -\int_{\Omega} \rho(x) ln\rho(x) dx \qquad (5)$$

which is usually called the Gibbs entropy. This concept can be generalized to quantum systems in a straightforward fashion, as shown by von Neumann (Von Neumann, 2018).

Random Phases Hypothesis [Pauli 1928]

In 1931, Von Neumann (1955) discovered the connection between two branches of physics: quantum mechanics and thermodynamics. Von Neumann defined the entropy of a density operator ρ by:

$$S_{VN} = -k_B Tr(\rho \ln \rho) \qquad (6)$$

The mechanical interpretation of thermodynamic irreversibility has been "solved" by Pauli in the quantum framework thanks to the use of an approximation (here, the hypothesis of "random phases").

The random phase hypothesis (Pauli, 1928) can be considered as a quantum replica of Boltzmann's molecular chaos hypothesis. According to this hypothesis, which is justified by the existence of uncontrollable external perturbations undergone by the system, the phase relations between its possible microstates are randomly distributed and are then very rapidly destroyed, in times very small on the scale of macroscopic evolution times (Vanchurin et al., 2021).

INTRODUCTION OF THE CONCEPT OF STATISTICAL INFORMATION IN PHYSICS

The terms `bit' and `qubit' denote the units of classical and quantum information, as well as classical and quantum systems, that are carriers of one information bit (qubit).

In modern classical computers, there are memory bits, which store information, and controllable bits in `circuits', which process information. In the magnetic memory of a computer, a bit is a magnetized region of a magnetic film: to two magnetization directions there correspond the `0' and `1' values of the information bit. The switching `0'!`1' or `1'!`0' requires overcoming the energy barrier between the two states of the film; it is the existence of the barrier that ensures the reliability of information storage.

The basis element of a quantum computer (the carrier of quantum information) is a quantum bit a qubit. In quantum communication systems, information is transmitted by the physical transfer of a qubit the information carrier or by teleportation of the quantum state of the qubit.

In this context, the algorithmic complexity of the qubit |Ψ as the length, in classical bits, of the smallest program of a quantum computer that could "produce" or "compute" exactly an approximated qubit |Φ of |Ψ, which would be "computable", while taking into account the fidelity of the computation, i.e., its degree of approximation which can be evaluated by the quantity $|\Psi|\Phi|^2$ (Vitányi, 2001).

For their part, Berthiaume, et al. (2001) propose to consider as a measure of the algorithmic complexity of a given sequence of qubits the length, measured in qubits, of the smallest quantum input of a quantum computer that "produces" or "computes" the sequence of qubits given initially, and always taking into account the fidelity of the computation.

The essential difference between these two approaches is that Vitányi (2001) proposes a measure of the quantum algorithmic complexity of a qubit in classical bits, as a measure of the length of a "classical" program running on a quantum computer, whereas Berthiaume et al. (2001) give a measure in terms of qubits, measuring the length of the quantum input of such a computer (Berthiaume et al., 2001).

However, these propositions raise the following essential question: what does it mean to "produce" or "calculate" a qubit? In the classical case, "calculating" a microstate of a physical system was tantamount to describing its space-time configuration, for example in terms of the positions and momentum of its constituents, i.e. describing what it really "is", independently of any other consideration such as our means of accessing knowledge of this state. "is" really, independently of any other consideration such as our means of access to the knowledge of this state. But, as we have noticed above, a quantum "state" cannot be considered as such. It is above all an abstract entity which encodes information relative to a given preparation and which preparation and which allows to predict the results of measurement that it would be possible to obtain by possible to obtain by measuring such or such observable.

QUANTUM PHYSICS AND COULOMB CRYSTAL

It is with trapped atomic ions that laser cooling was shown for the first time in the years 1970 (Scott et al, 1970; Wineland, 1979). But the decisive projection in this field intervened when it is realized that a chain "crystallized" ions could constitute the basic element for a computer quantum (Cirac et al, 2012). Since developments experimental based on crystals Coulomb led to one of the current achievements most elaborate of a quantum information system (Weckesser et al, 2021). What it is now advisable to call « quantum simulation » also became an important application: dynamics Hamiltonian of a complex system is simulated with through its implementation in another system like a Coulomb crystal: it is shown that the system obtained is then mathematically equivalent, while profiting from the possibilities from handling of the Coulomb crystal.

In this section, we have given an overview on strong correlation effects in quantum physics and the formation of Coulomb crystal, in particular Coulomb (Wigner) crystallization. Consequently, contrary to ionic crystals well known in solid state physics (such as Na^+Cl^-), coulomb crystals are composed of ions of the same charge. Their formation includes two essential steps essential steps: the trapping of the ions and their cooling. The trapping by an external action is necessary, since the ions strongly repel each other because of their charge (Willitsch et al., 2008).

At the experimental level, Willitsch et al. (2008) present the relevant energy levels used for Doppler laser cooling in Ca^+. It is shown that the Doppler laser cooling is performed on the 4s $^2S_{1/2}$-4p $^2P_{1/2}$ transition at a wavelength of 397 nm. The lowest translational temperature which can be achieved with this simple Doppler cooling scheme corresponds to: $T_{min,Doppler} = \dfrac{\hbar\Gamma}{2k_B}$ (where Γ is the linewidth of the laser-cooling transition) (Willitsch et al., 2008).

At the theoretical level, previous studies (Baiko et al., 2001) calculated the thermal thermodynamic functions (electrostatic and vibrational parts of the free energy, internal energy, and heat capacity) for *bcc* and *fcc* Coulomb crystals in the harmonic approximation. Baiko et al. present the calculations of thermodynamic functions for *bcc* and *fcc* lattices at any values of the quantum parameter $\theta = \dfrac{T_p}{T}$, where $T_p = \dfrac{\hbar\omega_p}{k_B}$ is the ion plasma temperature and $\omega_p = \sqrt{4\pi n_i Z^2 e^2 / M}$ is the ion plasma frequency (n_i, M, and Ze being the ion number *density*, *mass*, and *nuclear charge*, respectively) (Baiko et al, 2001).

In 2022, Baiko et al. performed calcul *ab initio* thermodynamics of one-component plasma for astrophysics of white dwarfs and neutron stars. They calculated energy of a crystal composed of atomic nuclei and uniform incompressible electron background in the temperature and density range, covering fully ionized layers of compact stellar objects, white dwarfs and neutron stars, including the high-density regime using path-integral Monte Carlo (PIMC) simulations of a bcc crystal composed of ions of a single sort and rigid uniform charge-compensating electron background (one-component plasma crystal) (Baiko et al., 2022).

On the other hand, Chabrier et al. (1992) suggested an approximate analytical model of the harmonic Coulomb crystal, which is widely used in astrophysics. However, precise numerical calculations of the thermodynamic functions, valid at any temperature T (Charbrier et al., 1992). The model of Chabrier et al. assumed a linear dispersion law for two acoustic (Debye-type) modes, $\omega_{\perp} = \alpha\omega q / q_B$, and an optical (Einstein-type) mode, $\omega_{\perp} = \alpha\omega q / q_B$. The known phonon spectrum moments of a Coulomb crystal are approximately reproduced with the choice $\alpha \cong 0.4$, $\gamma \cong 0.9$ (Chabrier et al., 1992).

ENTROPY AND COSMOLOGY

The second law of thermodynamics declares that entropy can only increase and that this principle extends, inevitably, to the universe as a whole (Eddington, 2022). John Wheeler was one of the first to examine more closely the notion of entropy in cosmology, which led him to transpose the problem to the physics of black holes (Wheeler, 2018). Bekenstein and Stephen Hawking came to conclude that a black hole has an entropy proportional to the area of the surface of its event horizon, the latter being itself proportional to the mass squared. This is the postulate of Bekenstein-Hawking entropy, then the black hole entropy, in dimensionless form, is given by: $S_{BH} = \frac{A}{4L_p^2} = \frac{c^3 A}{4G\hbar}$, where L_p stands for the Planck length $G\hbar/c^3$ *while* $G, \hbar$ *and* c denote, respectively, *Newton's gravity constant*, the *Planck-Dirac constant* and the *speed of light*. And A is given by: $A = 16\pi(GM/c^2)^2$, where M *is the black hole's mass* (Bekenstein, 2008). This evaporation process respects the laws of thermodynamics and produces an increase in the entropy of the Universe.

ENTROPY AND CHAOS

As a mathematical theory devoted to study dynamic systems strongly linked to entropy (Mussett, 2022), Chaos Theory presents special features including sensitivity to initial conditions, positive or close to zero Lyapunov exponents, statistics governed by gaussian or non-gaussian distributions. These features are unpredictable in the long run of time and space. This Theory were previously applied in different fields of science including philosophy (Sartre, 2022), mathematics (Xiong et al., 2021) phsycis (Ageev et al., 2021), chemistry (Dolomatov, 2021), geology (Essefi, 2009, 2013; Essefi et al., 2012), and biology (Mejia-Romero et al., 2021). Corresponding to dynamical chaotic systems, fractal features had been recently proven to be generated by the entanglement entropy (Ageev et al., 2021). Combining chaos with recent definition of entropies would give understanding to highly small systems at the microscopic levels of atoms and quantum (Fang et al., 2022) as well as highly big systems at the macroscopic level of galaxies and black holes (Giataganas, 2021).

CONCLUSION

Being a pivotal concept in the study of complicated systems, entropy (or entropies) has received a special care. In spite of the controversial definitions, entropy provides key answers for the study of the complicated systems. As a matter of fact, microscopic systems such as Coulomb crystals and quantum computers are governed by the evolution of the entropy. Macroscopic systems at the level of cosmological studies remain difficult to understand far from the entropy calculation.

REFERENCES

Ageev, D. S., Bagrov, A. A., & Iliasov, A. A. (2021). Deterministic chaos and fractal entropy scaling in Floquet conformal field theories. *Physical Review. B*, *103*(10), L100302. doi:10.1103/PhysRevB.103.L100302

Anta, J. (2021a). Can informational thermal physics explain the approach to equilibrium? *Synthese*, *199*(1-2), 1–24. doi:10.100711229-020-02967-8

Anta, J. (2021b). *A Philosopher against the Bandwagon*. Carnap and the Informationalization of Thermal Physics.

Arias-Gonzalez, J. R. (2021). Microscopically Reversible Pathways with Memory. *Mathematics*, *9*(2), 127. doi:10.3390/math9020127

Baiko, D. A., & Chugunov, A. I. (2022). Ab initio thermodynamics of one-component plasma for astrophysics of white dwarfs and neutron stars. *Monthly Notices of the Royal Astronomical Society*, *510*(2), 2628–2643. doi:10.1093/mnrastab3613

Baiko, D. A., Potekhin, A. Y., & Yakovlev, D. G. (2001). Thermodynamic functions of harmonic Coulomb crystals. *Physical Review. E*, *64*(5), 057402. doi:10.1103/PhysRevE.64.057402 PMID:11736150

Bekenstein, J. D. (2008). Bekenstein-hawking entropy. *Scholarpedia*, *3*(10), 7375. doi:10.4249cholarpedia.7375

Berthiaume, A., Van Dam, W., & Laplante, S. (2001). Quantum kolmogorov complexity. *Journal of Computer and System Sciences*, *63*(2), 201–221. doi:10.1006/jcss.2001.1765

Bychkov, E. G. (2022). An integrated approach for designing Joule-Thomson refrigerators operating with mixtures. *Applied Thermal Engineering*, *202*, 117837. doi:10.1016/j.applthermaleng.2021.117837

Chabrier, G., Ashcroft, N. W., & DeWitt, H. E. (1992). White dwarfs as quantum crystals. *Nature*, *360*(6399), 48–50. doi:10.1038/360048a0

Cirac, J. I., & Zoller, P. (2012). Goals and opportunities in quantum simulation. *Nature Physics*, *8*(4), 264–266. doi:10.1038/nphys2275

Dolomatov, M. Y. (2021). Theory of systems with chaos of chemical composition in nature and technology. *Butlerov Communications A*, *1*(2), 2.

Eddington, A. (2022). Entropy in Science and Metaphor. *Entropic Philosophy: Chaos, Breakdown, and Creation, 17*.

Essefi, E. (2009). *Multidisciplinary study of Sidi El Hani Saline Environment: The Geological History and the Climatic Variability*. Academic Press.

Essefi, E. (2013). *Wet Aeolian Sedimentology and Sequence Stratigraphy within the Terrestrial Analogues in Eastern Tunisia: Implications for Wet Aeolian Sedimentology and Sequence Stratigraphy on Mars* (Doctoral dissertation). National Engineering School of Sfax, Sfax, Tunisia.

Essefi, E., Tagorti, M. A., Touir, J., & Yaich, C. (2012). Modeling of the chaotic behaviors at Sidi El Hani discharge playa, eastern Tunisia: Contribution of the philosophy of causality to solve complex chaotic systems in geology and biology. *Tunis. J. Med. Plants Nat. Prod*, *7*, 116–128.

Fang, X., Gao, Q., Zhang, J., Wang, Y., Guo, X., & Guo, Y. (2022). Entropy Enhancement of Chaotic Laser via Quantum Noise. *Chinese Journal of Lasers*, *48*(21), 2112001.

Frank, M. P., & Shukla, K. (2021). Quantum Foundations of Classical Reversible Computing. *Entropy (Basel, Switzerland)*, *23*(6), 701. doi:10.3390/e23060701 PMID:34206044

Gibbs, J. W. (1902). *Elementary principles in statistical mechanics: Developed with especial reference to the rational foundations of thermodynamics*. C. Scribner's Sons.

Jalalzadeh, S., da Silva, F. R., & Moniz, P. V. (2021). Prospecting black hole thermodynamics with fractional quantum mechanics. *The European Physical Journal C, 81*(7), 1–13. doi:10.1140/epjc10052-021-09438-5

Jarzynski, C. (2011). Equalities and inequalities: Irreversibility and the second law of thermodynamics at the nanoscale. *Annual Review of Condensed Matter Physics, 2*(1), 329–351. doi:10.1146/annurev-conmatphys-062910-140506

Javier, A. (2021). *Historical and Conceptual Foundations of Information Physics.* Academic Press.

Júnior, P. F. (2021). On the Deduction of the Caratheodory's Axiom of the Second Law of Thermodynamics from the Clausius and Kelvin Principles. arXiv preprint arXiv:2110.02502.

Mejia-Romero, S., Lugo, J. E., Bernardin, D., & Faubert, J. (2021). Evaluation of Bio-movements Using Nonlinear Dynamics. In *Proceedings of International Conference on Data Science and Applications* (pp. 197-208). Springer. 10.1007/978-981-15-7561-7_16

Melkikh, A. V. (2021). Can Quantum Correlations Lead to Violation of the Second Law of Thermodynamics? *Entropy (Basel, Switzerland), 23*(5), 573. doi:10.3390/e23050573 PMID:34067021

Micadei, K., Peterson, J. P., Souza, A. M., Sarthour, R. S., Oliveira, I. S., Landi, G. T., Batalhão, T. B., Serra, R. M., & Lutz, E. (2019). Reversing the direction of heat flow using quantum correlations. *Nature Communications, 10*(1), 1–6. doi:10.103841467-019-10333-7 PMID:31165732

Mussett, S. M. (2022). *Entropic Philosophy: Chaos, Breakdown, and Creation.* Rowman & Littlefield.

Pauli, W. (1928). *Über das H-Theorem vom Anwaschsen der Entropie vom Standpunkt der neuen* [On the H-theorem of washing up entropy from the point of view of the new]. Quantenmechanik in Probleme des Modernen Physik.

Rajan, A. G. (2021). *A Pedagogical Approach to Obtain the Combined First and Second Law of Thermodynamics from Classical Statistical Mechanics.* Academic Press.

Rapf, M., & Kranert, M. (2021). Irreversible entropy to account for environmental impacts and sustainability. *Procedia CIRP, 98*, 601–606. doi:10.1016/j.procir.2021.01.161

Ribeiro, M., Henriques, T., Castro, L., Souto, A., Antunes, L., Costa-Santos, C., & Teixeira, A. (2021). The entropy universe. *Entropy (Basel, Switzerland)*, *23*(2), 222. doi:10.3390/e23020222 PMID:33670121

Sarkar, S., Hubbard, J. B., Halter, M., & Plant, A. L. (2021). Information thermodynamics and reducibility of large gene networks. *Entropy (Basel, Switzerland)*, *23*(1), 63. doi:10.3390/e23010063 PMID:33401415

Sartre, J. P. (2022). Entropy in Ancient Greek Thought. *Entropic Philosophy: Chaos, Breakdown, and Creation*, 35.

Scott, R. H., & Strasheim, A. (1970). Laser induced plasmas for analytical spectroscopy. *Spectrochimica Acta. Part B, Atomic Spectroscopy*, *25*(7), 311–332. doi:10.1016/0584-8547(70)80038-6

Seshadri, N., & Galperin, M. (2021). Entropy and information flow in quantum systems strongly coupled to baths. *Physical Review. B*, *103*(8), 085415. doi:10.1103/PhysRevB.103.085415

Shahsavar, A., Noori, S., Toghraie, D., & Barnoon, P. (2021). Free convection of non-Newtonian nanofluid flow inside an eccentric annulus from the point of view of first-law and second-law of thermodynamics. ZAMM-Journal of Applied Mathematics and Mechanics. *Zeitschrift für Angewandte Mathematik und Mechanik*, *101*(5), e202000266. doi:10.1002/zamm.202000266

Strasberg, P., & Winter, A. (2021). First and second law of quantum thermodynamics: A consistent derivation based on a microscopic definition of entropy. *PRX Quantum*, *2*(3), 030202. doi:10.1103/PRXQuantum.2.030202

te Vrugt, M. (2021). The five problems of irreversibility. *Studies in History and Philosophy of Science*, *87*, 136–146. doi:10.1016/j.shpsa.2021.04.006 PMID:34111817

te Vrugt, M. (2021). *The mereology of thermodynamic equilibrium*. arXiv preprint arXiv:2104.11140.

Thomson, W. (1851). Article. *Transactions of the Royal Society of Edinburgh*, *20*, 261–268, 289–298.

Tovbin, Y. K. (2021). Second Law of Thermodynamics, Gibbs' Thermodynamics, and Relaxation Times of Thermodynamic Parameters. *Russian Journal of Physical Chemistry A*, *95*(4), 637–658. doi:10.1134/S0036024421020266

Tributsch, H. (2021). Time-Neutrality of Natural Laws Challenged: Time Is Not an Illusion but Ongoing Energy-Driven Information Loss. *Journal of Modern Physics*, *12*(03), 300–327. doi:10.4236/jmp.2021.123023

Tu, F. Q., Sun, B., Wan, M., & Huang, Q. H. (2021). Evolution of the universe from the perspective of entropy and information. *Modern Physics Letters A*, *36*(16), 2150111. doi:10.1142/S021773232150111X

Ulyanov, S. V. (2021). Quantum Algorithm of Imperfect KB Self-organization Pt I: Smart Control-Information-Thermodynamic Bounds. *Artificial Intelligence Advances, 3*(2).

Valente, G. (2021). Taking up statistical thermodynamics: Equilibrium fluctuations and irreversibility. *Studies in History and Philosophy of Science*, *85*, 176–184. doi:10.1016/j.shpsa.2020.10.006 PMID:33966773

Vanchurin, V., Wolf, Y. I., Koonin, E. V., & Katsnelson, M. I. (2021). *Thermodynamics of Evolution and the Origin of Life*. arXiv preprint arXiv:2110.15066.

Vitányi, P. M. (2001). Quantum Kolmogorov complexity based on classical descriptions. *IEEE Transactions on Information Theory*, *47*(6), 2464–2479. doi:10.1109/18.945258

Von Neumann, J. (1955). Mathematical Foundations of Quantum Mechanics (R. T. Beyer, Trans.). Princeton University Press.

Von Neumann, J. (2018). *Mathematical foundations of quantum mechanics*. Princeton university press.

Wang, L. S. (2021). PROGRESS in entropy principle, as disclosed by nine Schools of thermodynamics, and its ecological implication. *Journal*, *16*(4), 359–372. doi:10.18280/ijdne.160403

Wang, X., Zhang, K., & Wang, J. (2021). *What can we learn about islands and state paradox from quantum information theory?* arXiv preprint arXiv:2107.09228.

Weckesser, P., Thielemann, F., Hoenig, D., Lambrecht, A., Karpa, L., & Schaetz, T. (2021). Trapping, shaping, and isolating of an ion Coulomb crystal via state-selective optical potentials. *Physical Review. A*, *103*(1), 013112. doi:10.1103/PhysRevA.103.013112

Weinstein, G. (2021). *Demons in Black Hole Thermodynamics: Bekenstein and Hawking*. arXiv preprint arXiv:2102.11209.

Wheeler, J. A. (2018). *Information, physics, quantum: The search for links*. CRC Press.

Willitsch, S., Bell, M. T., Gingell, A. D., & Softley, T. P. (2008). Chemical applications of laser-and sympathetically-cooled ions in ion traps. *Physical Chemistry Chemical Physics*, *10*(48), 7200–7210. doi:10.1039/b813408c PMID:19060963

Wineland, D. J., & Itano, W. M. (1979). Laser cooling of atoms. *Physical Review A*, *20*(4), 1521–1540. doi:10.1103/PhysRevA.20.1521

Xiong, P. Y., Jahanshahi, H., Alcaraz, R., Chu, Y. M., Gómez-Aguilar, J. F., & Alsaadi, F. E. (2021). Spectral entropy analysis and synchronization of a multi-stable fractional-order chaotic system using a novel neural network-based chattering-free sliding mode technique. *Chaos, Solitons, and Fractals*, *144*, 110576. doi:10.1016/j.chaos.2020.110576

Xu, X. (2021). *Gibbs Paradox in the View of Information Entropy*. arXiv preprint arXiv:2105.12566.

Chapter 10
Modelling of Active Magnetic Regenerative Refrigeration System Performance by New Approaches

Zina Meddeb
Institut Supérieur des Sciences Appliquées et Technologies de Gabès, Tunisia

ABSTRACT

This work aimed to study the coefficient of performance (COP) of an active magnetic regenerative refrigeration (AMRR) system by new analytical approaches of magnetic work Wm(B,x,y) and magnetocaloric effect MCE (T,B). Those approaches were applied to a permanent magnet magnetic refrigerator. The studied refrigeration system consisted of four regenerators, each of which was formed by parallel plates of gadolinium, a circulation pump, a rotating magnet, and two heat exchangers. The heat transfer fluids used were water and gallium. A resolution of the continuity equation, the amount of movement equation, and the heat equation were carried out in order to study the temperature profile in both the regenerator and the fluid. Furthermore, the authors deduced the temperatures at the inlet and the outlet of the heat exchangers in order to establish a thermal balance.

DOI: 10.4018/978-1-7998-8801-7.ch010

INTRODUCTION

Magnetic refrigeration is a cooling technology based on the magnetocaloric effect (MCE). The MCE describes the resulting change in temperature of a material due to the application of an external magnetic field. From an environmental point of view, this technology is so promising because it does not involve the use of greenhouse gases. The study of magnetic refrigeration has begun with the discovery of MCE (Lebouc et al., 2014, Allab et al. 2005) and then has been used in cryogenic refrigeration. It is maturely used in liquefaction of hydrogen and helium (Smaili et al., 2011, Utaki et al., 2007, Numazawa et al., 1993, Baker et al., 1978, Matsumoto et al., 2011, Park et al. 2014). The study of magnetic refrigeration of paramount importance in both the design and size of prototypes.

Several works on optimization were carried out (Bouchekara et al., 2008, Roy et al., 2016, Niknia et al., 2016). The majority of these works used a constant MCE, which is not proper. Similarly, previous works did not take into account magnetic work. Only Andrew Row proposed a model of magnetic work with several simplifying assumptions (Andrew 2012 Part I, Andrew 2012 Part II).

In this work, the originality of this work lies in the fact that the expression of the magnetic work (Wm) is distinct from those used in the literature. It is given by a new detailed analytical method. Furthermore, the MCE was used as a parameter variable which is an analytical expression as a function of the material nature, the local regenerator temperature and the applied magnetic field. It is a new important approach because in general this parameter is either considered as constant, which is not always the case, or determined by an approximate semi-empirical method.

The coefficient of performance (COP) is a dimensionless quantity that depicts the performance of the refrigeration cycle. The Variation of the COP, as a function of Qc, was studied.

STUDIED SYSTEM

The cooling system consisted of four generators constituted by parallel gadolinium plates, two heat exchangers, a pump and a rotating magnet (Figure 1) (Rosario et al 2011, Lebouc et al. 2014). The heat transfer fluid was water or gallium. The regenerator was magnetized and demagnetized periodically (Figure 2).

The system fluid passed through the hot heat exchanger to transfer heat Qh to the hot source. Then, the fluid passed through the demagnetized regenerator to provide heat. This cold fluid cooled the cold source by exchanging heat with cooling load Qc. The fluid was then heated by the magnetized regenerator, and it continued the cycle around.

Figure 1. A prototype used (proposed by G2Elab staff)

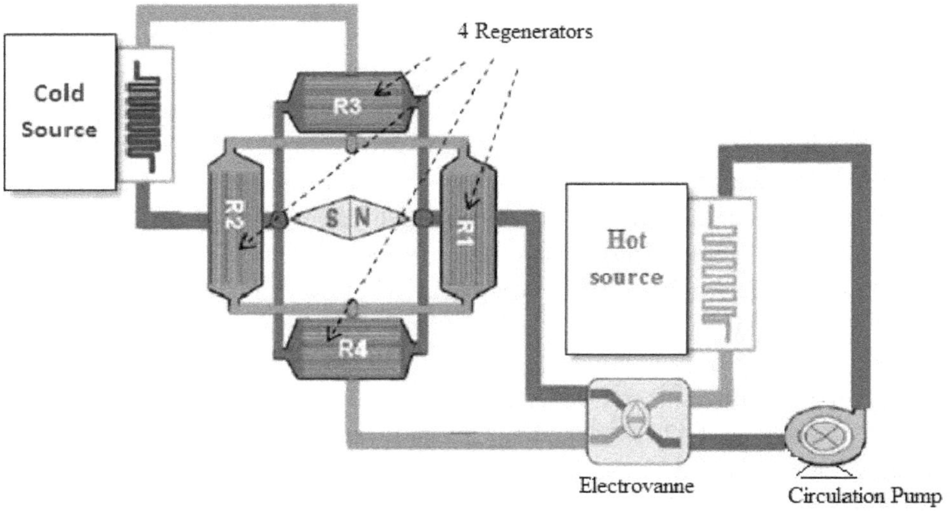

Figure 2. The steps of a period

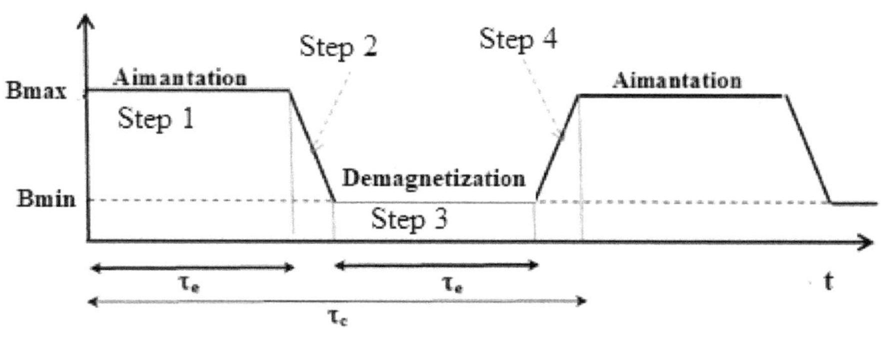

PERFORMANCE COEFFICIENT

Thermodynamic System

The existing energies in the system are defined as follow: Qc is the heat transferred with the cold source. Qh is the heat transferred with the hot source, and Qr is the heat transferred with regenerators 1 and 2 or 3 and 4. Qlosses is the lost energy by the system. Wp is the work of pump. We is the engine work of the magnet rotation and Wm is the work received by the material during magnetization.

The thermodynamic system is the fluid. It transfers heats and works with the hot source, the cold source and the circulation pump. Therefore, Wp embodies the thermodynamic cycle work.

The COP is calculated by the ratio of the cooling load (Qc) and the supplied work to the system (Ws).

So the COP is calculated by:

$$COP = \frac{Gain}{Supply} = \frac{Qc}{Ws} = \frac{Qc}{Wp + We + Wm} \tag{1}$$

Concerning *Wp* and *We*, we used two Classic formulas (Arques, 2009, Padet, 2008):

$$Wp = \frac{8\mu L}{\pi R^4} D_v{}^2.\tau_e \tag{2}$$

And

$$W_e = C_e.w.t_c \tag{3}$$

Concerning Wm, we proposed a New Approach which will be described below.

The heat transferred with the cold source (*Qc*) is:

$$Qc = \int_{Tfce}^{Tfcs} m_f Cp_f(T)dT \tag{4}$$

The heat transferred with the hot source (*Qh*) is:

$$Qh = \int_{Tfhe}^{Tfhs} m_f Cp_f(T)dT \tag{5}$$

The heat transferred with the regenerator (*Qr*) is:

$$Qr = \int_{Tf_c}^{Tf_h} m_f Cp_f(T)dT == \rho_m.e_m.l_{eq}.\Delta x \sum_{1}^{Nx-1} Cp_m(T).MCE(T,B) \tag{6}$$

The second focus of our work consisted in identifying the formulas of MCE (T,B), which was determined by an analytical new approach. Afterwards, T(x) was determined by a numerical calculation.

The New Approach to Calculate the Wm

The magnetization (M) is given by the following relation:

$$M = \chi_m \frac{B}{\mu_0} \tag{7}$$

The magnetic moment is:

$$dm = M.dV \tag{8}$$

dV is an elemental volume in the magnetized material. So

$$dm = \chi_m \frac{B}{\mu_0} dV \tag{9}$$

The elemental force acting on the material is given by eq (10):

$$\overrightarrow{dF} = \left(\overrightarrow{dm}.\overrightarrow{grad}\right).\vec{B} \tag{10}$$

If we use the expression proposed by (Tomas B. et al.), the average applied wave form as a function of the angle of rotation of the magnet can be expressed by:

$$B = B_{min} + B_{amp} \frac{\sqrt{2}}{2}\left(1 + \sin\theta\right) \tag{11}$$

Where B_{min} and B_{amp} represent respectively the minimum field and amplitude.

According to the calculation presented in Appendix 1, we can write the equation as follows:

$$F_x(x) = \int_0^x \left(\int_d^{d+lm} \frac{\chi_m}{\mu_0} \left(B_x.\frac{\partial B_x}{\partial x} + B_y.\frac{\partial B_x}{\partial y} \right) dy \int_0^{em} dz \right) dx \tag{12}$$

Figure 3 shows the variation of the relative permeability as a function of magnetic fields and temperature (Bouchard, 2005). As can be seen from this figure, it can be assumed that the relative permeability is constant for a non-weak magnetic field($\geq$ 0,1T).

Figure 3. Relative magnetic permeability for gadolinium (Bouchard, 2005)

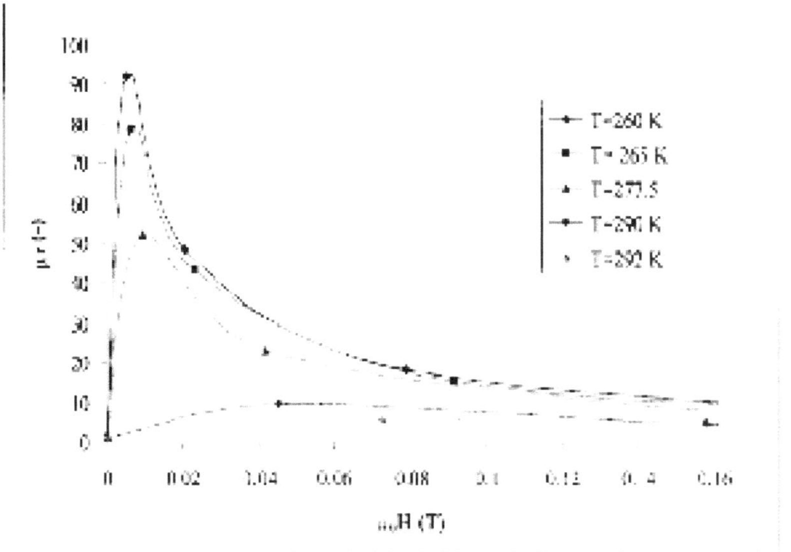

Hypothesis: $\mu_r \approx cst$ (13)

And since the magnetic susceptibility is given by the following relation:

$$\chi_m = \frac{\mu_r - 1}{\mu_r} \tag{14}$$

It may be assumed that:

$$\chi\text{m} = cst \tag{15}$$

$$F_x(x) = \frac{\chi_m}{\mu_0} \int_0^x \left(\int_d^{d+lm} \left(B_x . \frac{\partial B_x}{\partial x} + B_y . \frac{\partial B_x}{\partial y} \right) dy \int_0^{em} dz \right) dx \tag{16}$$

The same applies to *Fy*

$$F_y(y) = \frac{\chi_m}{\mu_0} \int_0^y \left(\int_0^{Lm} \left(B_x . \frac{\partial B_x}{\partial x} + B_y . \frac{\partial B_x}{\partial y} \right) dx \int_0^{em} dz \right) dy \tag{17}$$

So, the local magnetic work is:

$$dWm = F_x.dx + F_y.dy \tag{18}$$

$$Wm = \int_0^{Lm} F_x dx + \int_d^{d+lm} F_y dy \tag{19}$$

The integration is calculated numerically by MuPAD after replacing B_x and B_y by their expressions.

Or

$$d\vec{F} = \left(dm_x \frac{\partial}{\partial x} + dm_y \frac{\partial}{\partial y} + dm_z \frac{\partial}{\partial z} \right).\vec{B} \tag{20}$$

In effect

$$\begin{pmatrix} dm_x = \chi_m dV \dfrac{B_x}{\mu_0} \\ dm_y = \chi_m dV \dfrac{B_y}{\mu_0} \\ dm_z = \chi_m dV \dfrac{B_z}{\mu_0} \end{pmatrix} \tag{21}$$

So

$$\begin{pmatrix} dF_x = dm_x \dfrac{\partial B_x}{\partial x} + dm_y \dfrac{\partial B_x}{\partial y} + dm_z \dfrac{\partial B_x}{\partial z} \\ dF_y = dm_x \dfrac{\partial B_y}{\partial x} + dm_y \dfrac{\partial B_y}{\partial y} + dm_z \dfrac{\partial B_y}{\partial z} \\ dF_z = dm_x \dfrac{\partial B_z}{\partial x} + dm_y \dfrac{\partial B_z}{\partial y} + dm_z \dfrac{\partial B_z}{\partial z} \end{pmatrix} \tag{22}$$

Which gives:

$$\begin{pmatrix} dF_x = \chi_m dV \dfrac{B_x}{\mu_0}.\dfrac{\partial B_x}{\partial x} + \chi_m dV \dfrac{B_y}{\mu_0}.\dfrac{\partial B_x}{\partial y} + \chi_m dV \dfrac{B_z}{\mu_0}\dfrac{\partial B_x}{\partial z} \\ dF_y = \chi_m dV \dfrac{B_x}{\mu_0}.\dfrac{\partial B_y}{\partial x} + \chi_m dV \dfrac{B_y}{\mu_0}.\dfrac{\partial B_y}{\partial y} + \chi_m dV \dfrac{B_z}{\mu_0}\dfrac{\partial B_y}{\partial z} \\ dF_z = \chi_m dV \dfrac{B_x}{\mu_0}.\dfrac{\partial B_z}{\partial x} + \chi_m dV \dfrac{B_y}{\mu_0}.\dfrac{\partial B_z}{\partial y} + \chi_m dV \dfrac{B_z}{\mu_0}\dfrac{\partial B_z}{\partial z} \end{pmatrix} \quad (23)$$

$$B = B_{min} + B_{amp}\frac{\sqrt{2}}{2}\left(1 + \sin\theta\right) \quad (24)$$

$$B = B_{min} + B_{amp}\frac{\sqrt{2}}{2}\left(1 + \frac{x}{\sqrt{x^2 + y^2}}\right) \quad (25)$$

Then

$$\begin{pmatrix} B_x = Bcos\theta = B\dfrac{x}{\sqrt{x^2 + y^2}} \\ B_y = Bsin\theta = B\dfrac{y}{\sqrt{x^2 + y^2}} \\ B_z = 0 \end{pmatrix} \quad (27)$$

Hence

$$\begin{pmatrix} dF_x = \chi_m dV \dfrac{B_x}{\mu_0}.\dfrac{\partial B_x}{\partial x} + \chi_m dV \dfrac{B_y}{\mu_0}.\dfrac{\partial B_x}{\partial y} \\ dF_y = \chi_m dV \dfrac{B_x}{\mu_0}.\dfrac{\partial B_y}{\partial x} + \chi_m dV \dfrac{B_y}{\mu_0}.\dfrac{\partial B_y}{\partial y} \end{pmatrix} \quad (28)$$

$$dF_x = \frac{\chi_m}{\mu_0}\left(B_x.\frac{\partial B_x}{\partial x} + B_y.\frac{\partial B_x}{\partial y}\right)dV \quad (30)$$

$$dF_x = \frac{\chi_m}{\mu_0}\left(B_x.\frac{\partial B_x}{\partial x} + B_y.\frac{\partial B_x}{\partial y} \right) dxdydz \tag{31}$$

$$F_x(x) = \int_0^x \left(\int_d^{d+lm} \frac{\chi_m}{\mu_0}\left(B_x.\frac{\partial B_x}{\partial x} + B_y.\frac{\partial B_x}{\partial y} \right) dy \int_0^{em} dz \right) dx \tag{32}$$

The same applies to *dFy*

$$F_y(y) = \int_0^y \left(\int_0^{Lm} \frac{\chi_m}{\mu_0}\left(B_x.\frac{\partial B_y}{\partial x} + B_y.\frac{\partial B_y}{\partial y} \right) dx \int_0^{em} dz \right) dy \tag{33}$$

EQUATIONS OF MOVEMENTS AND ENERGIES

In this work, the effects of the applied magnetic field on the fluid were taken into consideration. The governing equations of the active magnetic regenerator (AMR) model equations are presented as follows (Allab et al., 2005; Mehrez, 2013).

Continuity equation

$$\frac{\partial u}{\partial x} + \frac{\partial v}{\partial y} = 0 \tag{34}$$

Motion quantity conservation equations

Following x: $$\rho\frac{\partial u}{\partial t} = -\frac{\partial p}{\partial x} + \mu\left(\frac{\partial^2 u}{\partial x^2} + \frac{\partial^2 u}{\partial y^2} \right) - \rho\sigma_e B^2 u \tag{35}$$

Following y: $$\rho\frac{\partial v}{\partial t} = -\frac{\partial p}{\partial y} + \mu\left(\frac{\partial^2 v}{\partial x^2} + \frac{\partial^2 v}{\partial y^2} \right) + \rho g \tag{36}$$

Energy Balance

$$\text{Fluid: } c_f\left(u\frac{\partial T_f}{\partial x}+v\frac{\partial T_f}{\partial y}\right)-\frac{\lambda_f}{\rho_f}\left(\frac{\partial^2 T_f}{\partial x^2}+\frac{\partial^2 T_f}{\partial y^2}\right)+c_f\frac{\partial T_f}{\partial t}=\frac{1}{m_f}hS\left(T_m-T_f\right) \tag{37}$$

$$\text{Material: } c_m\frac{\partial T_m}{\partial t}+\frac{\lambda_m}{\rho_m}\left(\frac{\partial^2 T_m}{\partial x^2}+\frac{\partial^2 T_m}{\partial y^2}\right)=\frac{1}{m_m}hS\left(T_f-T_m\right) \tag{38}$$

NUMERICAL CALCULATION

Two softwares were used. To perform the parametric study, we used Matlab. Then, once the geometrical parameters were optimized, the ANSYS Fluent software was used.

Resolution with MATLAB

To solve the above equation system, we used the finite difference method. Therefore, after discretization, and some developments, the energy balance can be written as:

The energy balance can be written as follows:

$$T_{f(i+1,j)}=A_{f1}T_{f(i,j)}+A_{f2}T_{f(i,j-1)}+A_{f3}T_{m(i,j)} \tag{39}$$

$$T_{m(i+1,j)}=A_{m1}T_{m(i,j)}+A_{m2}T_{f(i,j)} \tag{40}$$

$$Q_{(i+1,j)}=hS\left(T_{m(i+1,j)}-T_{f(i+1,j)}\right) \tag{41}$$

With

$$A_{f1}=\left(1-\left(v\frac{\Delta t}{\Delta x}+\frac{hS}{m_f C_f}\right)\Delta t\right) \tag{42}$$

$$A_{f2}=v\frac{\Delta t}{\Delta x} \tag{43}$$

$$A_{f3} = \frac{hS}{m_m c_m} \Delta t \tag{44}$$

$$A_{m1} = \left(1 - \frac{hS}{m_m c_m} \Delta t\right) \tag{45}$$

$$A_{m2} = \frac{hS}{m_m c_m} \Delta t \tag{46}$$

A Matlab code was developed. The calculation process encompassed four steps of AMR. The flowchart given in Figure 4 illustrates the various steps of the code.

Figure 4. Calculation process

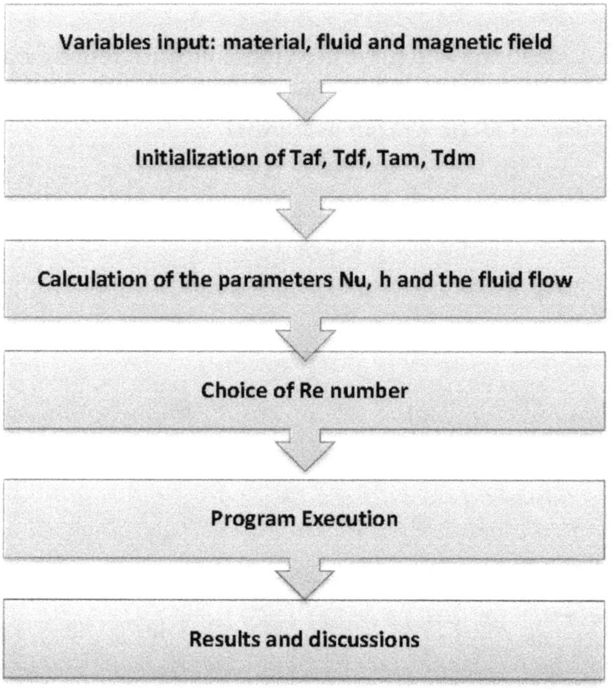

Resolution with ANSYS Fluent

Modeled Regenerator and Boundary Conditions

The boundary conditions of the modeled regenerator are presented in Figure 5.

Figure 5. The boundary conditions of the modeled regenerator

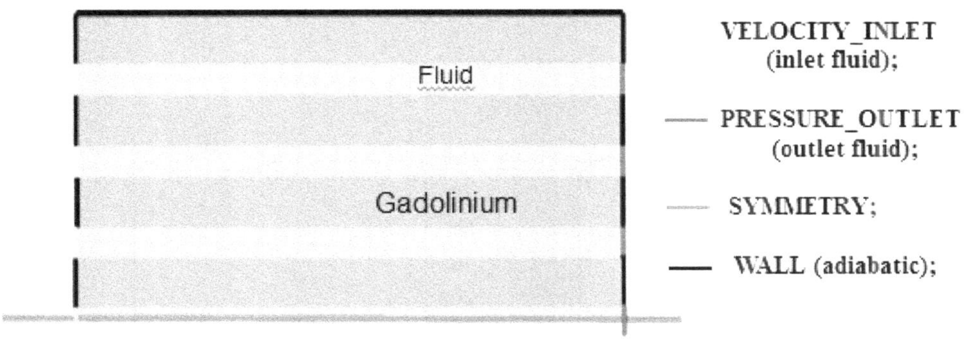

Initialization

After having introduced the boundary conditions and the first initialization, the initialization was repeated before each new half cycle. For this purpose, a function was defined:

Custom Field Function:

$$EMC(T,B)=\left(1,4+1,9e^{\left(-\frac{2(T-299)}{116}\right)}\right)B \tag{50}$$

Temperature Initialization:

$$\mathrm{T}=\mathrm{T}\pm \textit{Custom Field Function} \tag{51}$$

The program was run for each half cycle. Then the output was changed into input and the input into output.

RESULTS AND DISCUSSION

Parametric Study

Figure 6 shows the variation of the temperature profile along the regenerator at various times (from 1 to 60 s). Figures 6 (a) and (b) are for a length of the active material equal to 0.1 and 0.15m, respectively. It is noted that the temperature difference of the two ends of the material increases as a function of time. This difference increases if the length of the material is lower provided that the other parameters remain constant.

Figure 6. The effect of the material length

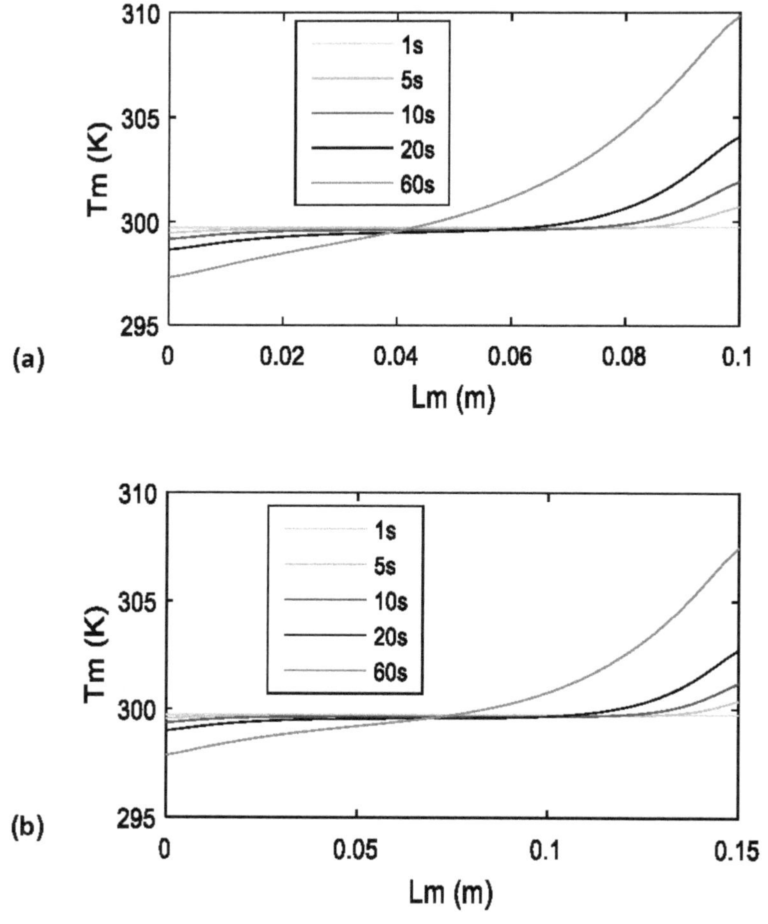

Figure 7 illustrates the variation of the temperature profile along the regenerator at different times. Figure 7(a) relates to a temperature difference between the hot and the cold sides equal to 10 and Figure 7(b) for a temperature difference between the hot and the cold sides equal to 20. It is noted that the temperatures of both the hot and the cold sources influence the temperature profiles and the duration of the transient regime. As can be noted from Figure 8, when the thickness of the fluid film increases, the medium of the material becomes a dead zone and a horizontal bearing can be observed. Therefore, if the thickness of the fluid film increases considerably, there will be an absence of transfer in the middle of the material.

Figure 7. The effect of temperature span

(a)

(b)

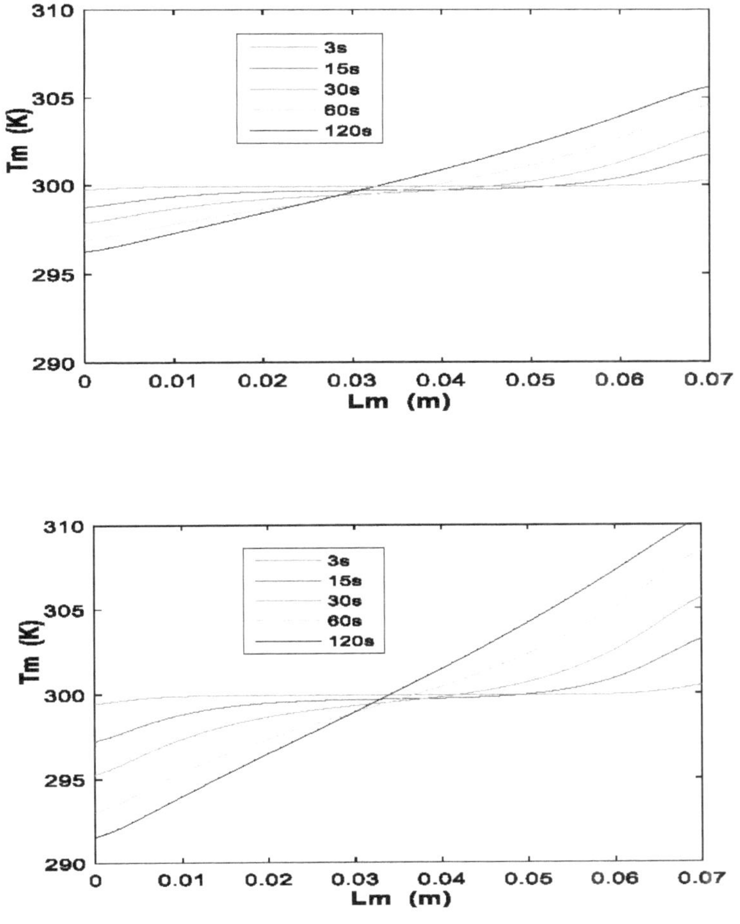

Figure 8. The effect of the fluid film thickness

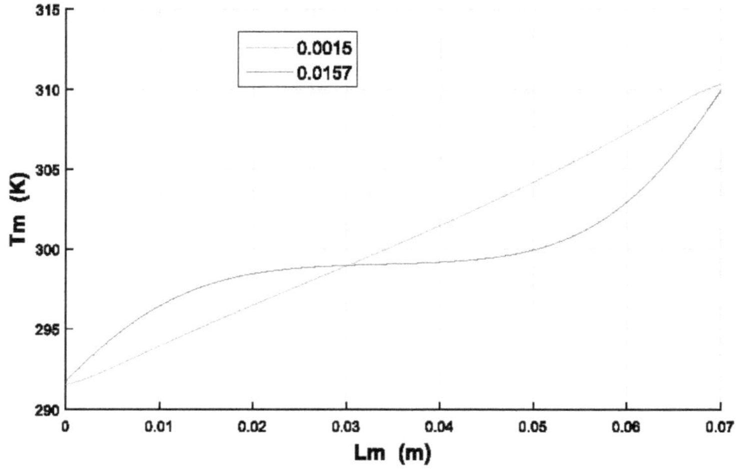

As the thickness of the material increased, the temperature of the material remained almost equal to the initial value (Figure 9) because the quantity of heat generated (Qr) is greater than that transferred. Figure 10 shows the temperature profiles along the material for two fluids of different properties. The blue curve is the temperature on the cold side and the red curve is the temperature on the warm side. Compared to water, heat transfer with gallium is superior because gallium is an electrically charged fluid. Figure 11 shows the temperature variation of two ends of the regenerator for several heat transfer fluid flow rates. It is deduced that there are several parameters affecting on the operating system.

Figure 9. The effect of the fluid film thickness

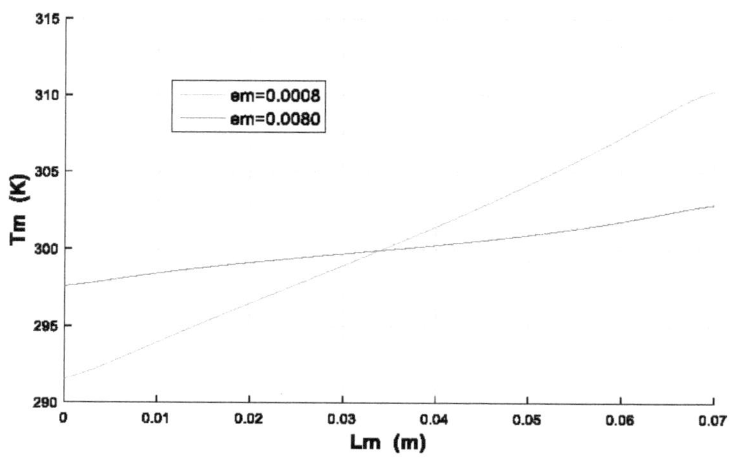

Figure 10. The nature of the heat transfer fluid effect

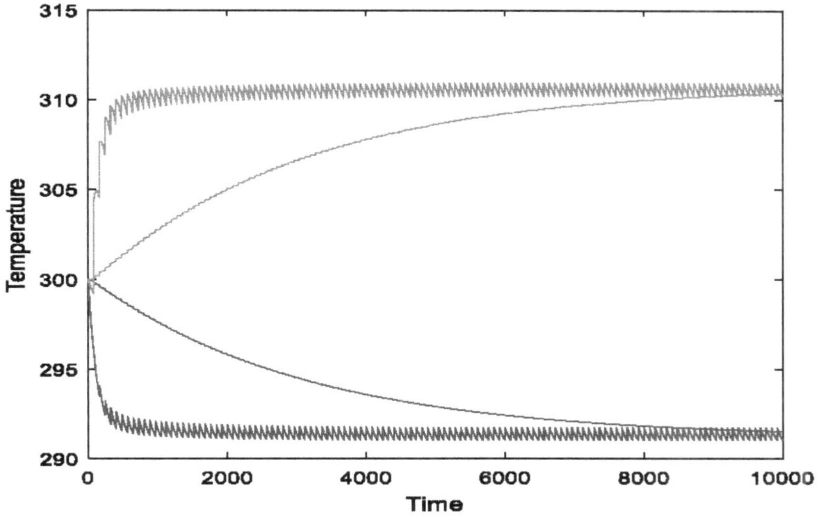

Figure 11. Temperature profiles for different flow rates

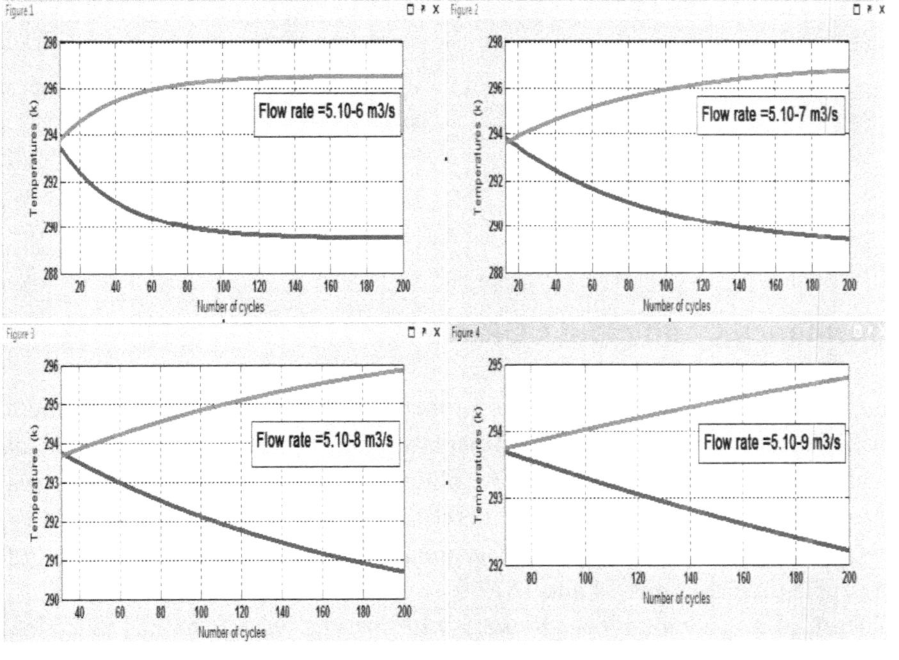

These results, therefore, show that it is important to optimize the operating parameters before discussing the performance of this cooling system. An optimization was achieved by DOE and FGOT software to determine the optimal operating parameters of the system. However, optimization can be possible only of we have a previous model of the response (Qr or Nc ....). Consequently, we used the experimental design method to do the modeling. Given that an experiment is equivalent to a simulation of the program with the chosen combination of independent parameters. The results are comparable to those found in the literature (Bouchekara, 2008, Roy, 2016,Niknia, 2016).

Figure 12 gives an example of the temperature profile along the regenerator T(x).

Figure 12. An example of a profile of the temperature along the regenerator

Performance Coefficient COP

Figure 13 shows the COP variation as a function of the period (τc). As can be seen from this figure, when the cycle period increases, the magnetization/demagnetization frequency decreases. Thus the heating / cooling frequency of the regenerator decreases which causes the COP to decrease. With the same amount of heat extracted from the cold source Qc, the COP at B = 1T was greater than the COP at B = 2T regardless of the cycle time (Figures 14 and 15).

Therefore it is advantageous to work with just the necessary magnetic field to increase the COP of the system. COP values are important in comparison with other cooling technologies. The COP can reach 11 as a value; whereas the COP of an absorption heat pump, for example, does not exceed 0.8. Figure 16 shows the COP variation as a function of the heat transfer fluid flow rate. It can be noted that if the flow of heat transfer fluid increases then the COP decreases. In fact, when the flow of fluid increases, the work of the circulation pump (Wp) increases. Therefore, the energy consumed increases, hence the decrease in the COP. The required mass increases when the power of the cold source increases (Figure 17).

Figure 13. The variation of COP in relation to the time cycle (for B =1T)

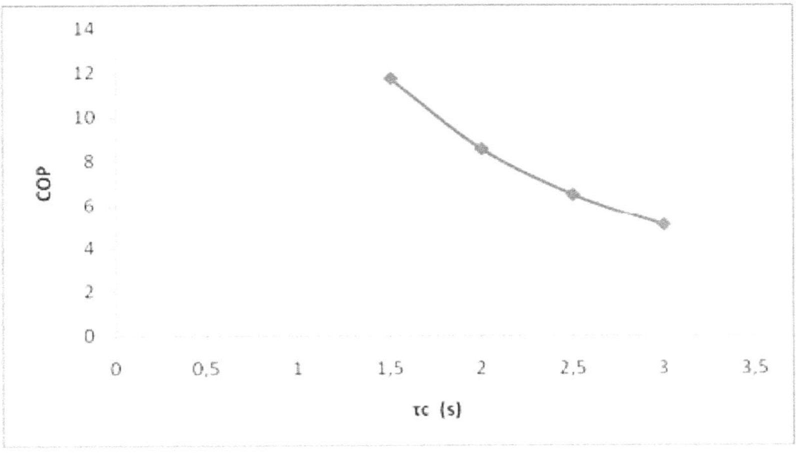

Figure 14. The variation of COP in relation to the Qc

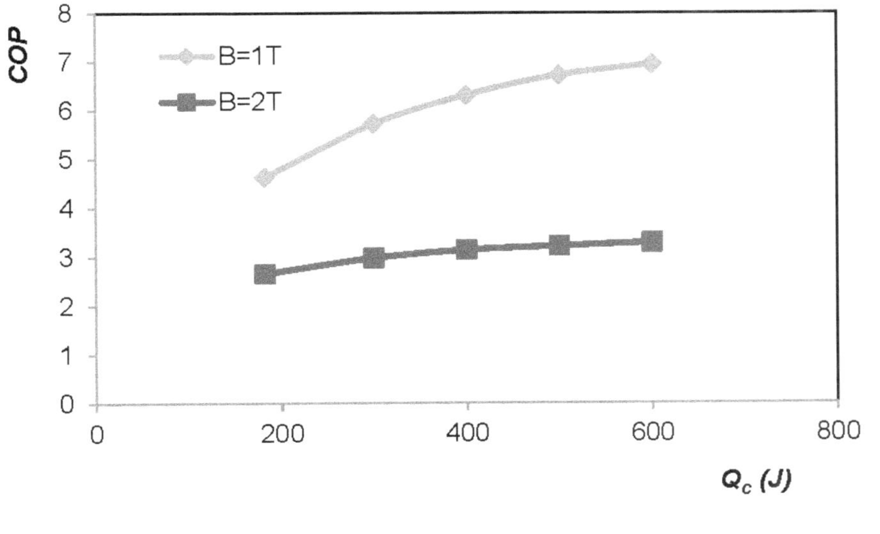

Figure 15. The variation of COP in relation to the cold power

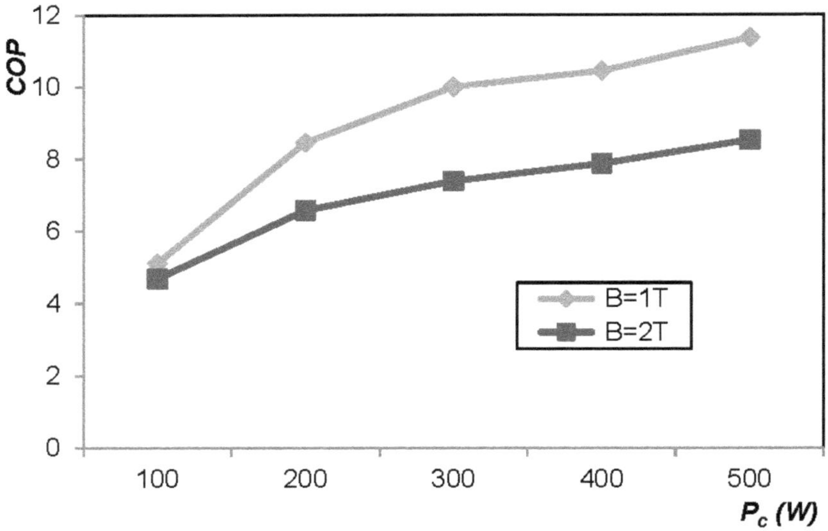

Figure 16. The variation of COP in relation to the fluid flow

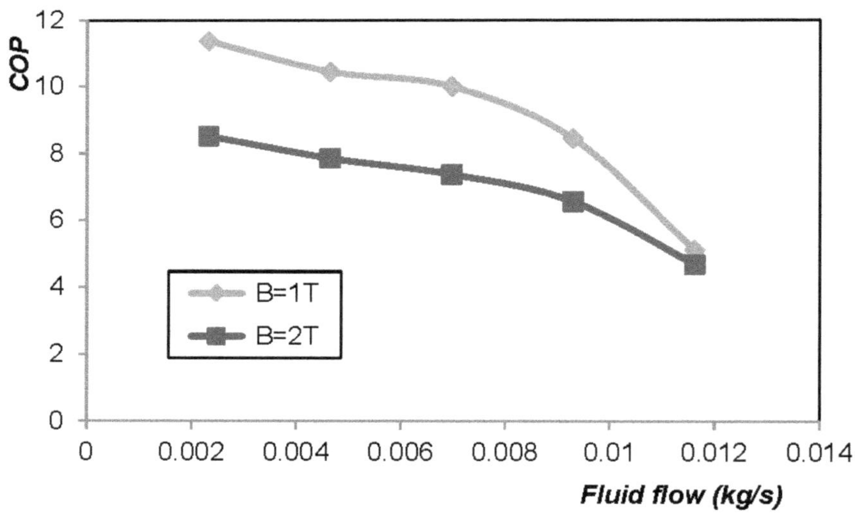

Since magnetic entropy is an extensive quantity, the regenerated heat increases under the effect of magnetization, if the quantity of the material increases. This increase is greater at B=1T than at B=2T. This can be explained by the fact that MCE decreases if the magnetic field decreases. Here, we proposed a combination between the COP and the mass required by a parameter α given by the following relation:

$$\alpha = \frac{a.COP}{b.m} \tag{52}$$

a and b are weighting coefficients according to the choice of the user.

Figure 17. The necessary mass of the regenerator in relation to Pc

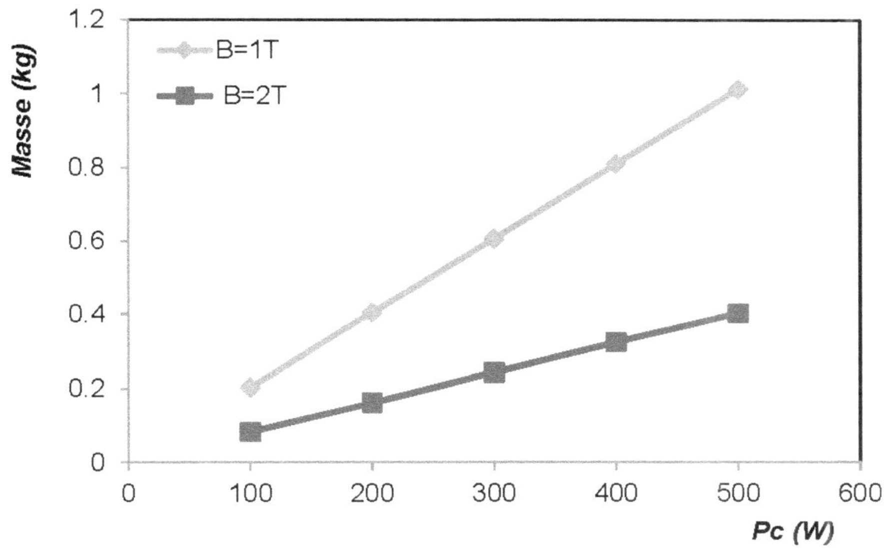

Figure 18. A Combination between COP-Mass in relation to Pc (a=b=1)

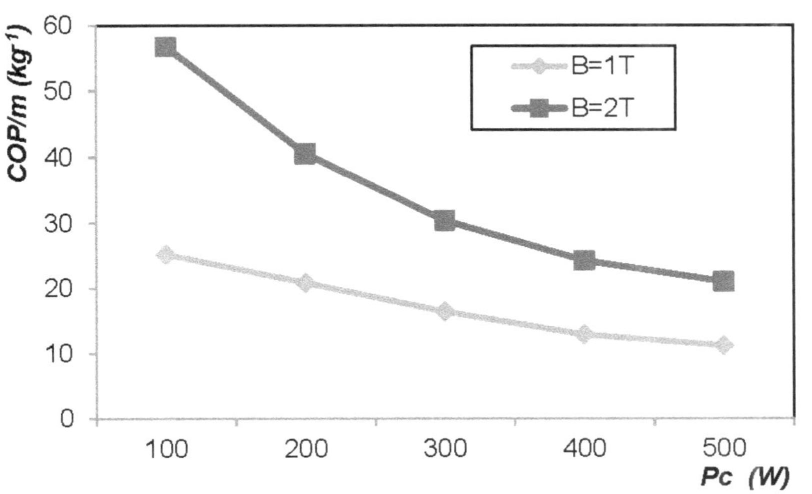

Figure 18 gives the COP-Mass combination as a function of the cold power. When α is high, then the choice is appropriate. Because α is, by definition, proportional to the COP and inversely proportional to the mass.

CONCLUSION

In this paper, we proposed a method to calculate the COP of a magnetic refrigerating system. To calculate the COP, we needed to analyze, size as well as optimize the operating parameters of the system in advance. Therefore, we studied the entire system and revealed that there existed an interaction among the parameters. In fact, we determined the temperature profile in the regenerator and in the fluid and deduced the exchanged heats and different works. Afterwards, Wm and MCE were determined by means of new approaches. Ultimately, the COP was deduced and its values are considered significant. A combination between the COP and the regenerator mass required by a parameter α was proposed to make the suitable choice. The new approaches to MCE and Magnetic work introduced in this work are important and can be useful for other works. A detailed energetic and economic analysis will be the aim of a future work.

REFERENCES

Allab, F., Kedous-Lebouc, A., Fournier, J. M., & Yonnet, J. P. (2005, October). (année). Numerical Modeling for Active Magnetic RegenerativeRefrigeration. *IEEE Transactions on Magnetics*, *41*(10), 3757–3759. doi:10.1109/TMAG.2005.854757

Arques, P. (2009). *Diagnostic prédictif et défaillances des machines*. Editions TECHNIP.

Baker, C. B. (1978). A Study of the Efficiency of Hydrogen Liquefaction. *International Journal of Hydrogen Energy*, *3*(3), 321–334. doi:10.1016/0360-3199(78)90037-X

Bouchekara, H., El-Hana, R., Kedous-Lebouc, A., Yonnet, J. P., & Chillet, C. (2014). Multiobjective optimization of AMR systems. *International Journal of Refrigeration*, *37*, 63–71. doi:10.1016/j.ijrefrig.2013.09.009

Bouchekara, H., Kedous-Lebouc, A., Dupuis, C., & Allab, F. (2008). Prediction and optimisation of geometrical properties of the refrigerant bed in an AMRR cycle. *International Journal of Refrigeration*, *31*(7), 1224–1230. doi:10.1016/j.ijrefrig.2008.02.007

Lebouc, A., Almanza, M., Yonnet, J. P., Legait, U., & Roudaut, J. (2014). Refrigeration magnetique Etat de l'art et developpements recents. *Symposium de Genie Electrique*, Cachan, France.

Matsumoto, K., Kondo, T., Ikeda, M., & Numazawa, T. (2011). Numerical analysis of active magnetic regenerators for hydrogen magnetic refrigeration between 20 and 77 K. *Cryogenics*, *51*(6), 353–357. doi:10.1016/j.cryogenics.2010.06.003

Mehrez, Z., Bouterra, M., El Gafsi, A., & Belghith, A. (2013). Heat transfer and entropy generation analysis of nanofluids flow in an open cavity. *Computers & Fluids*, *88*, 363–373. doi:10.1016/j.compfluid.2013.09.026

Niknia, I., Campbell, O., Christianse, T., Govindappa, P., Teyber, R., & Trevizoli, P. (2016). *A material screening technique for optimization performance of an AMR*. Seventh IIF-IIR International Conference on Magnetic Refrigeration at Room Temperature, Thermag VII, Torino, Italy.

Numazawa, T., Kimura, H., Sato, M., & Maeda, H. (1993). Carnot magnetic refrigerator operating between 1.4 and 10 K. *Cryogenics*, *33*(5), 547–554. doi:10.1016/0011-2275(93)90252-J

Padet, J. (2008). *Fluides en Écoulement Méthodes et modèles*. Seconde édition revue et augmenté, Reims.

Park, I., Kim, Y., Park, J., & Jeong, S. (2014). Investigation on the Two-stage Active Magnetic Regenerative Refrigerator for Liquefaction of Hydrogen. *Advances in Cryogenic Engineering*, *1573*, 74–81. doi:10.1063/1.4860685

Rosario, L., & Rahman, M. (2011). Analysis of magnetic refrigerator. *Applied Thermal Engineering*, *31*(6-7), 1082–1090. doi:10.1016/j.applthermaleng.2010.12.002

Row, A. (2012a). Therodynamics of active magnetic regenerators Part I. *Crygenics.*, *52*(2-3), 111–118. doi:10.1016/j.cryogenics.2011.09.005

Row, A. (2012b). Therodynamics of active magnetic regenerators: Part II. *Crygenics*, *52*(2-3), 119–128. doi:10.1016/j.cryogenics.2011.09.007

Roy, S., Poncet, R., & Sorin, M. V. (2016). *Multiobjective optimization of a reciprocating magnetic refrigerator using genetic algorithm.* Seventh IIF-IIR International Conference on Magnetic Refrigeration at Room Temperature, Thermag VII, Torino, Italy.

Smaili, A., Aït-Ali, S., & Chahine, R. (2011). Performance Predictions of First Stage Magnetic Hydrogen Liquefier. *International Journal of Hydrogen Energy*, *36*(6), 4169–4177. doi:10.1016/j.ijhydene.2010.09.006

Utaki, T., Kamiya, K., Nakagawa, T., Yamamoto, T. A., & Numazawa, T. (2007). Research on a Magnetic Refrigeration Cycle for Hydrogen Liquefaction. *International Cryocooler Conference, Inc*, Boulder, CO.

APPENDIX

Nomenclature

B ...... Magnetic field, T
C_{pf} ...... Specific heat of fluid, J.kg^{-1}.K^{-1}
C_{pm} ...... Specific heat of material, J.kg^{-1}.K^{-1}
L_m ...... Regenerator length, m
l_m ...... Regenerator width, m
M ...... Magnetisation, $A.m^{-1}$
m_f ...... Mass of fluid, kg
m_m ...... Mass of regenerator, kg
P_c ...... Cold power, W
Q_c ...... Heat transferred with the cold source, J
Q_h ...... Heat transferred with the hot source, J
Q_r ...... Heat regenerated, J
T_c ...... Transition temperature, K
W_e ...... Engine work, J
W_m ...... Magnetic work, J
W_p ...... Work of pump, J

Greek

α ...... A parameter
μ_0 ...... Magnetic permeability, H.m^{-1}
μ_r ...... Relative permeability, H.m^{-1}
τ_c ...... Cycle tim, s
τ_e ...... Exchange Time, s
θ...... Angle of rotation of the magnet, rd
χ_m ...... Magnetic susceptibility,

Chapter 11
Statistical Analysis and Linear Modeling of the Heat Exchangers Fouling in Phosphoric Acid Concentration Units

Saoussen El Aguel
Institut Supérieur des Sciences Appliquées et Technologies de Gabès, Tunisia

Zina Meddeb
Institut Supérieur des Sciences Appliquées et Technologies de Gabès, Tunisia

Mohamed Razak Jeday
National Engineering School of Gabes, Tunisia

ABSTRACT

Several factors influence the functioning of the heat exchangers in phosphoric acid concentration units and significantly affect their energy performance such as the cleanliness of the products, the operating conditions, the deposit of fouling on the walls. During the phosphoric acid concentration operation, fouling leads to a significant drop in the overall heat transfer coefficient, which is highly dependent on the thermal efficiency of the heat exchanger. This chapter presents the statistical study of the experimental database of two concentration units in order to study the variability of the system and identify outliers using principal component analysis. According to Hotelling's (T2) test, the authors identified 148 outliers for the two heat exchangers. Two reduced models of the thermal efficiencies were obtained by projection to latent structures (PLS) method. The application of the PLS regression method resulted in reliable correlation coefficients R2 equal to 0.9 for both configurations of heat exchangers.

DOI: 10.4018/978-1-7998-8801-7.ch011

INTRODUCTION

In almost all industrial activities, the supply of heat is an essential step in the production chains. It is performed by equipment such as heat exchangers, boilers and furnaces. To achieve this, two modes of heat transfer are possible: either by direct route, where two fluids exchange heat between them without any separation; or by indirect route where the hot fluid gives up its heat through a material, which separates it from the cold fluid. For these purposes, heat exchangers are widely used in industry in different configurations and sizes to best suit their various applications.

In this work, we are particularly interested in the enrichment of studies on the phenomenon of dross deposition on the heat exchange walls of heat exchangers, a phenomenon commonly known as "fouling". It refers to the accumulation of any unwanted deposit, i.e., crystalline, biological, particulate or chemical reaction product, on the surface of the heat exchanger representing additional thermal resistance, which leads to reducing heat transfer efficiency (Yamamoto et al., 2009).

Taborek (1995) traced the origin of the first industrial concern about fouling in the American electrical industry to 1880. The first mention of fouling in the literature is in the article presented by Orrok (1910). Fouling has been described both as: "The main unresolved problem in heat transfer" and "An almost universal problem in the design and operation of heat exchanger equipment". It could well be a result of the composition, temperature, pressure, contaminant concentration and rate of circulation of the fluid being treated (Eriksson et al., 2013, Nelson et al., 2006, Yamamoto et al., 2009). As an indication, in industrialized countries, according to Crittenden and Yang (2011), the additional costs due to fouling are of the order of 0.25% of their Gross Domestic Products (GDP) (Dunn, 2019).

According to Garret-Price et al., the conditions, which can influence the fouling factors in the heat exchanger (fouling resistance, thermal efficiency, etc.), can be subdivided into operating parameters, fouling fluid properties and design parameters of the heat exchanger (Hong, 2007).

In this context, the present work aimed to present the energy balances used to control the fouling behavior of the heat exchangers studied. The statistical study of the experimental database thus constituted was used to study the variability of the system, to detect outliers and to model the thermal efficiency of heat exchangers by the Projection to Latent Structures (PLS) regression method.

STATISTICAL ANALYSIS METHODS OF MULTIVARIATE DATA

General Information on "Chemometrics"

"Chemometrics" is defined as a discipline of chemistry that extracts relevant information from chemical data by mathematical and statistical tools. According to Gasteiger and Engel (2003), it is part of the larger field "chemoinformatics" which corresponds to the application of computational methods to solve chemical problems (Gasteiger & Engel, 2003). Some of the typical problems that can be successfully treated by chemo metric methods include:

1. Determining the concentration of a compound in a complex mixture (often from infrared data).
2. The classification of the origins of the samples (from analytical or spectroscopic chemical data).
3. Prediction of a property or activity of a chemical compound (from chemical structure data).
4. Assessment of the state of the chemical process.

The set of values collected in chemo metrics (Figure 1) are then databases in the form of matrices or spreadsheets composed of (n) rows, (m) columns and each cell containing a numerical value. Each row corresponds to an observation and each column corresponds to a particular variable. The size of a typical matrix is 20-1000 observations and 5-500 variables.

Figure 1. Modeling using chemometric methods (Varmuza &Filzmoser, 2009)

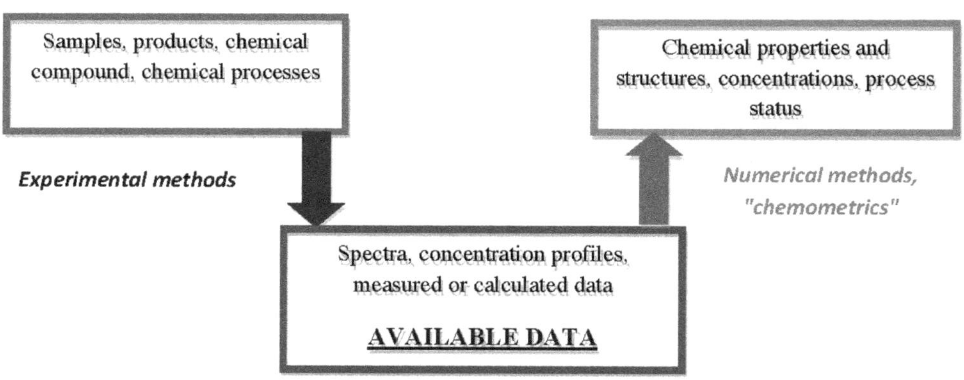

The purpose of statistical data analysis is finding groups of similar observations (forming clusters), outliers and similar (correlative) variables. The most important method for this is Principal Component Analysis (PCA), which allows visual inspection of the grouping of variables; other important (nonlinear) methods are hierarchical cluster analysis (dendrograms) and Kohonen maps. Other methods widely used in chemo metrics are Partial Least Squares Regression "Projection to Latent Structures" (PLS), Principal Component Regression (PCR), Linear Discriminant Analysis and Nearest Neighbor k classification. The application of artificial neural networks (ANN) is a non-linear approach to several problems in chemometrics aimed at modeling and predicting data. This type of data evaluation is called "supervised learning" (Varmuza & Filzmoser, 2009).

Data Preprocessing

The data to be analyzed by data mining techniques can be incomplete, noisy (containing errors or outliers) and inconsistent. There are many possible reasons for the presence of noisy data: Defective instruments, human or computer errors during data entry, technological limitations, and inconsistencies in naming conventions or data codes used. It is therefore essential to perform data transformation routines to fill in missing values, smooth noisy data, identify and / or remove outliers and resolve inconsistencies (Han et al., 2006).

In a database, the variables used can come from the same source (for example data from a single spectroscopic method), but can also have very different origins and sizes. Appropriate scaling or data preprocessing is required (Varmuza & Filzmoser, 2009).

The main stages of data preprocessing, according to figure 2, are data cleansing, integration, reduction and transformation. There are over 70 techniques for preprocessing data. Each technique has its own function and distinctive advantage. Its selection must be appropriate with the domain and data mining algorithm. Choosing the wrong technique will reduce data quality (Wen et al., 2009). The database preprocessing methods are (Varmuza & Filzmoser, 2009):

- Database transformation.
- Centering and scaling.

Database Transformation

It is often strongly recommended to start with the transformation of the data to approach a better symmetry. Unfortunately, this must be done for each variable separately, since it is not sure that the same transformation is useful for symmetrizing different variables. For skewed data, logarithmic "transformation" is often useful (this means taking the logarithm of the data values). More flexible is "power transformation" which uses an optimized power value p to transform x values into x^p.

Figure 2. Data preprocessing steps (Han et al., 2006)

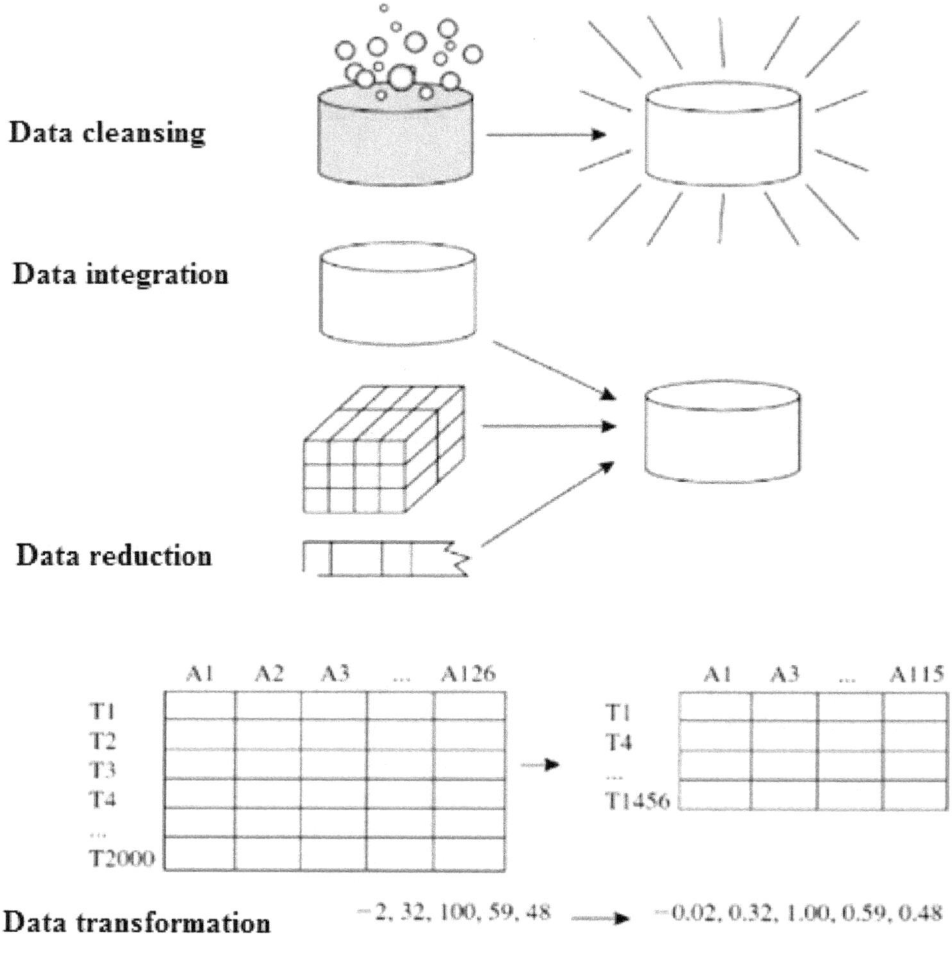

Centering and Scaling

Centering and scaling are typically applied after data transformation. They also refer to column-by-column manipulations of an X matrix with the aim that all columns have the mean of zero (centering) and the same variance (scaling). Let $\overline{x_j}$ be the mean and σ_j the standard deviation of a variable x_j. After centering and scaling, the variable has an average of zero; the data is shifted by x_j and the data center becomes the new origin.

Average centering simplifies many methods of multivariate data analysis and is expressed as follows:

$$x_{ij}(centred) = x_{ij}(original) - \overline{x_j} \quad (1)$$

Scaling standardizes each variable x_j by its standard deviation σ_j. It is generally combined with the average center of gravity to arrive at equation (2):

$$x_{ij}(centred + scaled) = \frac{x_{ij}(original) - \overline{x_j}}{\sigma_j} \quad (2)$$

The dataset then has a mean of zero and a variance (or standard deviation) of one, thus giving all variables equal statistical weight (Figure 3). The centering and scaling coupling is the most widely used pretreatment in chemometrics.

Figure 3. Graphical representation of mean centering and scaling for a two-variable X matrix (Varmuza & Filzmoser, 2009)

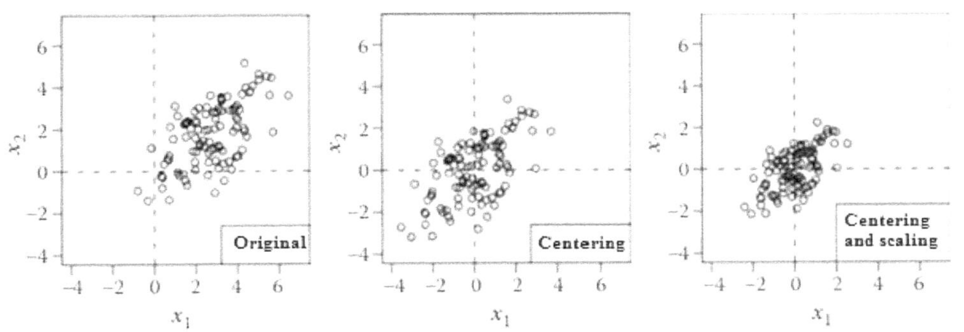

Statistical Analysis Methods

This section describes a set of statistical methods that are part of data analysis techniques. They are used to interpret the operation of a plant by analyzing its collected data: fault detection, parameter estimation and model discrimination. Although there are many mathematical and statistical methods in the literature, most of them only apply to installations atypical of normal operations or to situations where huge amounts of measurements are routinely processed (Varmuza & Filzmoser, 2009). We describe the following methods below:

1. The "CHAID" method (Chi-Square Automatic Interaction Detection).
2. The "PCA" method (Principal Component Analysis).
3. The "Kohonen Map" method.

Chi-Square Automatic Interaction Detection (CHAID)

Hartigan (1975) used a statistical method of dividing data into subsets called "Chi-Square Automatic Interaction Detection (CHAID)". The CHAID algorithm is sometimes used as a final modeling algorithm, but it has a number of drawbacks that limit its effectiveness as a multivariate preacher. It is more commonly used to reduce dimensionality. However, even here, there is a problem of bias, which can distort the interpretation of the responses (Breiman et al., 1984) (Blattberg et al., 2008, Nisbet et al., 2009).

Principal Component Analysis (PCA)

Principal Component Analysis (PCA) (Varmuza & Filzmoser, 2009) can be considered "the mother of all multivariate data analysis methods". It is a method that aims to calculate a new coordinate system formed by latent orthogonal variables, and where only the most informative dimensions are used. The PCA takes into account all the variables and the total structure of the data. It is an analysis method which can be applied to any X matrix.

The dimension reduction by PCA is mainly used for:

- Visualize multivariate data in the form of point clouds.
- Transform highly correlated variables into a smaller set of uncorrelated latent variables that can be used by other methods.
- Separate the relevant information (described by a few latent variables) from the noise.

Kohonen Map

Kohonen's maps are named after the Finnish mathematician Teuvo Kohonen who invented this method (Kohonen 1995). It is a non-linear method of representing high-dimensional data in a typically two-dimensional plot (map) - similar, for example, to the PCA score plot. Kohonen maps are advantageously used for exploratory data analysis in case of failure of the linear PCA method. The disadvantages of this approach are: The large number of adjustable parameters can be confusing; The results depend on the initial values; and For large databases, extensive memory requirements and long computing times (Kohonen, 1995).

DESCRIPTION OF PHOSPHORIC ACID CONCENTRATION UNITS

The concentration process is ensured by the evaporation under vacuum of the water contained in phosphoric acid at 28% by mass of P_2O_5 (Figure 4). The installation is mainly composed of the following equipment:

- Reboiler (RB),
- Barometric condenser (BC),
- Heat exchanger (HE),
- Basket filter (BF),
- Circulation pump (CP).

Figure 4. Schematic drawing of the process

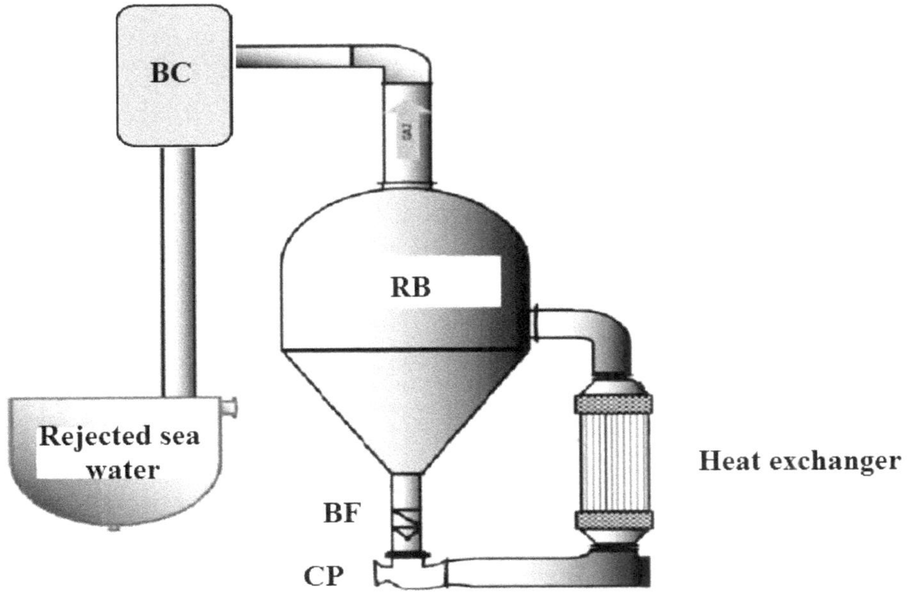

The heat exchanger ensures that the temperature of the acid is raised by low-pressure steam. The acid inlet and outlet temperatures are measured by thermocouples. In order to reach the concentration of 54% by mass of P_2O_5, the acid leaving the heat exchanger is subjected to evaporation at the boiling point in the reboiler (RB). The main function of the vacuum circuit is to reduce the amount of heat supplied.

EVALUATION OF THE THERMAL PERFORMANCE OF THE HEAT EXCHANGER

The thermal efficiency of the heat exchangers (η) is obtained by the following relationship:

$$\eta = \frac{U_{fouled}}{U_{clean}} \tag{3}$$

U_{fouled}: The heat transfer coefficient in fouling condition.
U_{clean}: The heat transfer coefficient in a clean condition.

We take the example of temporal evolution of the thermal efficiency (η) of the heat exchanger n°1 as represented in figure 5, for an arbitrary period of one year, chosen according to the availability of the greatest number of operating data. We take note that:

- After 1920 hours of operation, the thermal efficiency of the cross-flow graphite heat exchanger drops to 5%.
- A second drop in thermal efficiency of about 3% is observed after 4766 hours of operation to reach a value close to 64% after the cleaning operation.
- The third drop in thermal efficiency is recorded after 7252 hours of operation and is restored to nearly 60% after cleaning with industrial hot water.

Figure 5. Thermal efficiency of heat exchanger n°1

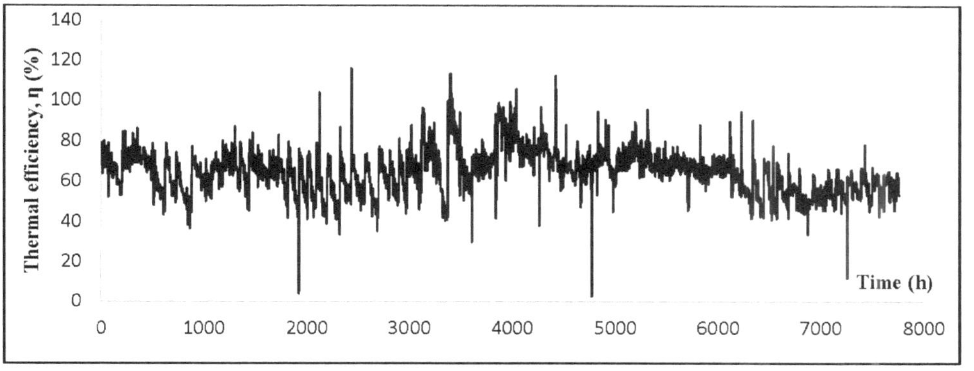

COLLECTION AND PREPROCESSING OF THE HEAT EXCHANGERS DATABASE

Database Collection

The increased pressure drop is a sign of clogging of the tubes. The operation must be stopped for washing the concentration loop. The washing is carried out in hot water (70-80 ° C) and / or acidulated (addition of concentrated sulfuric acid). The frequency of washing depends on the level of fouling. Note that in general, the average duration of a concentration cycle is four days.

The selection of operating periods is mainly related to the performance of the concentration units in terms of operating conditions (stability and continuity).

For the heat exchanger n°1, the data set consists of 73 operating cycles. Regarding the heat exchanger n°2, its database contains a total of 20 cycles. The variables are classified into three main categories:

- Heat exchanger operating variables.
- Variables of the concentration loop.
- Network operating variables.

Database Preprocessing

The preprocessing of all the data is essential before starting the modeling phase to obtain an appropriate form to analyze and a useful model without modifying the overall interpretation of the data.

Centering with the Median

The median can be expressed as follows:

$$median_j = \frac{x_{(v+0.5)} + x_{(v+1.5)}}{2} \quad (4)$$

Median values are calculated by relations (4) and then subtracted from each column of the data matrix (X).

Scaling

The way of scaling the data matrix (X) is such that each centered row is multiplied by its weight, which is the inverse of the interquartile range (1 / IIQ).

PRINCIPAL COMPONENT ANALYSIS (PCA)

The dimensionality reduction is an important preliminary step in the study of a basis containing several key parameters (22 parameters for the heat exchanger n°1 and 14 parameters for the heat exchanger n°2) (Yamamoto et al., 2009).

Taking into account all the data collected, the PCA allowed us to draw the graphs of:

- Eigen values and cumulative variability of the principal components.
- Observations.
- Correlations.

In the case of the heat exchanger n°1, the reduced plane is constructed by the PCA tool so that the first major component (P1) explains 25% of the variability of the raw data as shown in figure 6. The second principal component (P2) explains an additional 17% and is orthogonal to the direction of (P1). We notice from figure 7 corresponding to the heat exchanger n°2 that the first and second principal components represent 51% of the variability of the data matrix.

By simultaneously examining the graph of observations (Figure 8 and figure 9) of the first two principal components and the similar graphs between (P1) with the following principal components: P3. . . Pi (i = 7 for the heat exchanger n°1, i = 10 for the heat exchanger n°2), we can observe that no group is far from the origin. This proves the normal functioning of the two concentration units during the periods studied.

Figure 6. Eigen values and cumulative variability of the principal components for the heat exchanger n°1 database

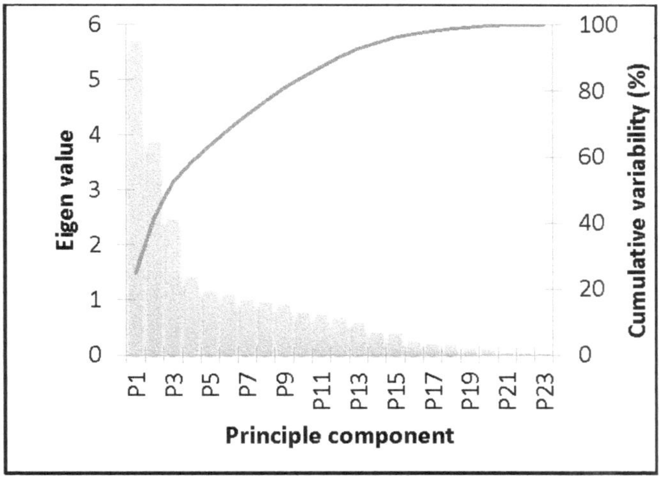

Figure 7. Eigen values and cumulative variability of the principal components for the database of the heat exchanger n°2

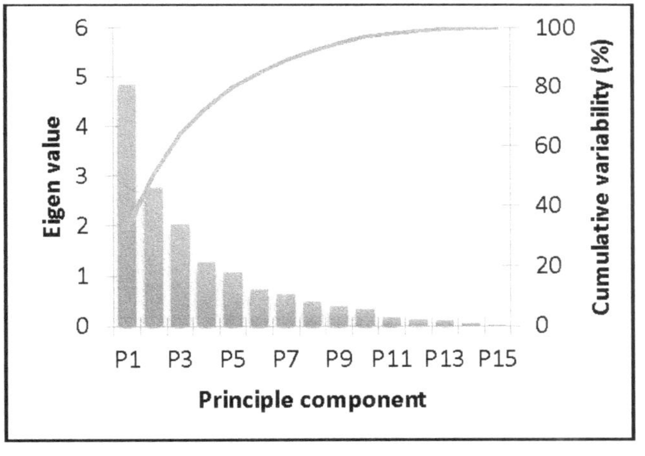

According to Hotelling's T^2 test (Figure 10), we identified 148 outliers for the heat exchangerS n°1 and n°2. Their elimination is therefore essential because they contribute to the rotation of the principal components due to their strong leverage effect (Nelson et al., 2006).

Figure 8. Diagram of the observations of the first two principal components (P1) and (P2) (Heat exchanger n°1)

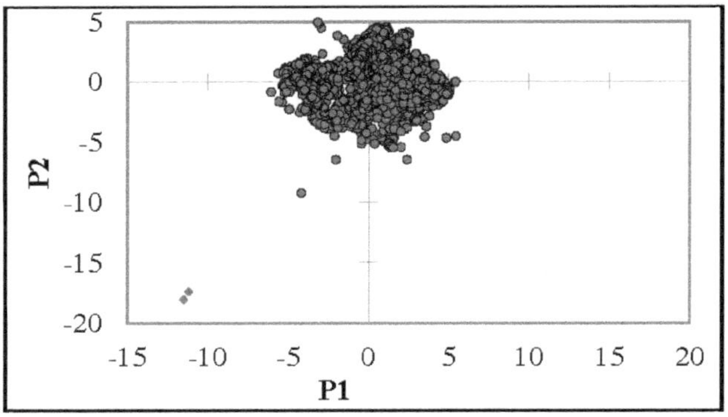

Figure 9. Diagram of the observations of the first two principal components (P1) and (P2) (Heat exchanger n°2)

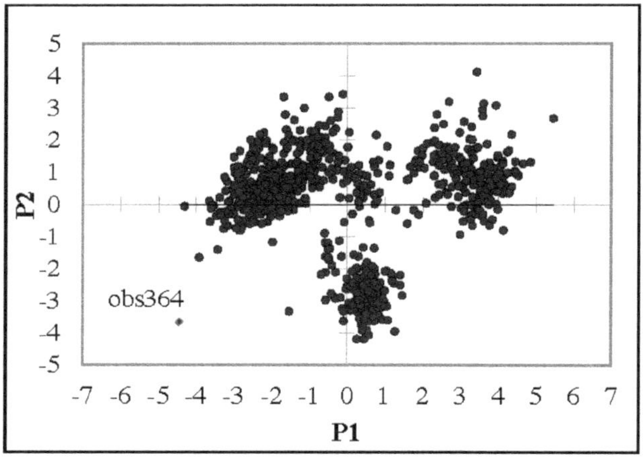

Figure 10. Example of the Hotelling Test T^2 (Heat exchanger n°1)

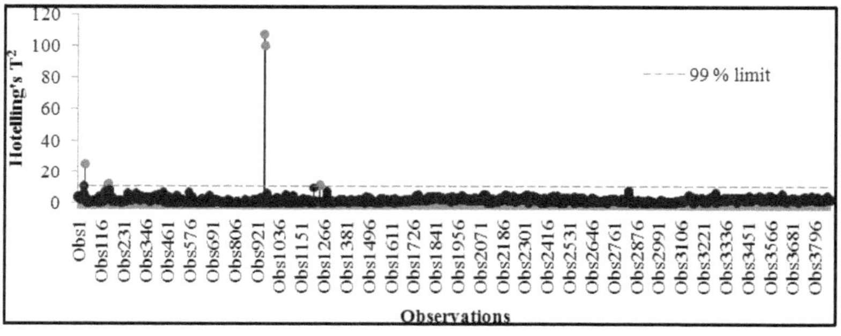

The examination of the correlation curves (Figure 11 and figure 12) reflects the reality of the interactions between the various operating parameters of the process. This can be deduced by:

- The negative correlation between the thermal efficiency (η) and the steam inlet temperature (T_v) located on either side of the second principal component (P2). Indeed, the increase in the temperature of the exchange surface promotes the crystallization of calcium sulphate due to its negative solubility in phosphoric acid.
- The temperature (T_v) and the pressure (P_v) of the heating steam are adjacent and positively correlated.

Figure 11. Correlation diagram for the first two principal components (P1) and (P2) (Heat exchanger n°1)

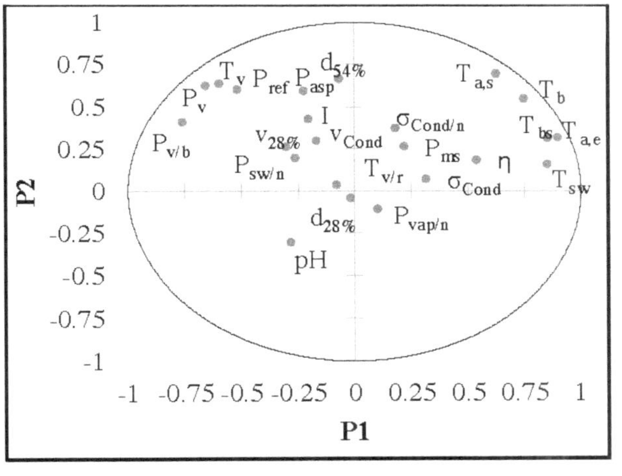

PROJECTION TO LATENT STRUCTURES (PLS)

Projection to latent structures (PLS) regression is a statistical method applied to large industrial databases since the early 1990s. It has powerful performance in the analysis and extraction of meaningful information from sets of data noisy and highly correlated. This statistical method is chosen for its predictive capacities and its stability. The "PLS" regression technique links one or more response variables (Y) to a set of predictor variables (x_j) (x_1, ..., x_j) by a linear model (Eriksson et al., 2013, Dunn, 2019, MacGregor, 2003, Sbárbaro & Del Villar, 2010).

Figure 12. Correlation diagram for the first two principal components (P1) and (P2) (Heat exchanger n°2)

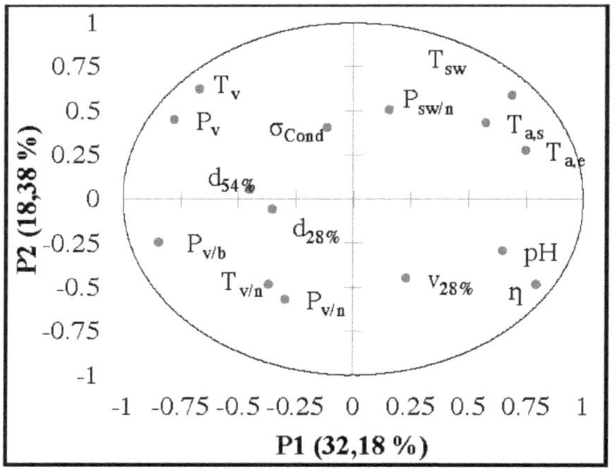

General Linear Model

According to the "PLS" method, the general linear model is made up of:

- The matrix (X) contains: 22 variables for the heat exchanger n°1 and 14 variables for the heat exchanger n°2.
- Thermal efficiency (η) is the answer (Y).

Heat Exchanger n°1

The coefficients of the general linear model corresponding to the cross-flow exchanger are given in Table 1.

Figure 13 shows the actual thermal efficiency versus the thermal efficiency predicted by the general linear model. All the observations form a cluster superimposed on a line at 45 °. This implies the good agreement of the values of the model proposed by the "PLS" method ($R^2 = 0.7$, Standard deviation = 0.4) with the collected results.

Heat Exchanger n°2

The coefficients of the general linear model for predicting the thermal efficiency of the heat exchanger n°2 are shown in Table 2. The corresponding precision measurements are $R^2 = 0.9$ and the standard deviation = 0.2. The good prediction of the model is further affirmed by the curve in figure 14, which represents a comparison of the actual thermal efficiency values to those calculated by the general linear model.

Table 1. Coefficients of the general linear model by the "PLS" method (Heat exchanger n°1)

Parameter	Coefficient
Constant	-1,2
$v_{28\%}$	0,5
T_b	0,2
$d_{54\%}$	0,2
$T_{a,e}$	-0,5
$T_{a,s}$	0,6
T_{bs}	-0,1
P_{ms}	0,1
P_{asp}	0,1
P_{ref}	-0,2
I	0
T_v	-0,5
P_v	-0,3
v_{Cond}	0,1
σ_{Cond}	0
Ph	-0,1
$P_{v/b}$	0
T_{sw}	0
$P_{v/n}$	-0,1
$T_{v/n}$	0
$\sigma_{Cond/n}$	0,1
$P_{sw/n}$	0
$d_{28\%}$	0

Reduced Linear Model

Heat Exchanger n°1

The scale model consists of two matrices (X) and (Y) formed respectively of 10 variables having the most important influences (according to the PCA) and of the thermal efficiency of the heat exchanger n°1.

Figure 13. Comparison between the actual thermal efficiency values and those predicted by the general linear model (Heat exchanger n°1)

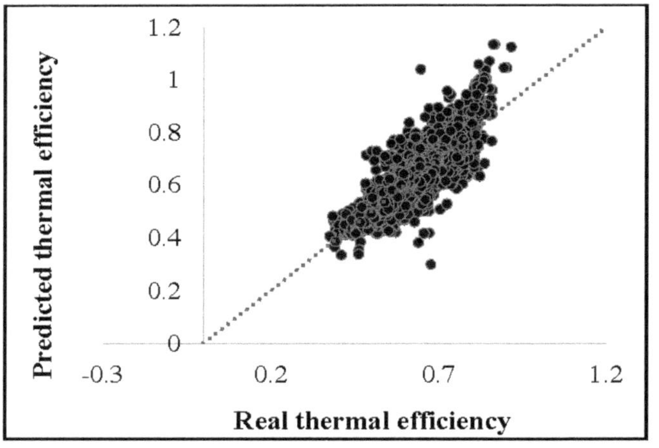

Table 2. Coefficients of the global linear model by the "PLS" method (Heat exchanger n°2)

Parameter	Coefficient
Constant	-1,6
P_v	0
$d_{54\%}$	0
$T_{a,e}$	0,2
$T_{a,s}$	0,1
$v_{28\%}$	0,1
T_v	-1
P_v	-0,2
σ_{Cond}	0,1
pH	0
$P_{sw/n}$	-0,1
T_{sw}	0
$P_{v/n}$	0
$T_{v/n}$	0
$d_{28\%}$	0

Figure 14. Comparison between the actual thermal efficiency values and those predicted by the general linear model (heat exchanger n°2)

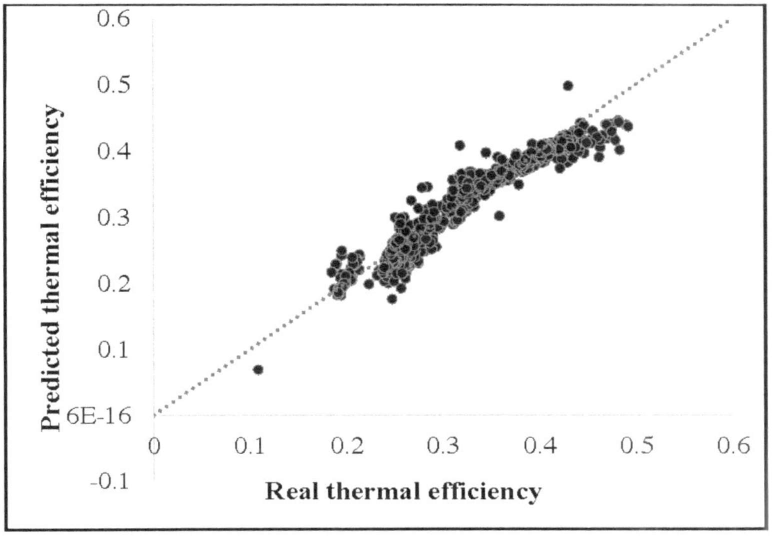

The reduced linear model designed for the prediction of the thermal efficiency of the heat exchanger n°1 is the sum of three functions F (v), G (T) and H (P):

$$\eta = F(v) + G(T) + H(P) \tag{5}$$

With:

$$F(v) = -2 + 0.3v_{28\%} \tag{6}$$

$$G(T) = 0.4T_{b} - 2.1T_{a,e} + 1.2T_{a,s} + 0.6T_{b,s} - 0.4T_{v} - 0.1T_{sw} \tag{7}$$

$$H(P) = -0.6P_{rev} - 0.2P_{v} - 0.02P_{b} \tag{8}$$

The examination of figure 15, which represents the thermal efficiency calculated by the reduced linear model as a function of the real thermal efficiency, reflects the good predictive capacity of the reduced linear model in the case of the heat exchanger n°1 with precision measurements: $R^2 = 0.9$ and standard deviation = 0.3 (the points are superimposed on the right: y = x).

Figure 15. Comparison between actual and predicted thermal efficiency by the reduced linear model (Heat exchanger n°1)

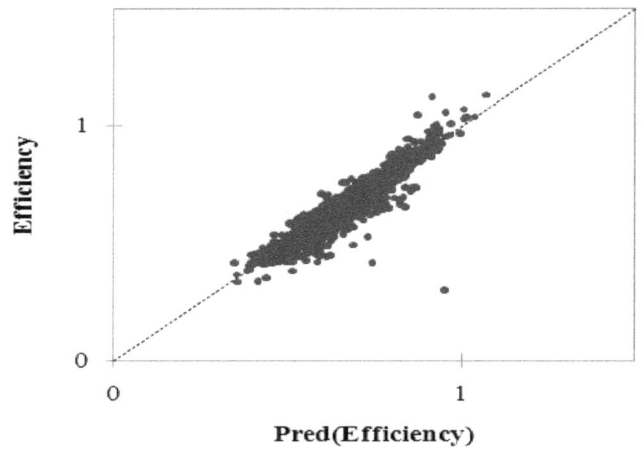

Heat Exchanger n°2

The reduced linear model equation for the heat exchanger n°2 is as follows:

$$\eta = -1,77 - 0,02P_v + 0,02d_{54\%} + 0,1T_{a,e} + 0,1T_{a,s} + 0,04v_{28\%} \\ -T_v - 0,05P_v + 0,04\sigma_{Cond} - 0,11pH + 0,05d_{28\%} \quad (9)$$

The precision measurements of the model obtained are equal to: $R^2 = 0.9$ and standard deviation = 0.1. Indeed, the superposition of the points of the comparison curve between the actual and predicted values of thermal efficiency on the equation line: y = x (Figure 16) reflects the good prediction capacity of the proposed model.

CONCLUSION

This work is devoted to the study of the thermal performance of the two operating modes of the heat exchangers used for the concentration of phosphoric acid, based on the history of the actual operating data of the units in the concentration plant. The statistical analysis by the "Principal Component Analysis" (PCA) method allowed to rule out outliers and to identify the system variables having the greatest impact on the modeling of the thermal efficiency of heat exchangers. The application of the "Partial Least Squares Regressions" (PLS) method to the databases provided by the "PCA" method led to the two reduced models having measures of precision greater than those of the two general models. The proposed mathematical models are used to predict a cleaning schedule for the heat exchangers and control operation of the phosphoric acid concentration plant.

Figure 16. Comparison between actual and predicted thermal efficiency by the reduced linear model (Heat exchanger n°2)

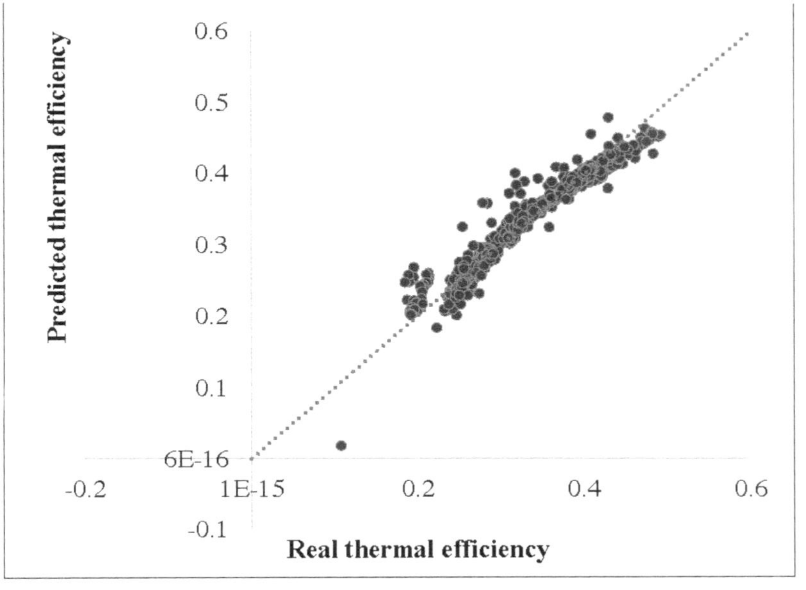

ACKNOWLEDGMENT

The authors would like to acknowledge the support of the Tunisian Chemical Group.

REFERENCES

Blattberg, R. C., Kim, B. D., & Neslin, S. A. (2008). *Database Marketing: Analyzing and Managing Customers*. Springer. doi:10.1007/978-0-387-72579-6

Dunn, K. (2019). *Process improvement using data*. Academic Press.

Eriksson, L., Johansson, E., Kettaneh-Wold, N., Trygg, J., Wikström, C., & Wold, S. (2013). *Multi- and Megavariate Data Analysis: Basic Principles and Applications* (3rd ed.). MKS Umetrics.

Gasteiger, J., & Engel, T. (2003). *Chemoinformatics: A Textbook*. Wiley-VCH. doi:10.1002/3527601643

Han, D. J., Pei, J., & Kamber, M. (2006). *Data Mining concepts and techniques. Southeast Asia Edition* (2nd ed.). Elsevier.

Hong, Y. (2007). Composite Fouling on Heat Exchanger Surfaces. Nova Science Books.

Kohonen, T. (1995). *Self-Organizing Maps*. Springer. doi:10.1007/978-3-642-97610-0

MacGregor, J. F. (2003). Multivariate statistical approaches to fault detection and isolation. *Fault Detection Supervision and Safety of Technical Processes*, *36*(5), 549–554. doi:10.1016/S1474-6670(17)36549-7

Nelson, P. R. C., MacGregor, J. F., & Taylor, P. A. (2006). The impact of missing measurements on PCA and PLS prediction and monitoring applications. *Chemometrics and Intelligent Laboratory Systems*, *80*(1), 1–12. doi:10.1016/j.chemolab.2005.04.006

Nisbet, R., Elder, J., & Miner, G. (2009). *Handbook of Statistical Analysis and Data Mining Applications*. Elsevier.

Sbárbaro, D., & Del Villar, R. (2010). *Advanced Control and Supervision of Mineral Processing Plants*. Springer. doi:10.1007/978-1-84996-106-6

Varmuza, K., & Filzmoser, P. (2009). *Introduction to multivariate statistical analysis in chemometrics*. Academic Press.

Wen, P., Li, Y., Polkowski, L., Yao, Y. Y., Tsumoto, S., & Wang, G. (2009). Rough Sets and Knowledge Technology. In *4th International Conference, Gold coast, Australia, Proceedings*. Springer.

Yamamoto, H., Yamaji, H., Abe, Y., Harada, K., Waluyo, D., Fukusaki, E., Kondo, A., Ohno, H., & Fukuda, H. (2009). Dimensionality reduction for metabolome data using PCA, PLS, OPLS, and RFDA with differential penalties to latent variables. *Chemometrics and Intelligent Laboratory Systems*, *98*(2), 136–142. doi:10.1016/j.chemolab.2009.05.006

KEY TERMS AND DEFINITIONS

Fouling: The phenomenon of accumulation of unwanted solid elements on the exchange surfaces.

Heat Exchanger: Equipment used in industrial installations to provide a large amount of heat to part of the system

Heat Transfer: It is a part of physics that deals with the mechanisms of the modes of heat propagation.

Modeling: A mathematical equation links a response to one or more parameters of the studied system.

PCA: A data analysis technique that allows, from n variables, to construct m other variables called principal components.

Phosphoc Acid: This product (H_3PO_4) is a deliquescent solid which readily absorbs moisture from the air. Phosphoric acid is commercially available in the form of aqueous solutions at various concentrations.

PLS: Tool for linear regression of a variable on other expicative variables which can be correlated with each other.

Chapter 12
Study of Foiling of Heat Exchangers in Phosphoric Acid Concentration Units

Zina Meddeb
Institut Supérieur des Sciences Appliquées et Technologies de Gabès, Tunisia

Saoussen El Aguel
National Engineering School of Gabes, Tunisia

Mohamed Razak Jeday
National Engineering School of Gabes, Tunisia

ABSTRACT

The purpose of this work is to establish mathematical models for monitoring the fouling of heat exchangers and programming shutdowns for the cleaning of this equipment. To achieve this objective, the authors have adopted an approach comprising essentially the identification of the operating parameters involved in the fouling of the heat exchangers used in the concentration of phosphoric acid, collection of the technical characteristics of the heat exchangers studied, collection and sorting of the operational data of the exchangers studied over a period of two and a half years, establishment of a database on the cycles of operation/shutdown, for the cleaning of the heat exchangers studied, and development of preliminary models for monitoring the fouling and to help the operator decide when to shut down for cleaning the heat exchangers.

DOI: 10.4018/978-1-7998-8801-7.ch012

INTRODUCTION

The rise in energy prices and the scarcity of fossil energy sources require manufacturers to achieve high energy performance. This constraint leads them to use optimized processes and high-performance technical equipment. This applies in particular to heat exchangers, which are essential elements of energy performance. The performance guarantee should apply not only to the profitability of the investments associated with the construction of a new installation, but also to the energy performance throughout its operation. Several effects tend to degrade the energy efficiency of a thermal installation, the most important of which is the fouling of the heat exchange walls. Manufacturers and operators of heat exchangers are constantly seeking to improve their knowledge in the operation, control and maintenance of these devices, with a view to optimizing their energy performance and reducing their downtime, in particular for cleaning and maintenance work. Controlling the fouling of heat exchangers is a very important point in their operation. Indeed, it is a preponderant phenomenon in the drop in the energy performance of these devices, which can have very serious consequences on the operating mode, the production capacity and on the economic competitiveness of the installations. A detailed knowledge of fouling mechanisms and kinetics is necessary to develop an optimal energy management strategy.

In this work we will try to study more closely, the fouling of the heat exchangers of the phosphoric acid concentration units of the Tunisian Chemical Group in Gabès. And, this by establishing a thermal balance on the heat exchangers in a first step and modeling the fouling factor (Global heat transfer coefficient) according to the operating parameters of each unit of acid concentration phosphoric.

CLOGGING OF EXCHANGERS

Thermal Fouling Resistance

The fouling resistance is defined as the difference between the overall thermal resistance of the exchanger in the dirty state, and that of the exchanger in the clean state. The expression of the overall heat transfer coefficient (Ug) generally includes five terms corresponding to the thermal resistances encountered on either side of the wall, which are responsible for the temperature difference between the two fluids. The fouling of heat exchangers (Figure 1) is defined as the deposit of materials, or unwanted substances, on the exchange surfaces (internal and external surfaces of the exchanger).

Figure 1. Schematization of the fouling of a wall.

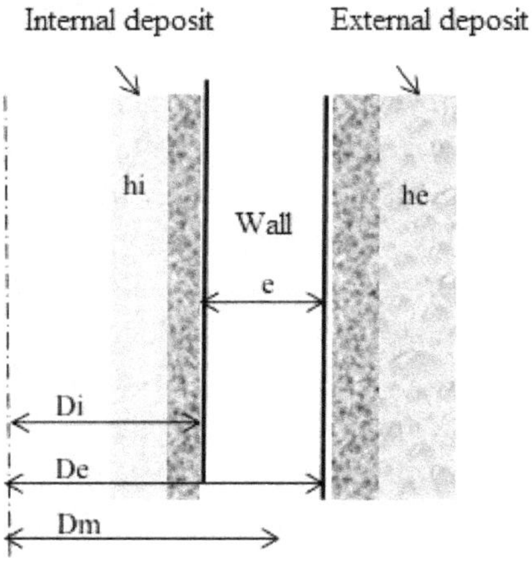

Different Types of Fouling

Fouling is divided into six categories (Thonon, 2007, VDI, 2010): (i) Particulate fouling: formation of a deposit linked to the presence of suspended solids in the heat transfer fluid. (ii) Fouling by scaling: formation of a deposit linked to the precipitation on the wall of the exchanger of the salts initially dissolved in the transfer fluid. (iii) Biofouling: deposition of organic layers including microorganisms as well as the growth and attachment of macroorganisms. (iv) Corrosion fouling: defined as the result of a chemical or electrochemical reaction between the metallic heat transfer surface and the flowing fluid. (v) Chemical fouling: deposit on the transfer surface by a chemical reaction (the material of the wall does not take part in this reaction). (vi) Fouling by icing: formation of ice, if the temperature of the wall is lower than the melting temperature of the fluid.

These different mechanisms interact, and/or overlap, in a more or less complex way. The most common types of fouling encountered in industry are particulate fouling and scaling (Epstein, 1988). The different types of fouling are controlled by several factors related to the fluids in contact or to the exchange surface. Among these factors, we can cite the configuration and mode of operation of the exchanger, the construction material of the exchanger and the operating conditions (Thonon, 2007, Finkbeiner et al., 1993).

Kinetics of Evolution of Fouling Resistance

According to the literature (Taborek, 1979, Bontemps, 1994, Bevevino, 1978), three forms of evolution of thermal fouling resistance can be distinguished as a function of time - as shown in Figure 2, namely: an evolution linear (1), an asymptotic rise (2) and a low-speed rise (3).

Figure 2. Evolution of fouling resistance as a function of time.

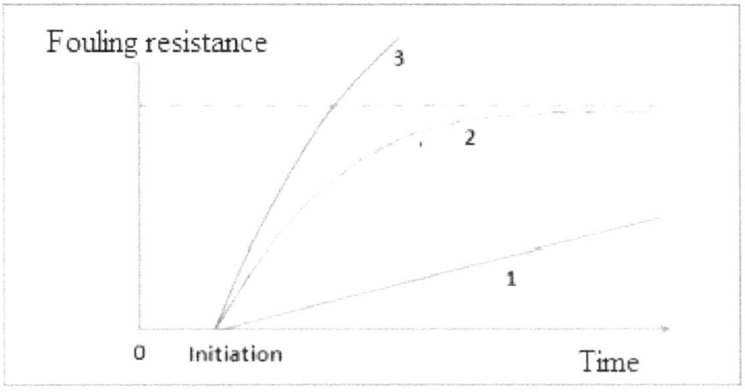

Linear Kinetics

The linear kinetics is represented by curve 1. The deposition rate is increasing and without re-entrainment. The deposits are hard and of the tartar or limestone type.

Asymptotic Kinetics

Asymptotic kinetics is one of the most frequent kinetics. The deposits are soft, hence the phenomenon of re-entrainment under the action of shear forces, linked to the speed of the fluid. The asymptotic fouling value is reached when the deposition rate is equal to the re-entrainment rate of the deposit. This kinetics is correlated by the Kern model (Kern & Seaton, 1959).

Increasing Kinetics

The kinetic evolution with increasing fouling rate, and without reaching a limit value, has been demonstrated in particular cases of particulate fouling in the gas phase, where the flow velocity is low.

Fouling Modeling Methods

Semi-Empirical Method

Kern is the first to have worked on the fouling of heat exchangers. The modeling assumption assumes that two processes act simultaneously:

- Deposition of particles, characterized by a constant flux (*Φd*), if the concentration is too.
- Re-entrainment of particles, characterized by a flux (*Φr*) which depends on the mass of particles (mp) deposited. The particle balance of the deposit is therefore formulated as follows:

$$\frac{dm_p}{dt} = \phi_d - \phi_r \tag{1}$$

Consider the following hypotheses:

- A single type of fouling;
- Homogeneous deposit;
- Failure to take into account the deposit initiation phase and the surface condition;
- Constant thermo-physical characteristics of the fluid and the deposit.

The transport phase of the particles at the wall controls the deposition process, while the shear stress controls the re-entrainment phase of the particles. Considering the proportionality of Φ_d as a function of the mass of the deposited particles, we can write:

$$\phi_d = k_p(C_b - C_w) \tag{2}$$

$$\phi_r = C_1 \tau_{wmp} \tag{3}$$

Where kp is the transport coefficient, Cb is the concentration of particles within the fluid, Cwest the concentration of particles at the wall, C1 is a dimensional constant, τw is the shear stress exerted by the fluid on the deposit

Equation 1 becomes:

$$\frac{d_{m_p}}{dt} = k_p(C_p - C_w) - C_1\tau_w m_p \tag{4}$$

The solution is:

$$m_p = \frac{k_p(C_p - C_w)}{(C_1 + \tau_w)}(1 - e^{-(C_1\tau_w t)}) \tag{5}$$

If we suggest:

$$m_p = \frac{1}{C_1\tau_w} \tag{6}$$

And

$$m_p^* = \frac{k_p(C_b - C_w)}{C_1\tau_w} \tag{7}$$

The mass of particles is then determined by the following equation:

$$m_p = m_p^*\left(1 - e^{-\frac{t}{\tau}}\right) \tag{8}$$

The Kern model therefore provides a mathematical description of fouling. This equation verifies curve 2 in Figure 2: asymptotic kinetics of the formation of a particulate deposit on the exchange surface.

Many efforts were made after the work of Kern, in 1959, to develop more precise fouling models: Watkinson (Watkinson, 1968), in 1968, Thomas (Thomas and Grigull, 1974) in 1974, Bowen (Bowen and Epstein, 1979) in 1979, Epstein (Epstein, 1988) in 1988, Wang (Wang et al., 2015) in 2015…all of which recognized the complexity of physical and semi-empirical modeling of the fouling problem.

To circumvent the difficulties of applying the previous methods, the development of another approach has been proposed: it is based on statistical modelling.

Methods of Statistical Analysis of Multivariate Data

Data analysis is the set of methods from which one collects, enters, organizes and studies data to draw conclusions. This section describes a set of statistical methods that are part of the data analysis techniques. They are used to interpret the operation of a plant by analyzing its collected data: detection of defects, estimation of parameters and discrimination of models. Although there are many mathematical and statistical methods in the literature, most of them only apply to installations atypical of normal operations or to situations where huge amounts of measurements are routinely processed. (Perry et al., 2008). We describe below the following methods:

- The "CHAID" method (Chi-Square Automatic Interaction Detection).
- The "PCA" method (Principal Component Analysis).
- The "Kohnen Map" method.

Chi-Square Automatic Interaction Detection (CHAID)

Hartigan (1975) used a statistical method of dividing data into subsets called "Chi-Square Automatic Interaction Detection (CHAID)". The CHAID algorithm is sometimes used as the final modeling algorithm, but it has a number of drawbacks that limit its effectiveness as a multivariate predictor. It is used more commonly to reduce dimensionality. But even here, there is a problem of bias that can distort the interpretation of responses (Breiman et al., 1984) (Blattberg, Nisbet, 2009).

Principal Component Analysis (PCA)

Principal Component Analysis (PCA) (Varmuza, 2009) can be considered "the mother of all multivariate data analysis methods". It is a method that aims to calculate a new coordinate system formed by orthogonal latent variables, and where only the most informative dimensions are used. PCA takes into account all the variables and the total structure of the data. It is an analysis method that can be applied to any X matrix.

Dimension reduction by PCA is mainly used for:

1. Visualize multivariate data in the form of scatter plots.
2. Transform highly correlated variables into a smaller set of uncorrelated latent variables that can be used by other methods.
3. Separate relevant information (described by a few latent variables) from noise.

Map of Kohnen

Kohonen charts are named after the Finnish mathematician Teuvo Kohonen who invented this method (Kohonen 1995). It is a non-linear method of representing high-dimensional data in a typically two-dimensional plot (map) - similar, for example, to the score plot in PCA. Kohonen maps are advantageously used for exploratory data analysis when the linear PCA method fails. The disadvantages of this approach are: (i) the large number of adjustable parameters can be confusing; (ii) The results depend on the initial values; and (iii) For large databases, large memory requirements and long computation times (Kohonen, 1995).

CONCENTRATION UNITS

The "Phosphoric Acid" (PA) plant of the Tunisian Chemical Group in Gabes includes three units for the concentration of phosphoric acid. The concentration of phosphoric acid consists in eliminating, by evaporation, part of the water contained in the incoming phosphoric acid at a mass concentration of approximately 28% in P2O5 up to the mass concentration of exit, at approximately 54% in P2O5. Note that each concentration unit using the "Rhône Poulenc" process, as shown in Figure 3, includes the following equipment: a filter, a circulation pump, a heat exchanger, a boiler and a condenser.

The Boiler

The boiler (B) (Figure 3) separates the evaporated water from the circulating acid. It works under vacuum, in order to lower the boiling point of phosphoric acid. The liquid phase, occupying between 25 to 33% of the total volume of the boiler, is introduced at the level of the upper part to avoid short-circuiting of the mixture (passage through the circulation line before its separation). The boiler is equipped with a deflector which ensures the dispersion of the phosphoric acid over the entire section, thus limiting the formation of foam which aggravates the entrainment of the phosphoric acid droplets with the outgoing gases. It should be noted that to have a good evaporation, the phosphoric acid must be allowed at a temperature slightly higher than its boiling point.

Circulation Pump

The type of circulation pump (P) (Figure 3) depends on the pressure drop across the heat exchanger. In the case of the phosphoric acid concentration units of the "AP" plant, the pumps installed are of the centrifugal type for the block heat exchangers (AP1 and AP2 units) and of the axial type for the shell and tube exchangers (unit AP3).

Figure 3. Diagram of the principle of operation of the concentration line.

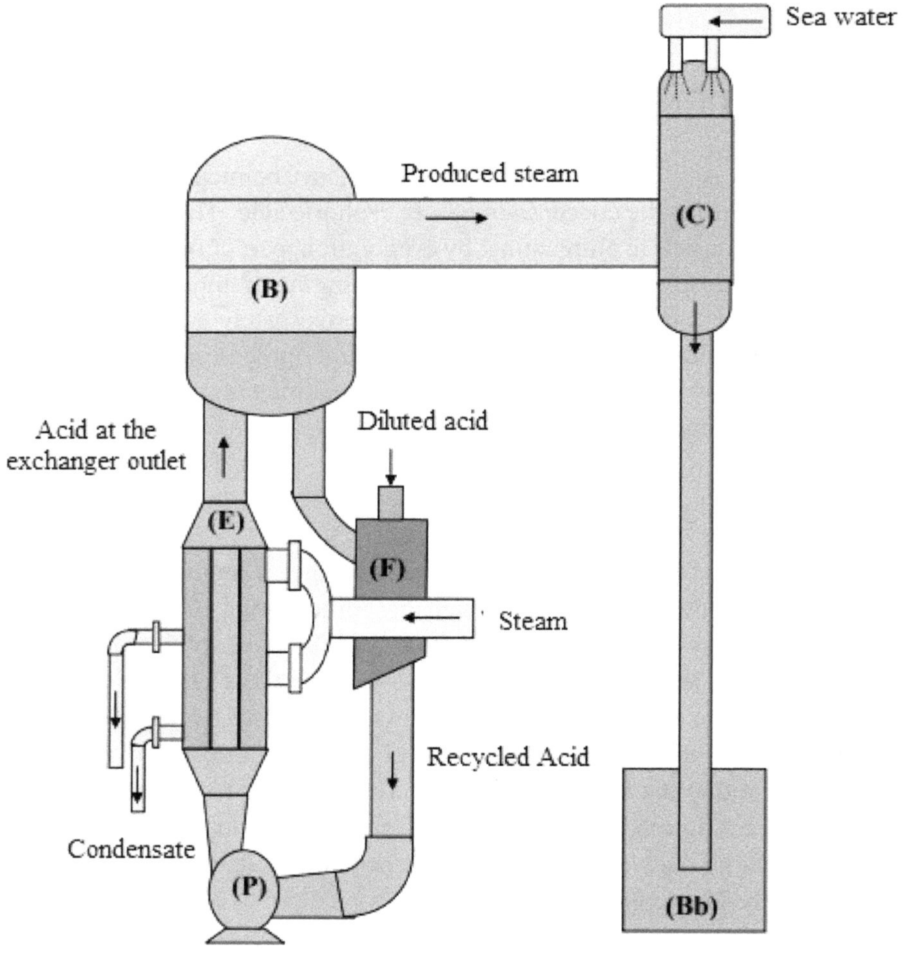

Vacuum Circuit (condenser)

The main function of the vacuum circuit ((C) and (Bb) in Figure 3) is to reduce the boiling temperature at the level of the boiler and to condense the amount of evaporated water.

Heat Exchangers

The Groupe Chimique Tunisien phosphoric acid plant in Gabes uses 3 types of heat exchangers (E) (In Figure 3) in the different concentration lines. These are:

1. Stainless steel tube bundle exchanger (unit AP1).
2. Graphite cylindrical block heat exchangers (unit AP2).
3. Graphite tube bundle heat exchangers (unit AP3).

DATABASE AND CALCULATION STEPS

Collection of Data

The collection of the operating parameters of the heat exchangers of the AP1, AP2 and AP3 units of the "AP" plant of the GCT-Gabès is spread over a period of two and a half years (first 6 months of 2009, the year 2010, first 6 months of 2013 and 6 months of 2014).

These data were used to form a database which will be used to establish fouling prediction models in order to help the operator choose when to trigger the washing operation of the heat exchanger concerned.

To monitor the operation of heat exchangers used in industrial phosphoric acid concentration units for the detection of fouling and the triggering of cleaning operations, we have retained the parameters recommended by the operator. These are:

The temperature of the steam from the power plant Tv, the temperature of the steam after the desuperheating Tev, the pH of the condensate at the outlet of the exchanger, the temperatures of the phosphoric acid, respectively, at the inlet and at the exchanger outlet Teac and Tsac, the temperature of the 54% acid in P2O5 at the boiler outlet Tsb, the condensate flow rate Dcond, the discharge pressure Pref and suction pressure Pasp of the circulation pump of the acid, the amperage of the pump I, the temperature of the sea water at the inlet of the condenser Tme, the relative pressure in the boiler read in relation to that absolute Pr, the loss of P2O5 in the sea water δ, the acid flow 28% in P2O5 D28%ac, the densities of the diluted acid and that concentrated in P2O5 respectively $\rho 28\%ac$ and $\rho 54\%ac$, the mass titles

$\chi 28\%ac$ and $\chi 54\%ac$ (% in P2O5), sulfate levels $\tau 28\%ac$ and $\tau 54\%ac$ (%) and solid matter concentrations $M28\%ac$ and $M54\%ac$.

For the collection of operating data, relating to the operation of the heat exchangers of three phosphoric acid concentration units (UAP1, UAP2 and UAP3) of the "AP" factory of the Tunisian Chemical Group in Gabès, we adopted an approach including the following steps:

- Setting the parameters for controlling the operation of the heat exchanger allowing the detection of clogging and the triggering of washing operations.
- Choice of the period for the collection of operating data.
- Collection and sorting of data and development of the database.

AP1 and AP2 Units

The AP1 and AP2 units use graphite cylindrical block heat exchangers and a stainless steel tube exchanger. Eleven measuring points are recommended for checking function and fouling (see Figure 4).

Figure 4. Block diagram of the concentration loop of the AP1 and AP 2 units.

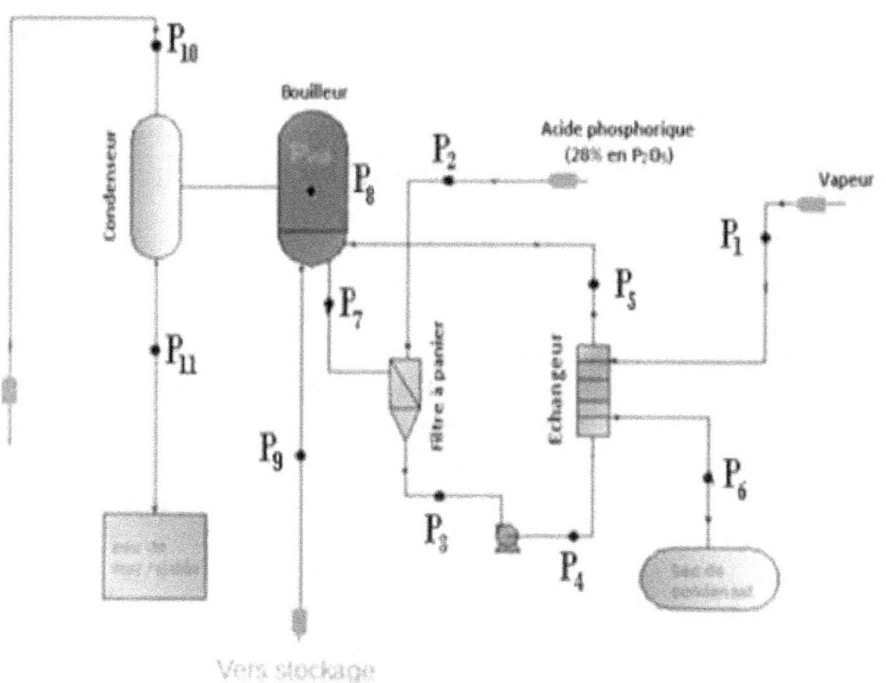

Remember that the AP1 and AP2 units have the same operating principle, the same measurement points. The only difference is in the type of heat exchanger used.

The parameters for each measurement point are shown in Table 1.

Table 1. The specifications of the measurement points of the AP1 and AP2 units.

Point	P1		P2					P3	
Parametre	T_e^v	P_v	$D_{28\%}^{ac}$	$\rho_{28\%}^{ac}$	$\chi_{28\%}^{ac}$	$\tau_{28\%}^{ac}$	$M_{28\%}^{ac}$	Pasp	I
Unite	°C	Bar	m³/h	kg/m³	%	%	mg/l	bar	A
P4		P5	P6		P7	P8	P9		
P_{ref}	T_e^{ac}	T_s^{ac}	D_{cond}	pH	T_{sb}	P_r	$P_{54\%}^{ac}$	$X_{54\%}^{ac}$	$T_{54\%}^{ac}$
bar	°C	°C	m³/h	-	°C	mmHg	kg/m³	%	%
P9	P10	P11							
$M_{54\%}^{ac}$	T_e^m	δ							
mg/l	°C	ppm							

Figure 5. Block diagram of the concentration loop of the AP3 unit.

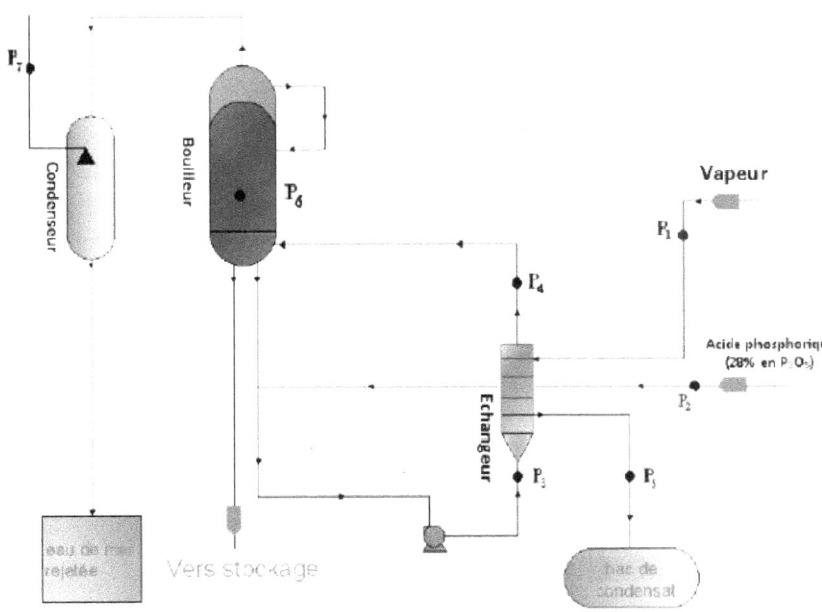

AP3 Unit

The heat exchanger installed in the AP3 unit is of the graphite shell and tube type. Figure 5 shows the block diagram of the phosphoric acid concentration loop of the AP3 unit. Its main specifications are listed in Table 2.

Table 2. The specifications of the measurement points of the AP3 unit.

Point	P1			P2		P3	P4	P5	P6	P7
Parametre	D_v	T_v	T_e^v	$D_{28\%}^{ac}$	$\rho_{28\%}^{ac}$	T_e^{ac}	T_s^{ac}	pH	P_r	T_e^m
Unit	$M^3/1$	°C	°C	m^3/h	kg/m^3	°C	°C	-	MmHg	°C

Establishing a database requires a fairly long period of sorting and data entry. Based on the data, recorded in written form, tables of parameters are made characterizing the concentration lines of phosphoric acid of the units AP1, AP2 and AP3 of the "AP" plant. Outliers are eliminated from the tables.

A graphical interface has been developed allowing the manipulator quick access to the recorded data.

Assumption and Calculation Steps

Hypotheses

The main objective of our work is to develop a mathematical model characterizing the drop in the overall heat exchange coefficient, due to fouling, as a function of the various operating parameters of the heat exchanger, the values of which are stored in the developed database.

The complexity of the fouling phenomenon imposes to admit certain hypotheses serving to facilitate the modelling. The main assumptions used are:

- Condensation inside the exchanger is total.
- The heat capacity of phosphoric acid is assumed to be constant over the operating temperature range.
- The flow of the two fluids within the graphite polyblock type heat exchanger is counter-current.
- Resistance to fouling is attributed solely to circulating phosphoric acid.
- Heat losses are zero (the heat exchanger is well insulated).

Calculation Steps

To perform the calculations necessary for the modelling, we adopted an approach based on the following steps:

Step 1: Determination of the circulation rate of phosphoric acid

The flow rate "Dv" is determined using the characteristic curve; the head as a function of the circulation flow. The manometric height is calculated according to the following formula:

$$H_m = \frac{P_{ref} - P_{asp}}{\rho^{ac}_{54\%} g} \tag{9}$$

Step 2: Calculation of the temperature difference "ΔT" and the logarithmic mean temperature difference "ΔTLM"

The calculation is done using the following equations:

$$\Delta T = T_s^{ac} - T_e^{ac} \tag{10}$$

$$\Delta TLM = \frac{(T_e^{ac} - T_e^{v}) - (T_s^{ac} - T_e^{v})}{\ln\left(\frac{T_e^{ac} - T_e^{v}}{T_s^{ac} - T_e^{v}}\right)} \tag{11}$$

Step 3: Calculation of the amount of heat exchanged

The heat Q is calculated from the following equations:

For UAP1 and UAP2 units:

$$Q = D_m C_p \Delta T \tag{12}$$

For UAP 3 unit:

$$Q = D_v L_v \tag{13}$$

Step 4: Calculation of the overall heat exchange coefficient Ug

The following relationship is used to calculate this coefficient:

$$U_g = \frac{Q}{S_{ech} \Delta TLM} \tag{14}$$

Step 5: Calculation of the fouling coefficient ΔUenc

The determination of the ΔUenc value is obtained by:

$$\Delta U_{emc} = U_{pr} - U_g \tag{15}$$

RESULTS

Modelization

Stainless Steel Tube Bundle Exchanger (unit AP1)

Depending on the operational data collected and after carrying out the sensitivity tests of the parameters, we selected the variables necessary for the development of the model, namely:

Tev: steam temperature (hot fluid) (°C).
ΔT: temperature difference between cold fluid inlet and outlet (°C).
Pref: delivery pressure of the cold fluid circulation pump (bar).
D28%ac: flow rate of 28% phosphoric acid in P2O5 (m3/h).
Pr: relative pressure read in relation to the absolute pressure ('vacuum pressure') (mmHg).

To improve the reliability of the model, we opted for a polynomial model, of order 4 whose expression is as follows:

$$\Delta U_{enc} = -40082549{,}813 + 8231{,}668\ \Delta T + 879782{,}357\ T_e^v + 20526086{,}298\ P_{ref} - 13786{,}581\ P_r - 23{,}544\ D_{28\%}^{ac} - 1960{,}254\ \Delta T^2 - 11521{,}492\ T_e^{v\,2} - 8784837{,}934\ P_{ref}^2 + 20{,}436\ P_r^2 - 4{,}051\ D_{28\%}^{ac\,2} + 195{,}989\ \Delta T^3 + 67{,}08\ T_v^3 + 1669535{,}53\ P_{ref}^3 - 1{,}009E\text{-}02\ P_r^3 + 0{,}311\ D_{28\%}^{ac\,3} - 7{,}159\ \Delta T^4 - 0{,}146\ T_v^4 - 118880{,}21\ P_{ref}^4 - 0.0054\ D_{28\%}^{ac\,4} \quad (16)$$

This model is recommended within the variation ranges of the parameters indicated in Table 3.

Table 3. Ranges of variation of the main operating parameters of the AP1 unit

Parameter	Minimum	Maximum
ΔT (°C)	4,7	9
T_v (°C)	112	119
P_{ref}(bar)	3,45	3,74
P_r(mmHg)	657	695
$D_{28\%}^{ac}$en P_2O_5 (m^3/h)	16	30

Figure 6. (ΔUenc) experimentally measured according to that calculated by the model (Unit AP1)

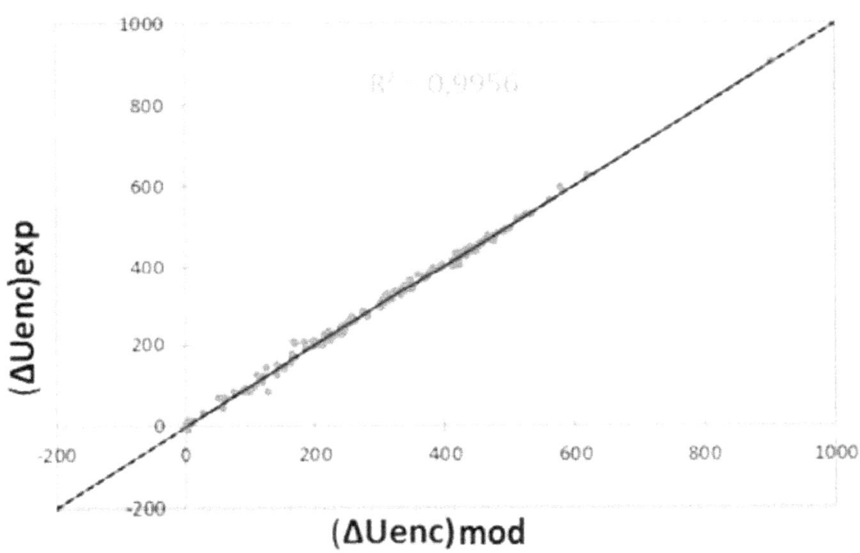

The values measured experimentally as a function of those calculated by the model of the heat exchange coefficient due to fouling, are represented in Figure 6. We note that this model is valid.

Cylindrical Polyblock Exchanger in Graphite (unit AP2)

By following the same approach as that described previously, the linear model proposed for the graphite heat exchanger of the poly cylindrical block type, is obtained from the linear regression of the data. The linear model expression for unit UAP2 takes the following form:

$$\Delta U_{enc} = -12072{,}29 + 37{,}254\, T_e^v + 880{,}84 P_{ref} + 8{,}689 P_r + 11{,}590\, D_{28\%}^{ac} - 263{,}058\, \Delta T \tag{17}$$

This model is recommended within the ranges of parameter variation, shown in Table 4.

Table 4. Ranges of variation of the main operating parameters of the AP2 unit

Parameter	Minimum	Maximum
ΔT (°C)	5	10
T_v (°C)	117	123
P_{ref}(bar)	4	4,38
P_r(mmHg)	690	700
$D_{28\%}^{ac}$ en P_2O_5 (m^3/h)	28	42

The experimentally measured values, as a function of those calculated by the model of the heat exchange coefficient due to fouling, are represented in Figure 7. We can see that this model is also valid.

Graphite Tube Bundle Heat Exchangers (unit AP3)

The AP3 unit at the GCT-Gabès "Phosphoric Acid" plant uses graphite exchangers of the tube and calender type. The proposed model will be established according to the following parameters: Steam temperature, temperature difference between inlet and outlet of the cold fluid, flow rate of the steam which feeds the exchanger, flow rate of the phosphoric acid and relative pressure read in relation to the absolute pressure.

Figure 7. (ΔUenc) measured experimentally, according to that calculated by the model (Unit AP2)

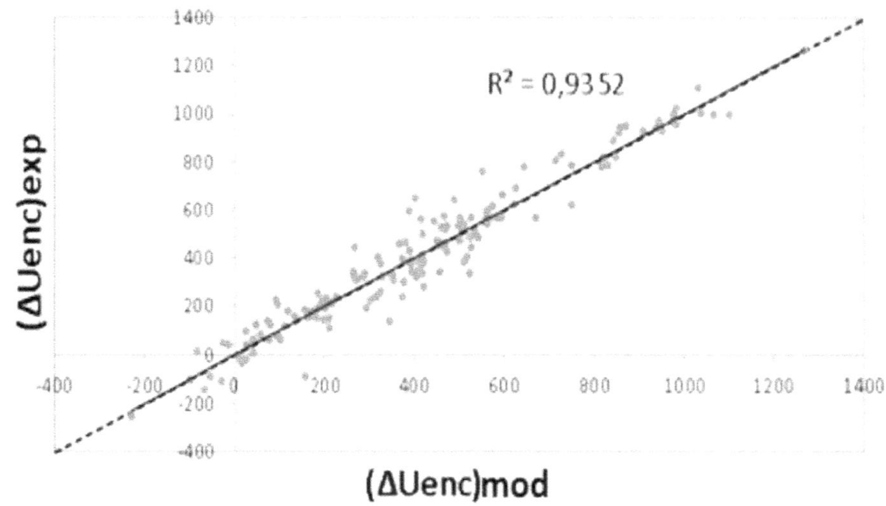

It is a polynomial model of order 5 whose expression is as follows:

$$\Delta U_{enc} = -652989{,}503 - 2339585{,}779\ D_v + 78245{,}454\ T_v + 683136{,}973\ \Delta T + 41689{,}861 P_r - 30376{,}597 D_{28\%} + 192686{,}639\ D_v^2 - 1016{,}075\ T_v^2 - 270931{,}976 \Delta T^2 - 62{,}381\ P_r^2 + 1302{,}856\ D_{28\%}^{ac\ 2} - 7898{,}521\ D_v^3 + 5{,}867\ T_v^3 + 53420{,}988 \Delta T^3 + 3{,}111E\text{-}02\ P_r^3 - 26{,}682\ D_{28\%}^{ac\ 3} + 161{,}203\ D_v^4 - 1{,}27E\text{-}02\ T_v^4 - 5236{,}979\ \Delta T^4 + 3{,}091E\text{-}17\ P_r^4 + 0{,}255\ D_{28\%}^{ac\ 4} - 1{,}311\ D_v^5 + 204{,}216\ \Delta T^5 - 8{,}793E\text{-}04\ D_{28\%}^{ac\ 5} \quad (18)$$

This model is recommended within the ranges of variation of the parameters indicated in Table 5.

Table 5. Ranges of variation of the main operating parameters of the AP2 unit

Parameter	Minimum	Maximum
$D_v (m^3/h)$	20	27
T_v (°C)	109	128,3
P_{ref} (bar)	4,2	6,1
P_r (mmHg)	654	680
$D_{28\%}^{ac}$ en P_2O_5 (m^3/h)	28	50

The values measured experimentally as a function of those calculated by the model of the heat exchange coefficient due to fouling are displayed in Figure 8.

Figure 8. (ΔUenc) measured experimentally according to that calculated by the model (Unit AP3)

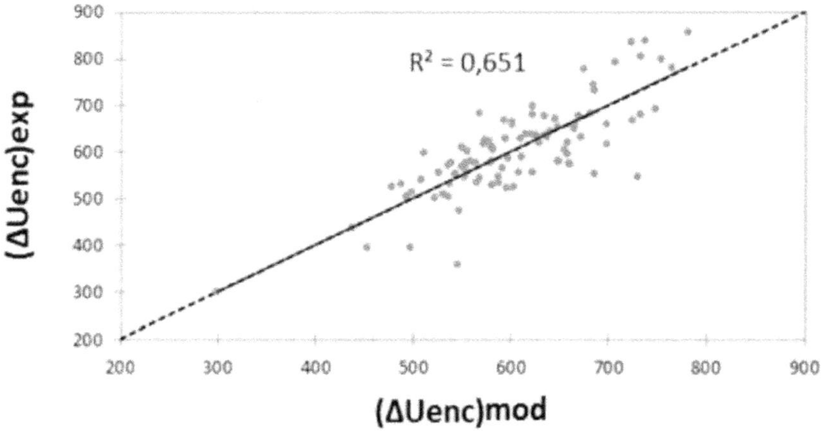

The lack of operating data and the absence of measurements for some operating parameters of the AP3 unit are the cause of the low reliability of the model developed for the unit concerned.

All the models developed are valid in well-defined operating parameter variation domains. Outside of these areas, the reliability of the models is suspect.

Parametric Study

We determined the effects of each parameter on the overall coefficient of thermal heat exchange (Ug). Figure 9 gives the effects of the parameters on Ug for the AP1 unit. Figure 10 shows the effects of the parameters on Ug for the AP2 unit. For unit AP3, we did not find definitive results. The two figures are roughly similar. We note that the steam quantity effect is negative for both figures. Indeed, the operators increase the quantity of steam during the cycle, to compensate for the decrease in the overall heat exchange coefficient. If the overall heat exchange coefficient is high, the heat transfer is also high, then the outlet temperature will increase, hence the positive effect of the exchanger outlet temperature on Ug.

The results are comparable to the results found in the bibliography [107-112]. If we compare the two figures we find that it is the difference between the inlet and outlet temperatures that has a significant effect and not the temperatures themselves.

Figure 9. Effects of parameters on the overall heat transfer conductance Ug (Unit AP1)

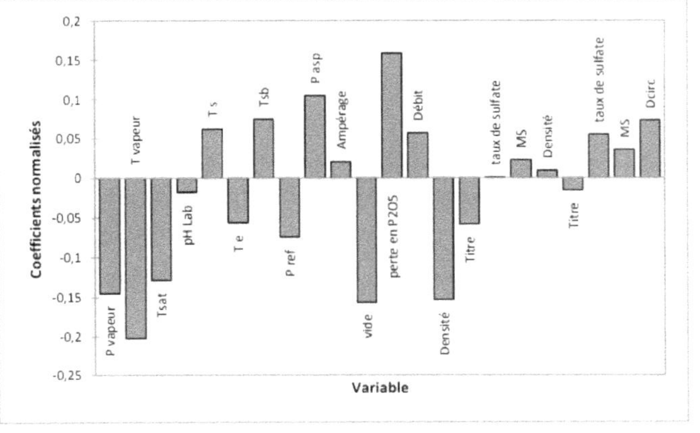

Figure 10. Effects of parameters on the overall heat transfer conductance Ug (Unit AP2)

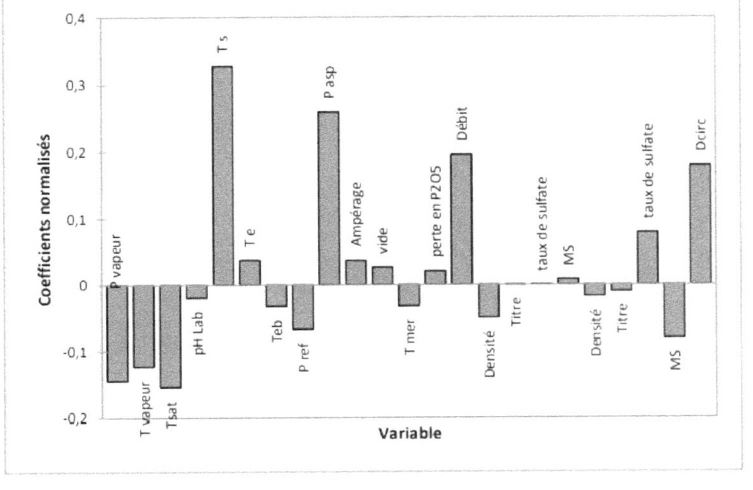

CONCLUSION

The main objective of our work is to establish mathematical models for monitoring the fouling of heat exchangers to, on the one hand, study the effects and interactions of operating parameters, on the other hand, to establish a good programming of the stops for the cleaning of this equipment automatically, by a regulation loop. To achieve this objective, we have adopted an approach comprising in particular the following steps: - Identification of the operating parameters involved in the fouling of the heat exchangers used in the concentration of phosphoric acid, and collection of the technical characteristics of the exchangers of heat studied. - Collection and sorting of operational data from the exchangers studied over a period of two and a half years. - Establishment of a database of on/off cycles for the cleaning of the heat exchangers studied. - Development of preliminary models for fouling monitoring and to help the operator decide when to shut down for cleaning the heat exchangers.

REFERENCES

Bevevino, J. W. (1978). Standards of Tubular Exchanger Manufacturers Association. *TEMA*.

Blattberg, R. C., Kim, B. D., & Neslin, S. A. (n.d.). *Database Marketing: Analyzing and Managing Customers. Springer.*

Bontemps, A., Garrigue, A., Goubier, C., Huetz, J., Marvillet, C., Mercier, P., & Vidil. (1994). *Échangeurs de Chaleur. Définitions et Architecture Générale* [Heat exchangers. Definitions and General Architecture]. Techniques de l'Ingénieur, Traité Génie Énergétique.

Bowen, B., & Epstein, N. (1979). Fine particle deposition in smooth parallel plate channels. *Journal of Colloid and Interface Science*, *72*(1), 81–87. doi:10.1016/0021-9797(79)90184-X

Epstein, N. (1988). General Thermal Fouling Models. In L. F. Melo, T. R. Bott, & C. A. Bernardo (Eds.), *Fouling Science and Technology. NATO ASI Series* (Vol. 145). Springer. doi:10.1007/978-94-009-2813-8_2

Finkbeiner, F., Gonard, T., & Filiol, B. (1993). Échangeurs Thermiques: Enjeux, Marchés, Technologie et Politique d'Innovation [Heat Exchangers: Challenges, Markets, Technology and Innovation Policy]. Éditions Européennes Thermique et Industrie (EETI).

Kern, D., & Seaton, R. (1959). A theoretical analysis of thermal surface fouling. British Chemical Engineering, 4(5), 258-262.

Kohonen, T. (1995). *Self-Organizing Maps*. Springer. doi:10.1007/978-3-642-97610-0

Nisbet, R., Elder, J., & Miner, G. (2009). *Handbook of Statistical Analysis and Data Mining Applications*. Elsevier.

Perry, R. H., & Green, D. W. (2008). *Perry's Chemical Engineers' handbook*. McGraw-Hill.

Taborek, J. (1979). Évolution of Heat Exchanger Design Techniques. *Heat Transfer Engineering*, *1*(1), 15–29. doi:10.1080/01457637908939546

Thomas, D., & Grigull, U. (1974). Experimental Investigation of the Deposition of Magnetite from the Fluid Flow in Steam Generating Tubes. *Brennstoff, Wärme, Kraft*, *26*(3), 109–115.

Thonon, B. (2007). *Guide de l'encrassement des échangeurs de chaleur* [Heat Exchanger Fouling Guide]. Editions greth France.

Varmuza, K., & Filzmoser, P. (2009). *Introduction to Multivariate Statistical Analysis in Chemometrics* (1st ed.). CRC Press. doi:10.1201/9781420059496

VDI. (2010). *Heat Atlas. Gesellschaft Verfahrenstechnik und Chemieingenieurwesen (GVC). s.l* (2nd ed.). Springer Verlag.

Wang, Y., Yuan, Z., Liang, Y., Xie, Y., Chen, X., & Li, X. (2015). A review of experimental measurement and prediction models of crude oil fouling rate in crude refinery preheat trains. *Asia-Pacific Journal of Chemical Engineering*, *10*, 607–625. doi:10.1002/apj.1895

Watkinson, A. P. (1968). *Particulate fouling of sensible heat exchangers (T)*. University of British Columbia. Retrieved from https://open.library.ubc.ca/collections/ubctheses/831/items/1.0059106

Chapter 13
Theoretical Analysis on a Household Heat Pump Water Heater With Immersed Condenser Coil

Sami Missaoui
École nationale supérieure d'ingénieurs de Tunis, Tunisia

Romdhane Ben Slama
Institut Supérieur des Sciences Appliquées et Technologies de Gabès, Tunisia

Bechir Chaouachi
National Engineering School of Gabes, Tunisia

ABSTRACT

In this study, a coupling model between the water tank with immersed condenser coil and the vapor compression system was developed in ANSYS fluent for simulating the heat transfer between the refrigerant in condenser and water in the storage tank. Further study was performed to analyze the effect of condenser coil location on the heating process. The results indicated that, when the condenser coil is placed in a lower part of the water tank, a higher water velocity can be observed. From the testing results, when the condenser coil is placed in the lower part of the water tank, the convective heat transfer is better than the other positions.

DOI: 10.4018/978-1-7998-8801-7.ch013

INTRODUCTION

Domestic hot water production accounts for a significant part of residential and commercial energy consumption in the world. Most conventional residential water heaters generate heat by directly consuming fossil fuels (wood, oil, gas water heaters…) or electric water heater (Ibrahim et al, 2013). With the demand of energy saving and environmental protection, heat pump water heater is a promising energy source and offers an economical solution by reducing the energy consumed for domestic hot water production and the Carbon dioxide emissions. For heat pump water heater with immersed condenser coil, optimizing the coil structure and the immersed helically coil position is important for improving the temperature stratification in the storage tank and the system performance. Many scholars have made experimental and numerical research on the air source heat pump water heater. (Jiang et al, 2006) conducted an experimental study on a modified air conditioner with a domestic hot water supply. Their results showed that the coefficient of performance of the modified air conditioner with a domestic hot water production is about 38.6% higher than that of the original unit. (Qu et al., 2014) studied the temperature and velocity field of the equal and variable diameter condensing coils at different positions of heat pump water storage tank with using ANSYS Fluent. Their results showed that the variable diameter condensing coils make the tank temperature raise stability, which is beneficial to the stability of the system operation. (Zhang et al, 2007) conducted an experimental and numerical research on air source heat pump water heater. From the results, it could be seen that the system performance (COP) could be improved obviously. (Dai et al, 2019) carried out an analysis on the charging and discharging process of a household heat pump water heater. They observed, in the water charging process, heat transfer coefficient and COP of variable diameter coil were 20% and 10.23% higher than that of fixed diameter coil, respectively. The effect of tank geometry on stratification was investigated by (Eames and Norton, 1998). According to the authors, low inlet jet velocities were found to have less cross sectional dependence on thermocline growth. The research also revealed that a single jet with a variable temperature inlet degrades the thermal stratification.

The effect of obstacles on stratification in hot water tanks was analyzed numerically by (Altuntop et al, 2005). According to the findings, tanks with obstacles and a hole in the middle tended to stratify better than tanks without obstacles. (Savicki et al, 2011) created and analyzed a three-dimensional model of a cylindrical storage tank, using it to predict the temperature profile and thermal stratification over time. These similarities are said to make solar collector and thermal storage tank modeling easier because they can provide reference data without the need for experiments in an experimental facility. In explaining quasi-steady warm up of a heat pump water heating system, (Shah and Hrnjak, 2014) proposed a related modeling method involving iteration between the CFD model of the water tank and the vapor-compression system model. His style, however, was for HPWHs with wrap-around coil condensers. (Park and Hrnjak, 2008) investigated the output of a microchannel condenser and a round-tube condenser experimentally, and found that the microchannel condenser's COP and heat flux were higher than the round-tube condenser's. (Wang, 2006) discovered that by reducing the condenser coil diameter and increasing the spiral diameter, the system's efficiency could be improved. The heat transfer efficiency of refrigerant in condenser coils could be slightly improved by adjusting the coil sectional structure, according to (Yang, Shao, and Zhang, 2014) despite the fact that the condenser coil layout has been altered, the heat transfer resistance has not been effectively decreased, and temperature stratification in the water tank continues to be evident. (Lu et al, 2014) studied the thermal-hydraulic properties of flowing fluid outside multi coils under various thermal boundary conditions. According to the findings, the heat transfer coefficients at constant wall temperature are similar to those at the fluid-to-fluid boundary. (Abolmaali et al., 2019) used a numerical simulation to establish Nusselt number and friction factor correlations for fluid flowing outside multi coils. They also discovered that the tube number in the first row has no effect on the Nusselt number or friction factor. For the shell-side of SWHE, (Wang et al, 2019) suggested a two-layer multi-objective optimization. According to the results, the Nusselt number was found to be significantly affected by the coiled radius. (Ali et al, 1998) investigated natural convection heat transfer from horizontally oriented uniformly heated helical tubes in air and suggested a coupled model to calculate the heat transfer coefficient. The effects of thermal radiation on convective heat transfer were investigated by (Zheng et al, 2000), who discovered that thermal radiation could increase the total heat transfer rate. For heating liquids, (Prabhanjan et al, 2002) demonstrated the advantages of using a helically coiled heat exchanger over a straight tube heat exchanger. The tests were carried out in the transitional and turbulent flow zones. (Prabhajan et al, 2004) investigated natural convection heat transfer from helical coiled tubes submerged in water in an experimental study. The outside Nusselt number was correlated to the Rayleigh number, and coil height was chosen as the best representation for a vertical coil. In the laminar regime, (Rennie et al, 2006) performed numerical analyses on a double pipe helical heat exchanger. Overall heat transfer coefficients for both parallel and counter flow were determined, and a correlation for annulus Nusselt number in terms of modified Dean number was discovered. (Kumar et al, 2006) conducted experimental and numerical investigations of heat transfer characteristics of tubes in tube heat exchangers in counter current mode operation. (Shokouhmand et al, 2007) investigated a helical coil heat exchanger with air and water as heat transfer fluids and suggested a correlation for calculating the heat transfer coefficient. (Conte et al, 2008) investigated forced laminar flow in rectangular coiled pipes with circular cross sections numerically. When a helical heat exchanger was used, (Xiaowen et al, 2009) found that the coefficient of performance of a domestic water cooled air conditioner improved. (Kharat et al., 2009) established a heat transfer coefficient connection for flow between concentric helical coils. (Chen et al, 2011) used the refrigerant R134a as the heat transfer fluid in horizontal helically coiled tubes to conduct heat flux experiments. According to the findings, the coil-to-diameter ratio is said to be more important than the length-to-diameter ratio, and a relationship is proposed to estimate heat flux. (Moawed et al, 2011) studied forced convection heat transfer from a helical coil tube with a constant heat flux. Variation of coil parameters was used to achieve a heat transfer coefficient correlation in terms of Reynolds number and geometric parameters. (Mahmoudi et al, 2017) used TiO2/ water nanofluid to investigate forced convection heat transfer and pressure drop in helically coiled pipes and found that the Dean number has a major impact on heat transfer for a given Reynolds number. (Naphon, 2007) investigated the thermal efficiency and pressure drop of a helical coil heat exchanger with and without helically crimped fins in an experimental setting.

In a turbulent flow, (Devanahalli et al, 2004) investigated natural convection heat transfer from helical coiled tubes in water. They used four coil sets with separate pitches of 47.4, 15.8, 13.5, and 40.5 mm and aspect ratios of 19.3, 19.3, 15.03, and 15.03. To investigate the heat transfer coefficient of a helically coiled tube, (Shah and Joshi, 1987) used the boundary conditions of constant wall temperature and constant heat flux. (Rennie et al, 2005; 2006a; 2006b) looked at experimental and numerical studies of the double pipe helically heat exchanger. (Guo et al. 2002) investigated the transient convective heat transfer in a helical tube subjected to pressure drop oscillations. (Choi, 1995) introduced the idea of nanofluids and stated that they have higher thermal conductivity than other traditional heat transfer fluids. (Xin et al, 1996) investigated two-phase flow in vertical helicoidal pipes in an experimental study. In their research, water and air were used as working fluids. Furthermore, eight coils were tested with various vertical helicoidal pipe configurations. (Patankar et al, 1974) looked at how the De number affected heat transfer in helically coiled tubes for developing and fully formed laminar flow. (Yang et al, 1995) examined laminar convective heat transfer in a helical pipe with completely formed laminar convection. (Pawar et al, 2014) investigated convective heat transfer in a helically coiled tube heat exchanger using experimental and CFD methods. The experimental results (Nui, hi, U, Q, T2, and Two) are compared to the CFD measurement results and found to match the CFD predictions fairly well. Whilst there are a large number of studies that have focused on the modified air conditioner with a domestic hot water production, only very few investigations have highlighted the effects of the condenser coil structure on the heating performance. Nevertheless, attempts have been made to analysis the effect of the condenser coil position on the temperature and velocity distribution and heat transfer characteristics during heating process. Therefore, the present study is an attempt to analyze the influence of condenser coil position on the water heating process.

SYSTEM DESCRIPTION

Figure 1 shows the schematic diagram of heat pump water heater with immersed condenser coil. The modified air conditioner with a domestic hot water production mainly consists of a vapor-compression system and a water heating unit, including water tank with immersed condenser coil, compressor, evaporator and expansion valve.

Figure 1. Schematic of the experimental apparatus

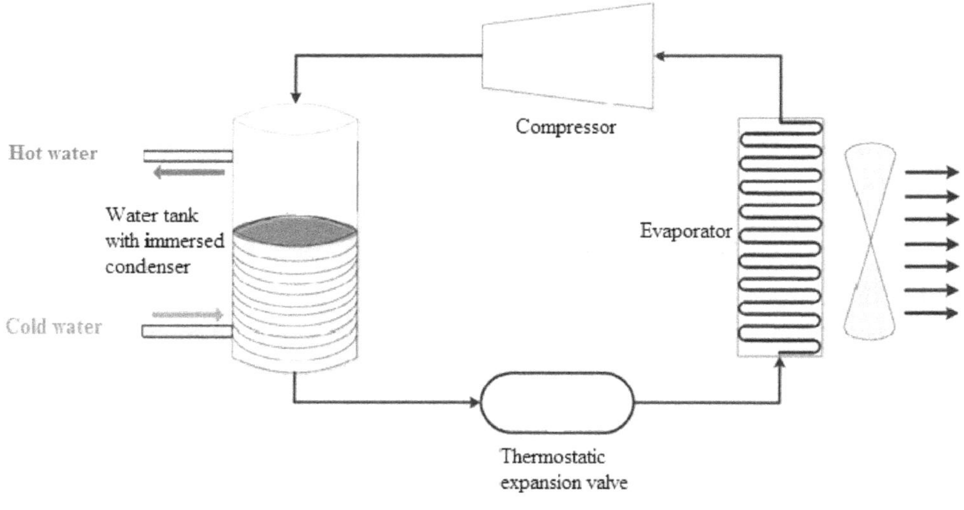

The stainless-steel water tank in this paper had a capacity of 80 liters. The tank's wall was made of aluminum, with a 30 mm thick layer of polyurethane thermal insulation. Cold water was directly connected to the tap water pipe at the bottom of the tank, while hot water was dispensed from the top, as shown in figure 2.

The main structural parameters of water tank and helically immersed copper coil are illustrated in table 1

For CFD simulation, REFPROP 7.0 is used to measure the thermodynamic properties of water and R22.

Figure 2. Schematic of the immersed helical coiled tube heat exchanger in the cylindrical water tank

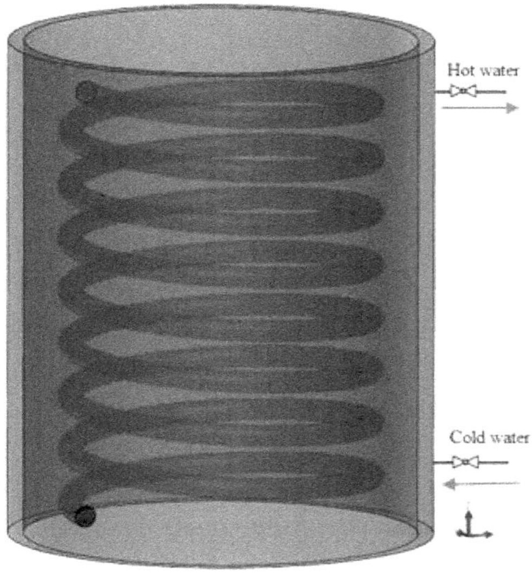

Table 1. Main structural parameters of water tank with immersed condenser coil. (Dai et al, 2018)

Component	Structural parameter	Symbol	Value
Water tank Condenser coil	Height of tank (m)	H_t	1.15
	Diameter (m)	D_t	0.38
	Volume (L)	V_t	80
	Number of turns (-)	N	36
	Total length (m)	L_c	19
	Coil diameter (m)	D_c	0.16
	Coil height (m)	H_c	0.70

Numerical Model

The water tank with immersed condenser coil model was simulated by discretizing the Navier-Stokes equations of mass, momentum and energy with using ANSYS Fluent software. In order to analyze the effects of condenser coil position on laminar water flow in the storage tank, a numerical investigation of natural convection from vertical helically coiled tubes for different condenser coil locations was carried out.

Boundary Conditions

The boundary conditions selected as main properties of water in computational fluid dynamics were: constant thermal expansion coefficient $\left(0.0002666\frac{1}{K}\right)$, constant dynamic viscosity (0.00098375 kg/ms), constant thermal conductivity (0.6104 W/mK), constant specify heat (4177.6 J/Kg•K) and the water density is the Boussinesq approximation (996.7 Kg/m^3). The heat flux of condenser coil surface was set as variable heat flux boundary conditions and input by Fluent User Defined Function (UDF) feature. In all the cases, the initial water temperature is set to be 15 °C. As for how to get the instantaneous heat flux, the waterside temperature and flow rate during the heating time were determined in CFD using an initial heat flux of the helical condenser coil. The variable heat flux $q(t)_i$ was measured using this waterside knowledge and a MATLAB code. The newly discovered $q(t)_i$ was then used as a heat flux boundary condition for the helically coil heat exchanger in CFD to obtain updated waterside data. Therefore, the following is the heat flux formulation for the entire heating time:

$$q(t) = -0.0008t^3 + 0.1187t^2 - 11.295t + 4376.4 \tag{1}$$

$R^2 = 0.9861$

Meshing

As shown in figure 3 the three-dimensional storage tank with immersed condenser coil was simplified to two-dimensional (2-D) axisymmetric geometry. The 2-D axisymmetric model can be assumed about the axial centerline of the cylindrical water tank. The mesh was refined near the condenser coil and tank wall to more precisely capture the water recirculation during the heating process. In this study, the structural quadrilateral mesh was used for CFD model of water tank.

Mathematical Formulation

The flow has been modeled using the continuity equation, momentum equation and energy equation:

The continuity equation is written as follows:

$$\frac{\partial u_x}{\partial x} + \frac{\partial u_y}{\partial y} = 0 \tag{2}$$

Figure 3. Geometric model of HPWH

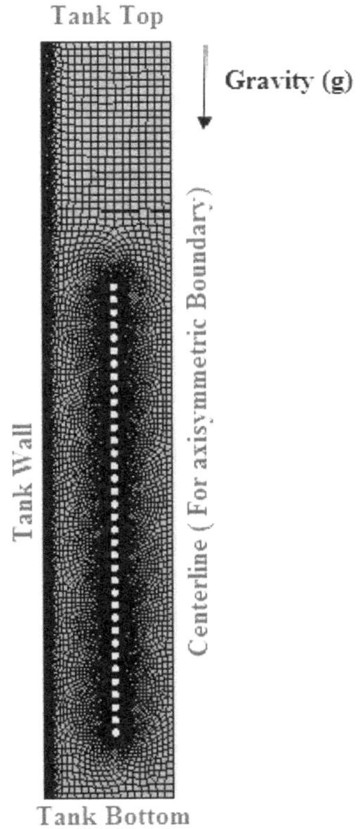

The momentum equation is written as follows:

$$\frac{\partial u_x}{\partial t}+u_x\frac{\partial u_x}{\partial x}+u_y\frac{\partial u_x}{\partial y}=-\frac{1}{\rho_{w0}}\frac{\partial P}{\partial x}+\nu_w\left[\frac{\partial^2 u_x}{\partial x^2}+\frac{\partial^2 u_x}{\partial y^2}\right] \tag{3}$$

$$\frac{\partial u_y}{\partial t}+u_x\frac{\partial u_y}{\partial x}+u_y\frac{\partial u_y}{\partial y}=-\frac{1}{\rho_{w0}}\frac{\partial P}{\partial y}+\nu_w\left[\frac{\partial^2 u_y}{\partial x^2}+\frac{\partial^2 u_y}{\partial y^2}\right]-g\beta\left(T_w-T_{w0}\right) \tag{4}$$

Energy equation is written as follows:

$$\frac{\partial T}{\partial t}+u_x\frac{\partial T}{\partial x}+u_y\frac{\partial T}{\partial y}=\lambda_w\left[\frac{\partial^2 T}{\partial x^2}+\frac{\partial^2 T}{\partial y^2}\right] \tag{5}$$

CFD Modeling

The experimental setup and the basic geometry used for numerical modeling are the same. The SIMPLE algorithm is used to solve velocity and pressure distribution using the coupling equations of continuity and momentum. The convection terms in the equations are treated using a second order upwind scheme. In a laminar flow state, a PRESTO scheme was used for pressure, and a second order upwind scheme was used for momentum and energy equations. Furthermore, for continuity, momentum and energy, the convergence criterion was $1.0e^{-6}$.

Model Validation

In order to validate the accuracy of the results, the two-dimensional axisymmetric model of HPWH with immersed condenser coil is modeled as per the literature of (Dai et al, 2018). For the present numerical study and that of the literature of (Dai et al, 2018), figure 4 shows the temperature distribution of water with height of tank for heating time is set to be 60 min. Figure 4 shows the comparison and it was seen that the predicted values are in good agreement with that of the results available in the literature.

Figure 4. Variation of water temperature versus height of tank at t= 60 min

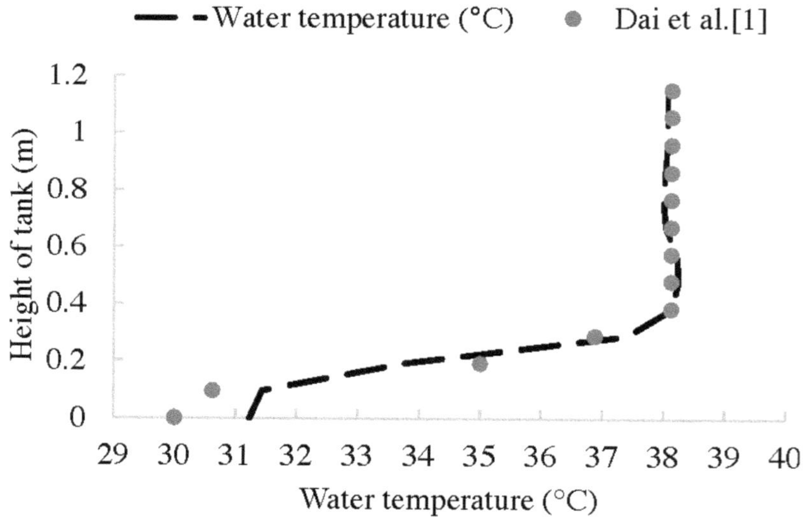

RESULTS AND DISCUSSION

In this paper, CFD simulation was conducted to study the effects of condenser coil position on the heating effects. The variation of the water velocity with height of tank at different positions of condenser coil is presented in figure 5. The water velocity in the tank was measured for three positions indicated by V1, V2 and V3. V1 represents the water velocity when the condenser coil is located in a position equal to 40 mm from the bottom. V2 represents the water velocity in the vertical direction of the tank center when the condenser coil is placed in a position equal to 100 mm from the bottom. Whereas, V3 indicate the water velocity when the condenser coil is located in a position equal to 160 mm from the bottom.

Effects of the Condenser Coil Position

In order to investigate the influence of condenser coil position on the heating effect, the structural parameters of the storage tank and the spiral condenser coil are kept constant. According to these results, it has been noted that the velocity is weak when the condenser coil is at the upper part, 160 mm from the bottom, of the tank. When the copper pipe is placed at the lower part of the tank (40 mm from the bottom), the water velocity reaches its maximum value.

When the helical coil is mounted in the lower part of the tank, as seen in this figure, the water velocity in the lower part of the tank is clearly high than the upper and middle parts, and the high velocity is concentrated near the condenser region in the tank's centerline direction. This is due to the helical coil's higher normal convection, which is lead to increase the water velocity. As a result of the increased water velocity, the water temperature rises.

Temperature Distribution

Figure 6 shows the temperature distribution of the water under different condenser coil position (40 mm, 100 mm and 160 mm). According to this study, it is clear that the water temperature in the storage tank increases with time taking variant thermal behavior depending on the condenser location. The effects of condenser coil position on water temperature variation were investigated in this case. According to this study, when the copper pipe is located at the bottom of the water tank, the thermal stratification improves. The water temperature of the lower part of the tank is also very high when the condenser coil is placed in the lower part of the tank, as opposed to the two other conditions. As a result, the effects of helical coil position on thermal characteristics were verified in this study.

Figure 5. Water velocity profiles with height of tank at t= 60 min

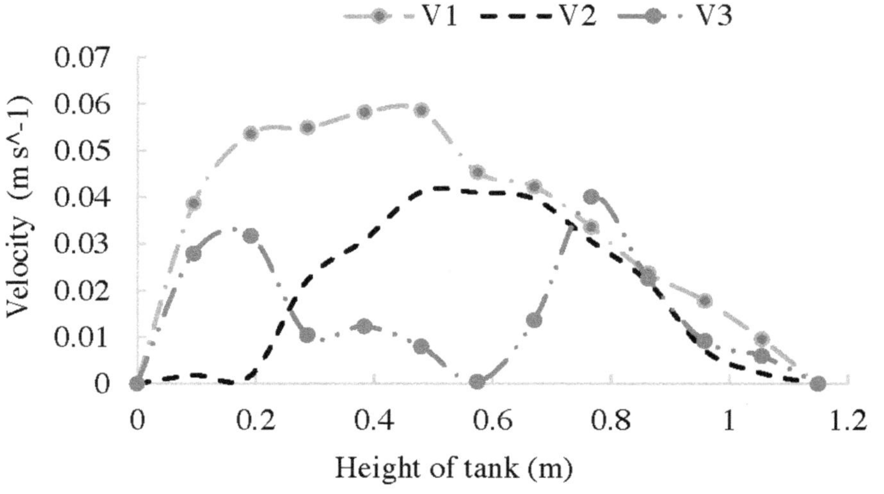

Figure 6. Temperature distribution of the water in the storage tank at t= 60 min

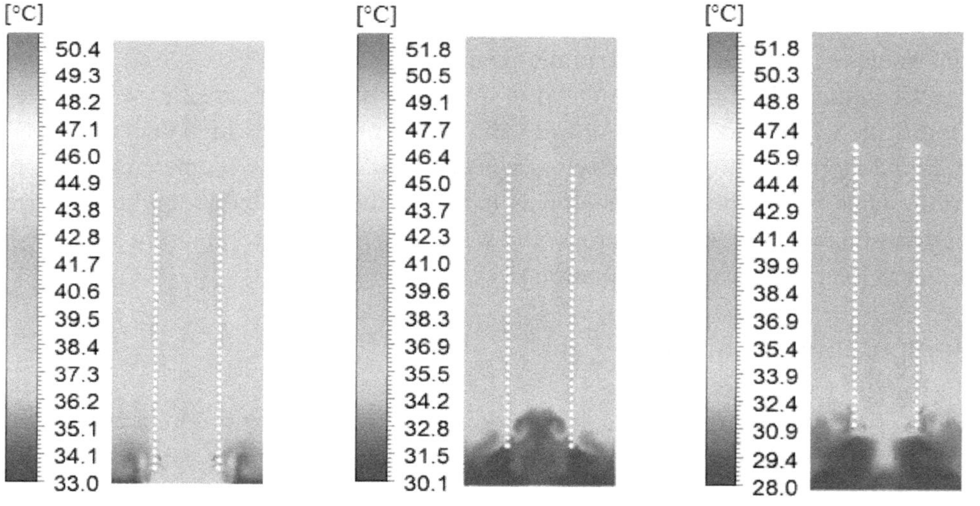

Velocity Distribution

Figure 7 shows the velocity distribution for different coil position equal to 40 mm, 100 mm, and 160 mm. From these results, it is clear that the velocity of the water becomes higher with decreasing the condenser coil distance from the bottom of the tank. In these conditions, the water velocity increases. Based on the CFD analysis, the effects of condenser coil position on water velocity is obvious. Based on this simulation, it is obvious that the water velocity reach's its maximum when the coil is at the lower part.

Figure 7. Velocity distribution of the water in the storage tank at t= 60 min

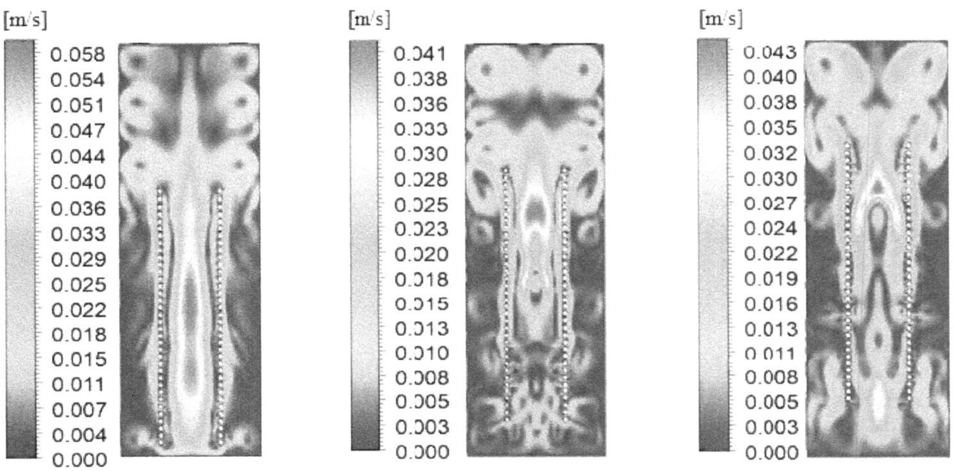

Static Pressure

Figure 8 shows the distribution of the static pressure for different condenser coil position equal to 40 mm, 100 mm and 160 mm. According to these results, it is obvious that the static pressure is on its maximum when the condenser coil located at the middle part of the storage tank. Compared with the two conditions, we can see that the static pressure in the upper part of the tank is higher than the middle and lower part of the storage tank and the higher-pressure area is concentrated near the tank top.

CONCLUSION

This paper presented a coupling model of household heat pump and storage tank with immersed condenser coil. Fluent software was used to study the effect of condenser coil location on the water velocity distribution in the tank. According to the results, the following conclusions can be drawn:

- When the condenser coil is placed in the lower part of the storage tank, the convective heat transfer is better than the others locations.
- When the condenser coil is placed in a lower part of the water tank, a higher water velocity can be observed.
- The static pressure increases in the upper part of the tank.
- The thermal stratification could be enhanced by reducing the distance between the condenser coil and the tank bottom.

Figure 8. Distribution of the static pressure in the storage tank

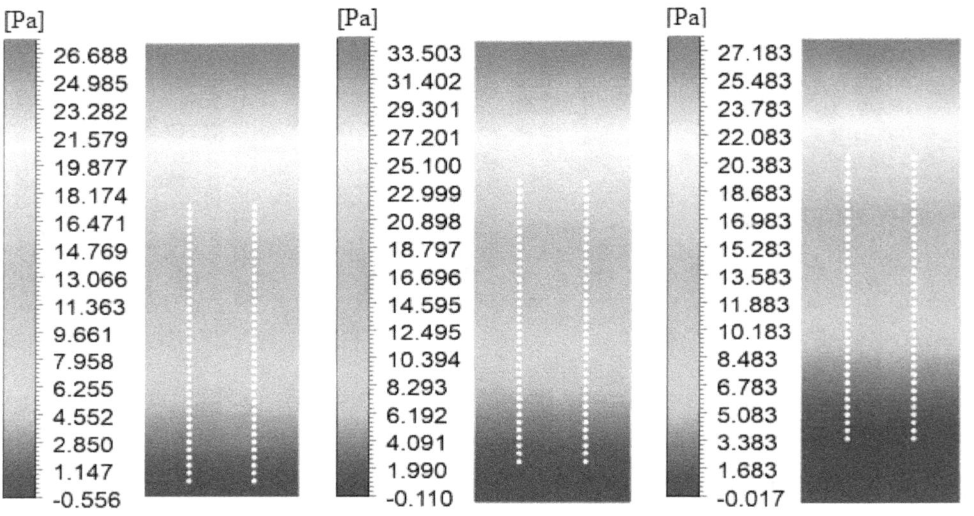

REFERENCES

Abolmaali, A. M., & Afshin, H. (2019). Development of Nusselt number and friction factor correlations for the shell side of spiral wound heat exchangers. *International Journal of Thermal Sciences*, *139*, 105–117. doi:10.1016/j.ijthermalsci.2019.01.038

Ali, M.E. (1998). Laminar natural convection from constant heat flux helical coiled tubes. *Int. J. Heat Mass Transf.*, *41*, 2175–2182. (97)00322-0 doi:10.1016/S0017-9310

Altuntop, N., Arslan, M., Ozceyhan, V., & Kanoglu, M. (2005). Effect of obstacles on thermal stratification in hot water storage tanks. *Applied Thermal Engineering*, *25*(14-15), 2285–2298. doi:10.1016/j.applthermaleng.2004.12.013

Chen, C. N., Han, J. T., Jen, T. C., Shao, L., & Chen, W. W. (2011). Experimental study on critical heat flux characteristics of R134a flow boiling in horizontal helically-coiled tubes. *International Journal of Thermal Sciences*, *50*(2), 169–177. doi:10.1016/j.ijthermalsci.2010.10.002

Choi, S. U. S. (1995). Enhancing thermal conductivity of fluids with nano particles. *ASME FED*, *231*, 99–103.

Conté, I., & Peng, X. F. (2008). Numerical investigations of laminar flow in coiled pipes. *Applied Thermal Engineering*, *28*(5-6), 423–432. doi:10.1016/j.applthermaleng.2007.05.009

Dai, N., & Li, S. (2018). Simulation and performance analysis on condenser coil in household heat pump water heater. *Sustainable Cities and Society*, *36*, 176–184. doi:10.1016/j.scs.2017.10.020

Dai, N., Li, S., & Ye, Q. (2019). Performance analysis on the charging and discharging process of a household heat pump water heater. *International Journal of Refrigeration*, *98*, 266–273. doi:10.1016/j.ijrefrig.2018.10.009

Devanahalli, G. P., Timothy, J. R., & Raghavan, G. S. V. (2004). Natural convection heat transfer from helical coiled tubes. *International Journal of Thermal Sciences*, *43*(4), 359–365. doi:10.1016/j.ijthermalsci.2003.08.005

Eames, P. C., & Norton, B. (1998). The effect of tank geometry on thermally stratified sensible heat storage subject to low Reynolds number flows. Int. *J. Heat Mass Tran.*, *41*(14), 2131–2142. doi:10.1016/S0017-9310(97)00349-9

Guo, L. J., Feng, Z. P., & Chen, X. (2002). Transient convective heat transfer of steam–water two-phase flow in a helical tube under pressure drop type oscillations. *International Journal of Heat and Mass Transfer*, *45*(3), 533–542. doi:10.1016/S0017-9310(01)00178-8

Ibrahim, O., Fardoun, F., Younes, R., & Louahlia-Gualous, H. (2013). Air source heat pump water heater: Dynamic modeling, optimal energy management and mini-tubes condensers. *Energy*, ●●●, 1–15.

Jiang, H., Jiang, Y., Wang, Y., Ma, Z., & Yao, Y. (2006). An experimental study on a modified air conditioner with a domestic hot water supply (ACDHWS). *Energy*, *31*(12), 1789–1803. doi:10.1016/j.energy.2005.07.004

Kharat, R., Bhardwaj, N., & Jha, R. S. (2009). Development of heat transfer coefficient correlation for concentric helical coil heat exchanger. *International Journal of Thermal Sciences*, *48*(12), 2300–2308. doi:10.1016/j.ijthermalsci.2009.04.008

Kumar, V., Saini, S., Sharma, M., & Nigam, K. D. P. (2006). Pressure drop and heat transfer study in tube-in-tube helical heat exchanger. *Chemical Engineering Science*, *61*(13), 4403–4416. doi:10.1016/j.ces.2006.01.039

Lu, X., Du, X., Zeng, M., Zhang, S., & Wang, Q. (2014). Shell-side thermal-hydraulic performances of multilayer spiral-wound heat exchangers under different wall thermal boundary conditions. *Applied Thermal Engineering*, *70*, 1216.

Mahmoudi, M., Tavakoli, M. R., Mirsoleimani, M. A., Gholami, A., & Salimpour, M. R. (2017). Experimental and numerical investigation on forced convection heat transfer and pressure drop in helically coiled pipes using TiO2/water nanofluid. *International Journal of Refrigeration*, *74*, 627–643. doi:10.1016/j.ijrefrig.2016.11.014

Moawed, M. (2011). Experimental study of forced convection from helical coiled tubes with different parameters. *Energy Conversion and Management*, *52*(2), 1150–1156. doi:10.1016/j.enconman.2010.09.009

Naphon, P. (2007). Thermal performance and pressure drop of thehelical-coil heat exchangers with and without helically crimped fins. *International Communications in Heat and Mass Transfer*, *34*(3), 321–330. doi:10.1016/j.icheatmasstransfer.2006.11.009

Park, C. Y., & Hrnjak, P. (2008). Experimental and numerical study on microchannel and round tube condensers in a R410A residential air-conditioning system. *International Journal of Refrigeration*, *31*(5), 822–831. doi:10.1016/j.ijrefrig.2007.10.007

Patankar, S. V., Pratap, V. S., & Spalding, D. B. (1974). Prediction of laminar flow and heat transfer in helically coiled pipes. *Journal of Fluid Mechanics*, *62*(03), 53–551. doi:10.1017/S0022112074000796

Pawar, S. S., & Sunnapwar, V. K. (2014). Experimental and CFD investigation of convective heat transfer in helically coiled tube heat exchanger. *Chemical Engineering Research & Design*, *92*(11), 2294–2312. Advance online publication. doi:10.1016/j.cherd.2014.01.016

Prabhanjan, D.G., Raghavan, G.S.V., & Rennie, T.J. (2002). Comparison of heat transfer rates between a straight tube heat exchanger and a helically coiled heat exchanger. *Int. Commun. Heat Mass Transf.*, *29*, 185–191. (02)00309-3 doi:10.1016/S0735-1933

Prabhanjan, D. G., Rennie, T. J., & Raghavan, G. S. V. (2004). Natural convection heat transfer from helical coiled tubes. *International Journal of Thermal Sciences*, *43*(4), 359–365. doi:10.1016/j.ijthermalsci.2003.08.005

Qu, Y., Wang, F., Wang, Y., Wang, P., Li, T., & Feng Meng, Z. (2014). Simulation Optimization and Experiment of R1234-ze on the Heat Pump Water Heater Storage Tank. *Advanced Materials Research*, *1051*, 828–831. doi:10.4028/www.scientific.net/AMR.1051.828

Rennie, T. J., & Raghavan, V. G. S. (2005). Experimental studies of a double-pipe helical heat exchanger. *Experimental Thermal and Fluid Science*, *29*(8), 919–924. doi:10.1016/j.expthermflusci.2005.02.001

Rennie, T. J., & Raghavan, V. G. S. (2006). Numerical studies of a double-pipe helical heat exchanger. *Applied Thermal Engineering*, *26*(11-12), 1266–1273. doi:10.1016/j.applthermaleng.2005.10.030

Rennie, T. J., & Raghavan, V. G. S. (2006a). Numerical studies of a double-pipe helical heat exchanger. *Applied Thermal Engineering*, *26*(11-12), 1266–1273. doi:10.1016/j.applthermaleng.2005.10.030

Rennie, T. J., & Raghavan, V. G. S. (2006b). Effect of fluid thermal properties on heat transfer characteristics in a double pipe helical heat exchanger. *International Journal of Thermal Sciences*, *45*(12), 1158–1165. doi:10.1016/j.ijthermalsci.2006.02.004

Savicki, D. L., Vielmo, H., & Krenzinger, A. (2011). Three-dimensional analysis and investigation of the thermal and hydrodynamic behaviors of cylindrical storage tanks. *Renewable Energy*, *36*(5), 1364–1373. doi:10.1016/j.renene.2010.10.011

Shah, R. K., & Joshi, S. D. (1987). Convective heat transfer in curved ducts. In Handbook of single phase convective heat transfer. New York: Wiley Inter science.

Shah, T., & Hrnjak, P.S. (2014). *Linked Modelling of Heat Pump Water Heater Vapor Compression System and Water Tank*. Academic Press.

Shokouhmand, H., & Salimpour, M. R. (2007). Entropy generation analysis of fully developed laminar forced convection in a helical tube with uniform wall temperature. *Heat and Mass Transfer*, *44*(2), 213–220. doi:10.100700231-007-0235-x

Wang, D. D. (2006). *Research on Temperature Field and Flow Field in Tank of ASHPWH*. Beijing Institute of Civil Engineering and Architecture Doctor of Philosophy.

Wang, G., Wang, D., Deng, J., Lyu, Y., Pei, Y., & Xiang, S. (2019). Experimental and numerical study on the heat transfer and flow characteristics in shell side of helically coiled tube heat exchanger based on multi-objective optimization. *International Journal of Heat and Mass Transfer*, *137*, 349.

Xiaowen, Y., & Lee, W. L. (2009). The use of helical heat exchanger for heat recovery domestic Water cooled air-conditioners. *Energy Conversion and Management*, *50*(2), 240–246. doi:10.1016/j.enconman.2008.09.039

Xin, R. C., Awwad, A., Dong, Z. F., Ebadian, M. A., & Soliman, H. M. (1996). An investigation and comparative study of the pressure drop in air–water two-phase flow in vertical helicoidal pipes. *International Journal of Heat and Mass Transfer*, *39*(4), 735–743. doi:10.1016/0017-9310(95)00164-6

Yang, G., Dong, F., & Ebadian, M. A. (1995). Laminar forced convection in a helicoidal pipe with finite pitch. *International Journal of Heat and Mass Transfer*, *38*(5), 853–862. doi:10.1016/0017-9310(94)00199-6

Yang, L., Shao, L. L., & Zhang, Z. C. (2014). Modeling and optimization of air source heat pump water heaters using wrap-around micro-channel condenser. *J. Refrig.*, *35*, 66–70.

Zhang, J., Wang, R. Z., & Wu, J. Y. (2007). System optimization and experimental research on air source heat pump water heater. *Applied Thermal Engineering*, *27*(5-6), 1029–1035. doi:10.1016/j.applthermaleng.2006.07.031

Zheng, B., Lin, C.X., & Ebadian, M.A. (2000). Combined laminar forced convection and thermal radiation in a helical pipe. *Int. J. Heat Mass Transf.*, *43*, 1067–1078. (99)00214-8 doi:10.1016/S0017-9310

Chapter 14
Thermodynamics of Lithium and Separation Processes From Natural Brine:
Finding a Needle in a Haystack

Elhoucine Essefi
University of Gabes, Tunisia

soumaya Hajji
https://orcid.org/0000-0002-6371-7223
LR3E.ENIS, University of Sfax, Tunisia

Hassan Khlissa
Institut Supérieur des Sciences Appliquées et Technologies de Gabès, Tunisia

ABSTRACT

Lithium has worldwide been proven of great energetic interest. One of the origins of lithium is salt lakes brine. Tunisia is marked by the presence of many saline systems containing lithium ranging from 20 mg/L to 50 mg/L. Nonetheless, extracting lithium from natural brine is really finding a needle in a haystack. This difficulty is basically due to the nuclear, electronic, and thermodynamic features of lithium as well as the other ions present in the brine. In this chapter, the authors study the technical and thermodynamic methods leading to the enrichment of lithium in brine, which in turn leads to its easy extraction. For this reason, techniques of extraction and their limitation were reviewed. In addition, the evaporation process of the brine is thermodynamic methods leading to an enrichment of with lithium in the brine due to the extraction of water molecules. Also, the precipitation of minerals including ions representing a noise of the lithium, especially magnesium, paves the way for easy extraction within lithium window.

DOI: 10.4018/978-1-7998-8801-7.ch014

INTRODUCTION

Alkali metals are also called the lithium family (Quintero et al., 2021). Like other alkali metals sodium (Na), potassium (K), rubidium (Rb), cesium (Cs), and francium (Fr), lithium has a single valence electron which is easily yielded to form a cation. For this reason, lithium is a good conductor of heat (Wong et al., 2022) and electricity (Anokhina et al., 2021) as well as a highly reactive element (Park et al., 2021), although it is the least reactive of the alkali metals. The low reactivity of lithium is due to the proximity of its valence electron to its nucleus (the two remaining electrons are in the 1s orbital, much less energetic and do not participate in chemical bonds). Molten lithium is significantly more reactive than its solid form (Harvey et al, 2021).

Figure 1. Ion structure of lithium

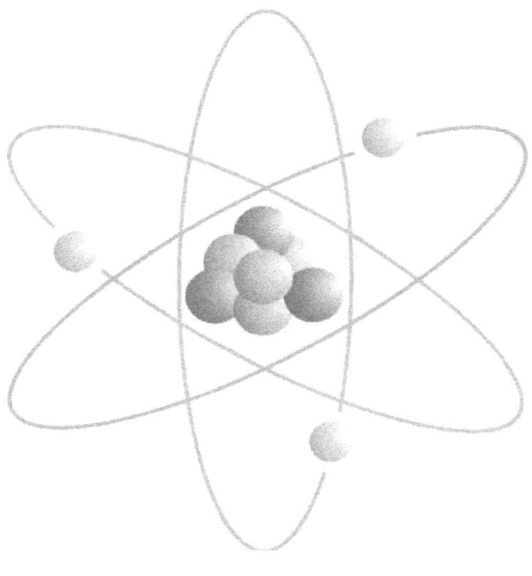

Lithium metal is soft enough to be cut with a knife. When cut, it has a silvery white color that quickly turns gray when oxidized to lithium oxide. Its melting point of 180.50 °C (453.65 K; 356.90 °F) and its boiling point of 1342 °C (1615 K; 2448 °F) which is the highest of all the alkali metals while its specific gravity of 0.534 is the lowest.

Lithium has a very low density (0.534 g / cm^3), comparable to pine wood. It is the least dense of all solid elements at room temperature; the next lightest solid element (potassium, at 0.862 g / cm^3) is 60% denser. Aside from helium and hydrogen, as a solid it is less dense than any other element in liquid form, being only two-thirds as dense as liquid nitrogen (0.808 g / cm^3). Lithium can float on the lightest hydrocarbons and is one of only three metals that can float on water, the other two being sodium and potassium.

The coefficient of thermal expansion of lithium is twice that of aluminum and almost four times that of iron. Lithium is superconducting below 400 μK at standard pressure and at higher temperatures (over 9 K) at very high pressures (> 20 GPa). At temperatures below 70 K, lithium, like sodium, undergoes phase change transformations without diffusion. At 4.2K, it has a rhombohedral crystal system (with a repeat spacing of nine layers); at higher temperatures, it changes to face-centered cubic and then to body-centered cubic. At liquid helium temperatures (4K), the rhombohedral structure is predominant; multiple allotropic forms have been identified for lithium at high pressure.

Lithium has a specific heat capacity of 3.58 kilojoules per kilogram-kelvin, the highest of all solids. For this reason, lithium metal is often used in coolants for heat transfer applications. These physicochemical properties create analytical difficulties for the analysis and extraction of lithium. In this chapter, we are really going to analyze and extract lithium is really looking for a needle in a haystack.

GLOBAL LITHIUM RESOURCES ESTIMATION AND FIELDS OF APPLICATION

Recently, the United States Geological Survey (USGS, 2021) estimated lithium resources at about 80 Million Tons (MT) distributed on 23 countries (12 major owners and 11 minor owners). Resources of major owners (Fig.2a) like Bolivia and USA are above 1 MT. On the other hand, resources of minor owners (Fig.2b) such as Peru and Spain are less than 1 MT. As a matter of fact, the production of lithium is not proportional to the existing reserves. Instead, other parameters including difficulty of extraction process make major owners not located in the production record.

Lithium is used in manufacturing of various products (Fig.3) such as ceramics, refrigerants glass, enamels and greases. It is also usedinotherfields including aerospace engineering and medicinal products. The majority of the global Li resources is used in Li-ion batteries industry.

Lithium Extraction Methods

In the middle of the 19th century, studies on the extraction of lithium from ores or aqueous media began. Currently, several experimental protocols have been proposed for efficient and inexpensive extraction. In fact, we have six main methods;

Figure 2. Lithium resources (Mt) and major countries distribution

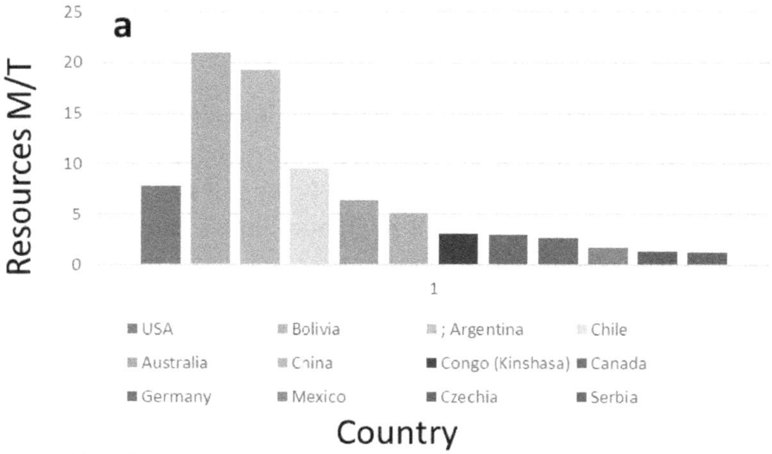

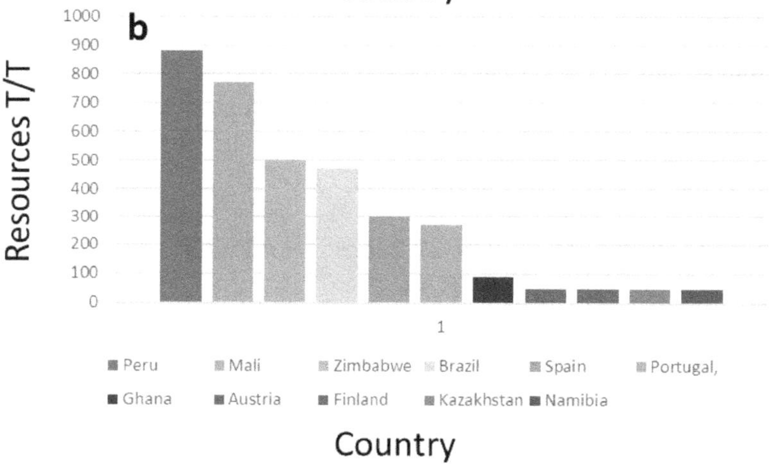

Figure 3. Lithium sectors of application

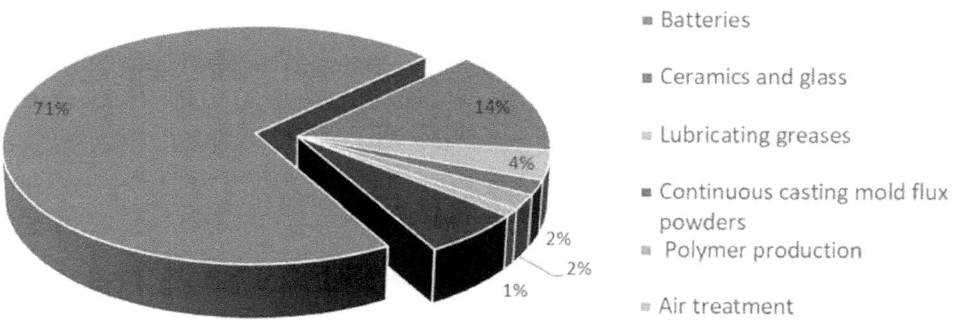

Separation of Lithium with an Organic Solvent

Based on the difference in solubility of lithium chloride and alkali and alkaline earth chlorides, in organic media, this method uses different reagents; (1) Solvents with a high boiling point such as pyridine (b.p. = 115.5 ° C) (Kahlenberg, 1958). (2) Aliphatic alcohols, with carbon number ranging from 3 to 8, branched or cyclic, have been used for the extraction of lithium chloride (Caley, 1942). For aliphatic monoalcohols, sodium and potassium salts have a more rapidly decreasing solubility than that of lithium chloride when changing from methyl to amyl alcohol. Organophosphoric acid compounds have a high selectivity for lithium ions such as di-2-ethylhexyphosphoric acid (D2EHPA). The latter was used for the extraction of lithium from geothermal waters. By synergistic effect, tributylphosphate (TBP) improves the efficiency of lithium extraction (Hano, 1992). (3) The depivaloylmethane (DPM) chelate is selective to lithium ions under alkaline conditions with a yield of 97%. (4) The lithium Corporation of America used a series of mixed solvents to optimize the extraction of lithium in the form lithium chloride (LiCl) or carbonate (Li_2CO_3).

Separation of Lithium by Absorption and Extraction

This method is based on the in-situ absorption of lithium by precipitated aluminum hydroxide, followed by extraction of LiCl with an appropriate organic solvent.

Separation of Lithium by Precipitation and use of Resins

The difference in solubility of carbonates of magnesium and calcium compared to that of lithium hydrogencarbonates is at the origin of Hering's work on the separation of lithium (Hering, 1952).

Separation of Lithium by Ion Exchange Resin

This procedure is based on the differences in the stability of the different complexes formed by the chelates of the resins with the metal ions. It is recommended for separating multivalent ions in quantities from alkali salts. Specific resins then make it possible to separate the alkalis.

Separation of Lithium by Evaporation, Crystallization and Thermal Decomposition

The evaporation of sea or salt lake waters favors the precipitation of magnesium chlorideMgCl2.6H2O. As result of this crystallization, the lithium will be split between the formed solid and the liquid phase (Flint, 1971). The thermal decomposition of this latter will lead to the formation of hydrogen chloride HCl (gaseous) and the magnesium oxide (insoluble). The separation of the obtained solid-liquid mixture will enable the extraction of LiCl-rich solution that would, after evaporation or carbonation, would give a lithium salt.

Membrane Methods

Membrane methods are an emerging and promising technology to extract lithium from salt lake brine with high energy efficiency and continuous operation. Depending on the driving force, these methods can be classified as (1) nanofiltration (pressure driven) or (2) electrodialysis (electrical potential driven). (3) The bipolar membrane and the membrane capacitivedeionization system is considered trending modern compared to electrodialysis since it offers improvements (Sun, 2020). Numerous membrane structures have been developed to extract and separate lithium from salt lake brines with high Mg/Li ratios. Membrane materials with high selectivity, low energy consumption, and good cycling performance are essential for industrial application.

CASE STUDY: LITHIUM EXTRACTION FROM THE BOUJMEL SEBKHA BRINE

The Boujmel lagoon is located 10 km from sebkha El Maleh in Zarzis southeast Tunisia (Fig.5). An area known by an arid climate with an annual precipitation rates around 200mm/year. It covers nearly 48 km^2. It is considered as ''evaporitic extension'' of Bhiret El Bibane. The sebkha is connected to seawater and supplied by the Makhada river.

To extract lithium from brines of sebkhas, many methods have been studied such as the use of membranes (Zante, 2019), adsorption (Xue, 2019), electrodialysis (Melnikov, 2017) and solvent extraction (Yu, 2019). Considered to be the most promising and recyclable method, solvent extraction has received special attention. In fact, tri-n-butyl phosphate (TBP) has been considered the most typical extracting agent and the most feasible method of extracting lithium from brines (Shi, 2019). On the other hand, from an analytical point of view, following the use of strong acids and alkalis in the extraction price (Ji, 2016a), TBP causes severe corrosion of equipment. For this reason, the new methods used milder products such as N, N-bis (2-ethylhexyl) acetamide (Shi, 2018) N, butyl acetate (Zhu, 2011) and N-bis (2-ethylhexyl) -3-oxobutanamide (Ji, 2016b). Regardless of the extractor used, and in order to ensure extraction efficiency, TBP was always added to the extraction system. Despite the varied research on lithium extraction by TBP, the thermodynamics and kinetics of this process remain poorly understood to date. In fact, many mechanisms such as diffusion resistance and chemical reaction rate and mass transfer merit further study, in order to clarify the mechanism of lithium extraction in the extraction process. In this context, the kinetics of extraction manage to solve these problems. Currently, the constant interfacial area cell is widely used to obtain extraction kinetics data due to its stability and workability (Xue, 1996). In the present work, the kinetics of extraction of lithium from salt lake brine by TBP was investigated using the cell with constant interfacial surface area and the thermodynamic function was calculated by the extraction experiments at different temperatures. Given its communication with the sea, the sebkha of Boujmel should in principle contain lithium. The total amount of lithium is a sort of average of the amount brought in from seawater, the amount brought in by hydrology and the amount brought in by hydrogeology.

Figure 4. Membrane developed by Toray Company that might be used in lithium extraction

Figure 5. Location of the case study Boujmal sebkha southeast Tunisia (Jouadi et al., 2016).

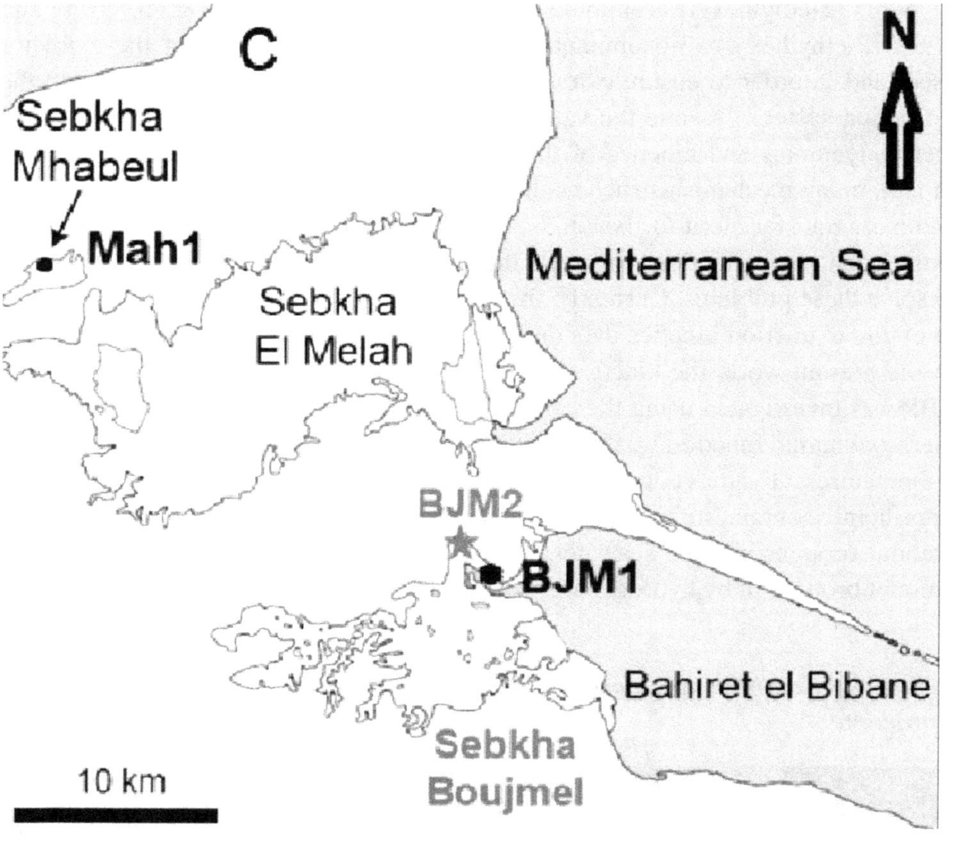

Hydrology of an Exoreic Wetland: Low and Continuous Supply of Lithium

Exoreic wetlands always have direct or indirect communication with seawater. In fact, each wetland has a hydrological or hydrogeological watershed. In the case of a wetland without communication with the sea, we can delimit the hydrological watershed while the hydrogeological watershed (if it exists) is still difficult to delimit. In the case of coastal wetlands, the contribution of marine water adds to hydrogeology and hydrology to give a complex hydrodynamic model. The balance sheet for lithium can be read:

Total Lithium = LiQr + LiQf + LiQsw + LiQgd + LiQgs - LiQe

Qr (L3 / T) is the average precipitation (L3 / T). Qf is the quantification of the average water slice covering the wetland (L3 / T). Qgs is the shallow groundwater input to the wetland (L3 / T); Qgd is the deep groundwater input (L3 / T). These last two components relating to the groundwater contribution (Qgs and Qgd) are usually deduced from this equation (1). They would be separately estimated. But due to a complicated hydrogeological situation, it would be better to treat (Qg = Qgd + Qgs) (L3 / T) as the average groundwater input).

Brine Geochemistry

The study of lithium in brine necessitates the study of the whole brine geochemistry. Three samples were collected from the sebkha of Boujmel (C1, C2 and sub). The geochemical relatedness of the facies is also evident with parallel lines on the Schoeller-Berkalof diagram (Fig.6). The marine contribution, especially after evaporation in a lower arid climate, results in a saltier sample. The aquifer looks smoother when mixed with the contrition of hydrology and hydrogeology results in a less salty sample. Since the marine contribution of lithium is known (2 mg / L), the contribution of hydrogeology and hydrology in lithium can provide enrichment. For this, we will choose the sample recovered after coring (Boujmal sub) to do the thermodynamic modeling and 'progressive evaporation.

Following the evaporation of 1 liter of the brine, the salts formed are recovered (Tab.1). On evaporation of 50%, 102 g (55%) are obtained. Then, at 100% evaporation, the remaining 45% is precipitated. This non-linear evolution of evaporation shows that the precipitated salts are not of the same nature. However, only thermodynamic modeling and mineralogical characterization give us an idea about the minerals formed.

Thermodynamic Modeling of Brine

The thermodynamic modeling of Boujmel's brine between the initial solution and 100% evaporation converges towards a solution which is stable to the precipitation of evaporitic minerals (Fig.7). This step covers the precipitation of calcite ($CaCO_3$) in a continuous and weak way. Pure sulfated minerals such as gypsum ($CaSO_4$: $2H_2O$) and anhydrite ($CaSO_4$) precipitate more strongly. The other minerals appear sequentially. Concerning the geochemistry of the brine, which is already the most interesting from the point of view of analysis and lithium extraction, we have a decrease in Na^+, Ca^{2+}, and K^+. While Cl^- and Mg^{2+} maintain high values.

Figure 6. Schoeller-Berkalof diagram of three samples of sebkha Boujmel

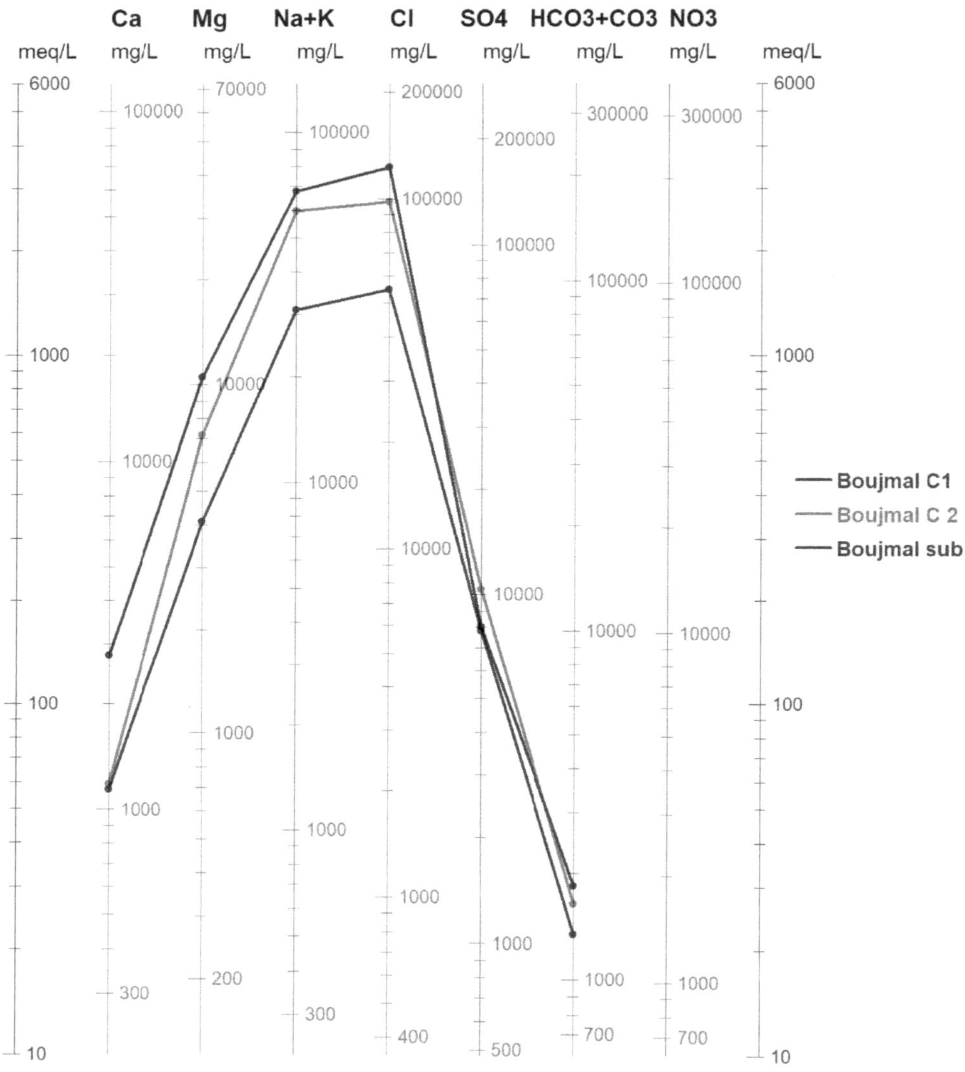

Table 1. Dry residues of Boujmel brines following gradual evaporation

Evaporation rate (%)	Weight of salt (g)	Percentage (%)
50	102.1	55
100	84	45
Cumulative	186.1	100

Figure 7. Modelingon thermodynamics of Boujmel brine

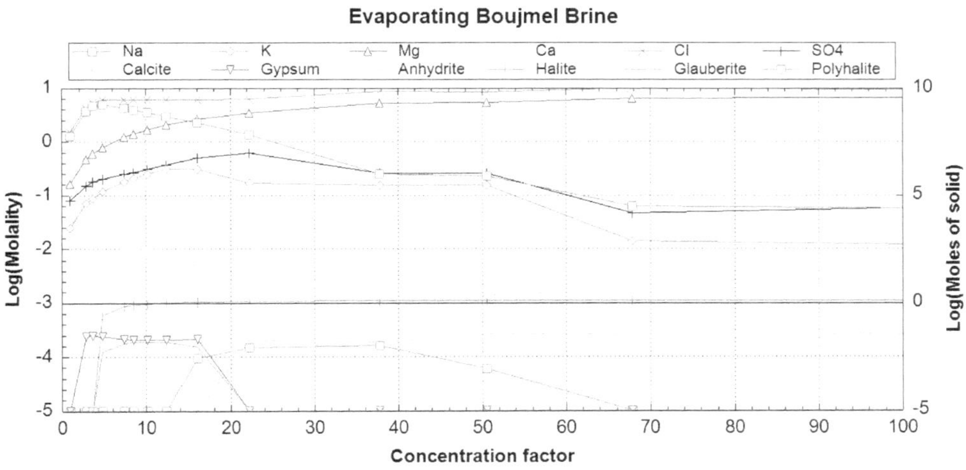

Lithium Detection by Progressive Evaporation

Following evaporation, lithium has two choices: either it passes to the solid phase of the salt formed or it concentrates in the residual brine. Our goal in this work is the localization of the phase and the time of this exceptional concentration: that we can call it by analogy with the petroleum industry: the Lithium Window In fact, the monitoring of this window requires a multidisciplinary approach to detect elemental lithium signal by geochemistry or mineralogical signal combined with infrared or X-ray.

Elementary Geochemistry

The elementary geochemistry of the salts formed following gradual evaporation shows a decrease in lithium contents (Fig.8). This decrease in the solid phase is controlled by a whole thermodynamics of the fractional crystallization of evaporitic minerals. The average value is around 7 mg / Kg.

In the liquid phase, an increase in lithium values was observed. The concentration of lithium in brines as a result of evaporation is a good indicator for reaching classification concentrations for a deposit. Values of 17 mg / L are close to salt lakes in the world.

Boujmel sebkha contains lithium in considerable proportions; the seawater feed ensures continuous pumping of lithium at low concentrations. Intensive evaporation governed mainly by an arid climate favors the enrichment of brines by lithium reaching 17 mg / L.

Figure 8. Evolution of lithium in Boujmel salts during progressive evaporation

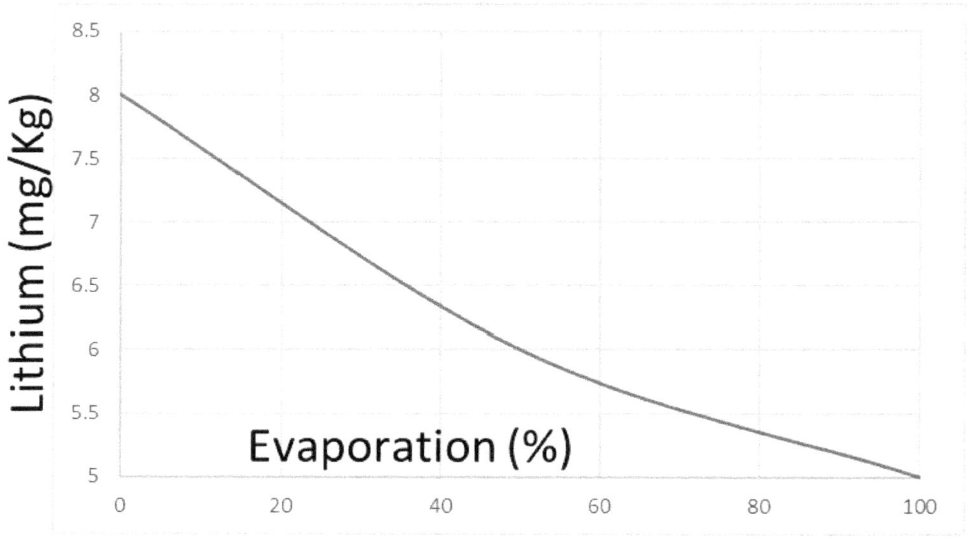

Figure 9. Evolution of lithium in Boujmel brines during progressive evaporation

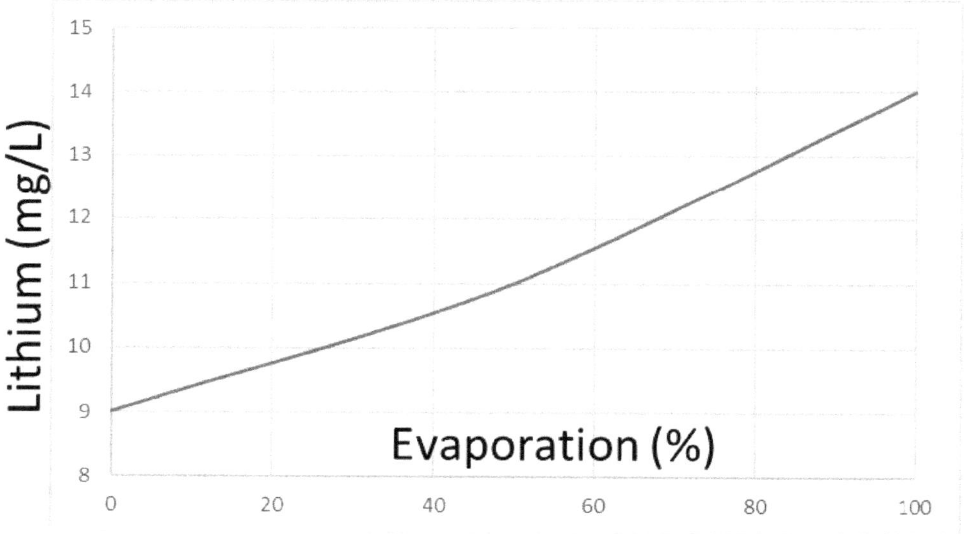

CONCLUSION AND PERSPECTIVES

It is difficult to reveal the relations of the nearby crystallization areas in the complex systems by using only the experimental results in the diagrams. For that reason, experimental work and thermodynamic modeling are usually combined to complete the description of phase equilibrium for the brine systems. Although the thermodynamic models are mainly empirical, they are very convenient for the thermodynamic property calculation and for other research. Moreover, the construction of the thermodynamics model must be affirmed with the experimental results (Meng, 2015). Recently, the world is witnessing many advances in the field of lithium separation and extraction from salt lake brines. Such as adsorption may be with electrochemical method, membrane method and the reaction-coupled separation technology, etc (Sun, 2020). The main objective of the new technologies is to extract efficiently the maximum amount of pure lithium from brines by the mean of solvents/reagents with high selectivity for lithium and while using less energy and materials.

REFERENCES

Anokhina, I. A., Animitsa, I. E., Voronin, V. I., Vykhodets, V. B., Kurennykh, T. E., Molchanova, N. G., Vylkov, A. I., Dedyukhin, A. E., & Zaikov, Y. P. (2021). The structure and electrical properties of lithium doped pyrochlore Gd2Zr2O7. *Ceramics International*, *47*(2), 1949–1961. doi:10.1016/j.ceramint.2020.09.025

Caley, E., & Axilrod, H. (1942). Separation of Lithium from potassium and sodium. *Industrial & Engineering Chemistry. Analytical Edition*, *14*(3), 242–244. doi:10.1021/i560103a018

Flint, G. (1971). Great Salt Lake Minerals and Chemicals Corporation (2nd ed.). Academic Press.

Hano, T. M. (1992). *Solvent extraction and ion exchange. Process Metallurgy*, 7(B).

Harvey, J. P., Singh, S., Oishi, K., Acheson, B., Turcotte, R., Pilon, D., Lavoie, J., & Grange, B. (2021). Quantification of the chemical reactivity of molten nitrate salts with heat treatable aluminum alloys. *Materials & Design*, *198*, 109293. doi:10.1016/j.matdes.2020.109293

Hering, H. (1952). Séparation quantitative de trace de lithium contenues le calcium. *Analytica Chimica Acta*, *6*, 340–350. doi:10.1016/S0003-2670(00)86954-8

Ji, Hu, Y., Li, L., Shi, D., Li, J., Nie, F., Song, F., Zeng, Z., Sun, W., & Liu, Z. (2016a). Lithium extraction with a synergistic system of dioctyl phthalate and tributyl phosphate in kerosene and FeCl3. *Hydrometallurgy*, *162*, 71–78. doi:10.1016/j.hydromet.2016.02.018

Ji, Li, L., Shi, D., Li, J., Liu, Z., Xu, D., & Song, X. (2016b). Extraction equilibria of lithium with N, N-bis (2-ethylhexyl)-3-oxobutanamide and tributyl phosphate in kerosene and FeCl 3. *Hydrometallurgy*, *164*, 304–312. doi:10.1016/j.hydromet.2016.06.022

Kahlenberg, L., & Krauskopf, F. C. (1958). A new method of separating Lithium chloride from the chlorides of the other alkalis and the chloride of barium. *Journal of the American Chemical Society*, *30*(7), 1104–1120. doi:10.1021/ja01949a008

Melnikov, S., Sheldeshov, N., Zabolotsky, V., Loza, S., & Achoh, A. (2017). Pilot scale complex electrodialysis technology for processing a solution of lithium chloride containing organic solvents. *Separation and Purification Technology*, *189*, 74–81. doi:10.1016/j.seppur.2017.07.085

Meng, L. M. S., Gruszkiewicz, M. S., Deng, T., Guo, Y., & Li, D. (2015). Isothermal Evaporation Process Simulation Using the Pitzer Model for the Quinary System LiCl−NaCl−KCl−SrCl2−H2O at 298.15 K. *Industrial & Engineering Chemistry Research*, *54*(33), 8311–8318. doi:10.1021/acs.iecr.5b01897

Park, H., Tamwattana, O., Kim, J., Buakeaw, S., Hongtong, R., Kim, B., Khomein, P., Liu, G., Meethong, N., & Kang, K. (2021). Probing lithium metals in batteries by advanced characterization and analysis tools. *Advanced Energy Materials*, *11*(15), 2003039. doi:10.1002/aenm.202003039

Quintero, M. A., Hao, S., Patel, S. V., Bao, J. K., Zhou, X., Hu, Y. Y., Wolverton, C., & Kanatzidis, M. G. (2021). Lithium Thiostannate Spinels: Air-Stable Cubic Semiconductors. *Chemistry of Materials*, *33*(6), 2080–2089. doi:10.1021/acs.chemmater.0c04651

Shi, D., Cui, B., Li, L., Peng, X., Zhang, L., & Zhang, Y. (2019). Lithium extraction from low-grade salt lake brine with ultrahigh Mg/Li ratio using TBP – kerosene – FeCl3 system. *Separation and Purification Technology*, *211*, 303–309. doi:10.1016/j.seppur.2018.09.087

Shi, D., Zhang, L., Peng, X., Li, L., Song, F., Nie, F., Ji, L., & Zhang, Y. (2018). Extraction of lithium from salt lake brine containing boron using multistage centrifuge extractors. *Desalination*, *441*, 44–51. doi:10.1016/j.desal.2018.04.029

Sun, M. X. (1996). The method of solvent extraction kinetics—The interfacial area. *Chem. Bulletin.*, *7*, 50–52.

Sun, Y., Wang, Q., Wang, Y., Yun, R., & Xiang, X. (2021). Recent advances in magnesium/lithium separation and lithium extraction technologies from salt lake brine. *Separation and Purification Technology*, *256*, 117807.

Wong, H., Liu, N., & Reichmanis, E. (2022). Single-Pot Fabrication of Cellulose-Reinforced Solid Polymer Lithium-Ion Conductors. *ACS Applied Polymer Materials*, *4*(3), 1948–1955. doi:10.1021/acsapm.1c01789

Xue, F., Wang, B., Chen, M., Yi, C., Ju, S., & Xing, W. (2019). Fe3O4-doped lithium ion-sieves for lithium adsorption and magnetic separation. *Separation and Purification Technology*, *228*, 115750. doi:10.1016/j.seppur.2019.115750

Yu, X., Fan, X., Guo, Y., & Deng, T. (2019). Recovery of lithium from underground brine by multistage centrifugal extraction using tri-isobutyl phosphate. *Separation and Purification Technology*.

Zante, G., Boltoeva, M., Masmoudi, A., Barillon, R., & Trébouet, D. (2019). Lithium extraction from complex aqueous solutions using supported ionic liquid membranes. *Journal of Membrane Science*, *580*, 62–76. doi:10.1016/j.memsci.2019.03.013

Zhu, H. F. (2011). *The new process study of extracting lithium from salt lake brine by solvent extraction*. Academic Press.

Compilation of References

Abbassi, Y., Ehsan, B., & Hossein, A. (2017). Comparative performance analysis of differ-ent solar desiccant dehumidification systems. *Energy and Building*, *150*, 37–51. doi:10.1016/j.enbuild.2017.05.075

Abolmaali, A. M., & Afshin, H. (2019). Development of Nusselt number and friction factor correlations for the shell side of spiral wound heat exchangers. *International Journal of Thermal Sciences*, *139*, 105–117. doi:10.1016/j.ijthermalsci.2019.01.038

Agee, C. B., Wilson, N. V., McCubbin, F. M., Ziegler, K., Polyak, V. J., Sharp, Z. D., Asmerom, Y., Nunn, M. H., Shaheen, R., Thiemens, M. H., Steele, A., Fogel, M. L., Bowden, R., Glamoclija, M., Zhang, Z., & Elardo, S. M. (2013). Unique Meteorite from Early Amazonian Mars: Water-Rich Basaltic Breccia Northwest Africa 7034. *Science*, *339*(6121), 780–785. Advance online publication. doi:10.1126cience.1228858 PMID:23287721

Ageev, D. S., Bagrov, A. A., & Iliasov, A. A. (2021). Deterministic chaos and fractal entropy scaling in Floquet conformal field theories. *Physical Review. B*, *103*(10), L100302. doi:10.1103/PhysRevB.103.L100302

Aguerre, R. J., Suarez, C., & Viollaz, P. E. (1986). Enthalpy-entropy compensation in sorption phenomena: Application to the prediction of the effect of temperature on food isotherms. *Journal of Food Science*, *51*(6), 1547–1549. doi:10.1111/j.1365-2621.1986.tb13856.x

Ahrens, C., Cataldo, V., & Leone, G. (2021). Volcanic Eruptions on Mars, Lava Flow Morphology, and Thermodynamics. In *Mars: A Volcanic World* (pp. 71–94). Springer. doi:10.1007/978-3-030-84103-4_4

Ait Mohamed, L., Kouhila, M., Lahsasni, S., Jamali, A., Idlimam, A., Rhazi, M., Aghfir, M., & Mahroue, M. (2004). Equilibrium moisture content and heat of sorption of prickly pear seeds. *Journal of Stored Products Research*, *41*, 199–209. doi:10.1016/j.jspr.2004.03.001

Ali, M.E. (1998). Laminar natural convection from constant heat flux helical coiled tubes. *Int. J. Heat Mass Transf.*, *41*, 2175–2182. (97)00322-0 doi:10.1016/S0017-9310

Ali, M., Vukovic, V., Muhammad Ali, H., & Ahmed Sheikh, N. (2018). Performance analysis of solar-assisted desiccant cooling system cycles in world climate zones. *Journal of Solar Energy Engineering*, *140*(4), 041009. doi:10.1115/1.4039426

Ali, M., Vukovic, V., Sheikh, N. A., & Ali, H. M. (2015). Performance investigation of solid desiccant evaporative cooling system configurations in different climatic zones. *Energy Conversion and Management*, *97*, 323–339. doi:10.1016/j.enconman.2015.03.025

Alkasassbeh, M., Omar, Z., Mebarek-Oudina, F., Raza, J., & Chamkha, A. J. (2019). Heat transfer study of convective fin with temperature-dependent internal heat generation by hybrid block method. *Heat Transfer, Asian Research*, *48*(4), 1225–1244. doi:10.1002/htj.21428

Allab, F., Kedous-Lebouc, A., Fournier, J. M., & Yonnet, J. P. (2005, October). (année). Numerical Modeling for Active Magnetic RegenerativeRefrigeration. *IEEE Transactions on Magnetics*, *41*(10), 3757–3759. doi:10.1109/TMAG.2005.854757

Al-Salem, K., Oztop, H. F., Pop, I., & Varol, Y. (2012). Effects of moving lid direction on MHD mixed convection in a linearly heated cavity. *International Journal of Heat and Mass Transfer*, *55*(4), 1103–1112. doi:10.1016/j.ijheatmasstransfer.2011.09.062

Altuntop, N., Arslan, M., Ozceyhan, V., & Kanoglu, M. (2005). Effect of obstacles on thermal stratification in hot water storage tanks. *Applied Thermal Engineering*, *25*(14-15), 2285–2298. doi:10.1016/j.applthermaleng.2004.12.013

Anderson, R.C., Dohm, J.M., Haldemann, A.F.C., Pounders, E., Golombek, M., & Castano, A. (2006). Centers of tectonic activity for the eastern hemisphere. *Icarus*. (submitted for publication)

Andrews-Hanna, J., Phillips, R., & Zuber, M. (2007). Meridiani Planum and the global hydrology of Mars. *Nature*, *446*(7132), 163–168. doi:10.1038/nature05594 PMID:17344848

Anokhina, I. A., Animitsa, I. E., Voronin, V. I., Vykhodets, V. B., Kurennykh, T. E., Molchanova, N. G., Vylkov, A. I., Dedyukhin, A. E., & Zaikov, Y. P. (2021). The structure and electrical properties of lithium doped pyrochlore Gd2Zr2O7. *Ceramics International*, *47*(2), 1949–1961. doi:10.1016/j.ceramint.2020.09.025

Ansan, V., & Mangold, N. (2004). Impact crater paleolakes in Hellas and Thaumasia areas, Mars. In Early Mars Conference, Jackson Hole, USA.

Ansan, V., Loizeau, D., Mangold, N., Le Mouélic, S., Carter, J., Poulet, F., Dromart, G., Lucas, A., Bibring, J.-P., Gendrin, A., Gondet, B., Langevin, Y., Masson, P., Murchie, S., Mustard, J. F., & Neukum, G. (2011). Stratigraphy, mineralogy, and origin of layered deposits inside Terby crater, Mars. *Icarus*, *211*(1), 273–304. doi:10.1016/j.icarus.2010.09.011

ANSYS Inc. (2021). ANSYS Fluent Theory Guide. ANSYS Inc.

Anta, J. (2021a). Can informational thermal physics explain the approach to equilibrium? *Synthese*, *199*(1-2), 1–24. doi:10.100711229-020-02967-8

Anta, J. (2021b). *A Philosopher against the Bandwagon.* Carnap and the Informationalization of Thermal Physics.

Appelbaum, J., & Flood, D. J. (1990). Solar radiation on Mars. *Solar Energy*, *45*(6), 353–363. doi:10.1016/0038-092X(90)90156-7

Appelo, C. A. J., & Postma, D. (2005). *Geochemistry, groundwater and pollution* (2nd ed.). Balkema.

Appelo, C. A. J., & Postma, D. (2005). *Groundwater and pollution.* Balkema.

Arasteh, H., Mashayekhi, R., Ghaneifar, M., Toghraie, D., & Afrand, M. (2020). Heat transfer enhancement in a counter-flow sinusoidal parallel-plate heat exchanger partially filled with porous media using metal foam in the channels' divergent sections. *Journal of Thermal Analysis and Calorimetry*, *141*(5), 1669–1685. doi:10.100710973-019-08870-w

Arias-Gonzalez, J. R. (2021). Microscopically Reversible Pathways with Memory. *Mathematics*, *9*(2), 127. doi:10.3390/math9020127

Arkani-Hamed, J. (2004). Timing of the martian core dynamo. *Journal of Geophysical Research*, *109*(E3), E03006. Advance online publication. doi:10.1029/2003JE002195

Arques, P. (2009). *Diagnostic prédictif et défaillances des machines.* Editions TECHNIP.

Arslan, N., & Togrul, H. (2006). The fitting of various models to water sorption isotherms of tea stored in a chamber under controlled temperature and humidity. *Journal of Stored Products Research*, *42*(2), 112–135. doi:10.1016/j.jspr.2005.01.001

Arvidson, R. E., Poulet, F., Morris, R. V., Bibring, J.-P., Bell, J. F. III, Squyres, S. W., Christensen, P. R., Bellucci, G., Gondet, B., Ehlmann, B. L., Farrand, W. H., Fergason, R. L., Golombek, M., Griffes, J. L., Grotzinger, J., Guinness, E. A., Herkenhoff, K. E., Johnson, J. R., Klingelhöfer, G., ... Wolff, M. (2006). Nature and origin of the hematitebearing plains of Terra Meridiani based on analyses of orbital and Mars Exploration Rover data sets. *Journal of Geophysical Research*, *111*(E12). Advance online publication. doi:10.1029/2006JE002728

Assmus, M., & Koehl, M. (2012). Experimental investigation of the mechanical behavior of photovoltaic modules at defined inflow conditions. *Journal of Photonics for Energy*, *2*(1), 022002. doi:10.1117/1.JPE.2.022002

Baiko, D. A., & Chugunov, A. I. (2022). Ab initio thermodynamics of one-component plasma for astrophysics of white dwarfs and neutron stars. *Monthly Notices of the Royal Astronomical Society*, *510*(2), 2628–2643. doi:10.1093/mnrastab3613

Baiko, D. A., Potekhin, A. Y., & Yakovlev, D. G. (2001). Thermodynamic functions of harmonic Coulomb crystals. *Physical Review. E*, *64*(5), 057402. doi:10.1103/PhysRevE.64.057402 PMID:11736150

Baker, C. B. (1978). A Study of the Efficiency of Hydrogen Liquefaction. *International Journal of Hydrogen Energy*, *3*(3), 321–334. doi:10.1016/0360-3199(78)90037-X

Baker, L. L., Agenbroad, D. J., & Wood, S. A. (2000). Experimental hydrothermal alteration of a martian analog basalt: Implications for martian meteorites. *Meteoritics & Planetary Science*, *35*(1), 31–38. doi:10.1111/j.1945-5100.2000.tb01971.x

Banks, D. (2010). *How Wind Load Studies Will Impact the Solar Industry*. Academic Press.

Barbieri, R. (2012). *Continental evaporites from arid systems and their microfossils: a promising target in the search for traces of Martian life*. 3rd Conference on Terrestrial Mars Analogues, Marrakech, Morocco.

Barbieri, R., & Stivaletta, N. (2012). Halophiles, continental evaporites and the search for biosignatures in environmental analogues for Mars. In Life on Earth and other Planetary Bodies. Springer. doi:10.1007/978-94-007-4966-5_3

Barbieri, R., & Stivaletta, N. (2011). Continental evaporites and the search for evidence of life on Mars. *Geological Journal*, *46*(6), 513–524. doi:10.1002/gj.1326

Barlow, N. G. (1988). Crater size-frequency distributions and a revised Martian relative chronology. *Icarus*, *75*(2), 285–305. doi:10.1016/0019-1035(88)90006-1

Barlow, N. G., & Perez, C. B. (2003). Martian impact crater ejecta morphologies as indicators of the distribution of subsurface volatiles. *Journal of Geophysical Research*, *108*(E8), 5085. Advance online publication. doi:10.1029/2002JE002036

Basak, T., Kumar, P., Anandalakshmi, R., & Roy, S. (2012). Analysis of entropy generation minimization during natural convection in trapezoidal enclosures of various angles with linearly heated side wall (s). *Industrial & Engineering Chemistry Research*, *51*(10), 4069–4089. doi:10.1021/ie201107f

Battler, M. M., Osinski, G. R., & Neil, B. R. (2012). Mineralogy of saline perennial cold springs on Axel Heiberg Island, Nunavut, Canada and implications for spring deposits on Mars. *Icarus*.

Bedoui, S., Essefi, E., Besser, H., Benour, A., Ayadi, Y., Khelifi, F., & Hamed, Y. (2022). Geochemical, Mineralogical and Isotopic Fracturing during the Progressive Evaporation of Chott Djerid Brine (Southern Tunisia). In International Journal of Engineering Research in Africa (Vol. 58, pp. 107-125). Trans Tech Publications Ltd.

Bejan, A. (1977a). *The concept of irreversibility in heat exchanger design: counterflow heat exchangers for gas-to-gas applications*. Academic Press.

Bejan, A. (1978b). General criterion for rating heat-exchanger performance. *International Journal of Heat and Mass Transfer*, *21*(5), 655–658. doi:10.1016/0017-9310(78)90064-9

Bejan, A. (1980c). Second law analysis in heat transfer. *Energy*, *5*(8-9), 720–732. doi:10.1016/0360-5442(80)90091-2

Bejan, A. (1982). *Entropy Generation through Heat and Fluid Flow*. Wiley.

Bejan, A. (1996). *Entropy Generation Minimization*. CRC Press.

Bejan, A. (2016d). *Advanced engineering thermodynamics*. John Wiley & Sons. doi:10.1002/9781119245964

Bekenstein, J. D. (2008). Bekenstein-hawking entropy. *Scholarpedia*, *3*(10), 7375. doi:10.4249cholarpedia.7375

Belguith, Meddeb, & Ben Slama. (2020). Performance analysis of desiccant cooling systems in a hot and dry climate. *Euro-Mediterranean Journal for Environmental Integration*, *6*(2).

Belguith, Slama, Chaouachi, & Meddeb. (2020). Performance analysis of hybrid solid desiccant - vapor compression air conditioning system: Application and comparative study. *2020 11th International Renewable Energy Congress (IREC) Hammamet, Tunisia, 2020*, 1-6.

Bendaraa, A., Charafi, M. M., & Hasnaoui, A. (2021). Numerical and experimental investigation of alumina-based nanofluid effects on double-pipe heat exchanger thermal performances. *SN Applied Sciences*, *3*(2), 1–10. doi:10.100742452-021-04195-2

Benlahcen, A. (1996). *Modélisation de saumures carbonatees par le traitement multivariable en hydrogéochimie à la mine Niobec de Saint-Honoré (Québec)* [Modeling of carbonate brines by multivariate treatment in hydrogeochemistry at the Niobec mine in Saint-Honoré, Quebec]. Université du Québec à Chicoutimi.

Berthiaume, A., Van Dam, W., & Laplante, S. (2001). Quantum kolmogorov complexity. *Journal of Computer and System Sciences*, *63*(2), 201–221. doi:10.1006/jcss.2001.1765

Bevevino, J. W. (1978). Standards of Tubular Exchanger Manufacturers Association. *TEMA*.

Bibring, J. P., Langevin, Y., Mustard, J. F., Poulet, F., Arvidson, R., Gendrin, A., Gondet, B., Mangold, N., Pinet, P., Forget, F., Berthé, M., Bibring, J.-P., Gendrin, A., Gomez, C., Gondet, B., Jouglet, D., Poulet, F., Soufflot, A., Vincendon, M., ... Neukum, G. (2006). Global Mineralogical and Aqueous Mars History Derived from OMEGA/Mars Express Data. *Science*, *312*(5772), 400–404. doi:10.1126cience.1122659 PMID:16627738

Bishop J. L., Zent A. P., & Pieters, C. M. (2002). A model for formation of dust, soil, and rock coatings on Mars: Physical and chemical processes on the Martian surface. *Journal of Geophysical Research, 107*(5097). . doi:10.1029/2001JE001581

Blattberg, R. C., Kim, B. D., & Neslin, S. A. (2008). *Database Marketing: Analyzing and Managing Customers*. Springer. doi:10.1007/978-0-387-72579-6

Blattberg, R. C., Kim, B. D., & Neslin, S. A. (n.d.). *Database Marketing: Analyzing and Managing Customers. Springer.*

Bontemps, A., Garrigue, A., Goubier, C., Huetz, J., Marvillet, C., Mercier, P., & Vidil. (1994). *Échangeurs de Chaleur. Définitions et Architecture Générale* [Heat exchangers. Definitions and General Architecture]. Techniques de l'Ingénieur, Traité Génie Énergétique.

Bouchekara, H., El-Hana, R., Kedous-Lebouc, A., Yonnet, J. P., & Chillet, C. (2014). Multiobjective optimization of AMR systems. *International Journal of Refrigeration*, *37*, 63–71. doi:10.1016/j.ijrefrig.2013.09.009

Bouchekara, H., Kedous-Lebouc, A., Dupuis, C., & Allab, F. (2008). Prediction and optimisation of geometrical properties of the refrigerant bed in an AMRR cycle. *International Journal of Refrigeration*, *31*(7), 1224–1230. doi:10.1016/j.ijrefrig.2008.02.007

Boudhrioua, N., Bahloul, N., Kouhila, M., & Kechaou, N. (2008). Sorption isotherms andisosteric heats of olive leaves (Chemlali variety): Experimental and mathematical investigations. *Food and Bioproducts Processing*, *86*(3), 167–175. doi:10.1016/j.fbp.2007.10.010

Bourdoukan, P., Wurtz, E., & Joubert, P. (2010). Comparison between the conventional and recirculation modes in desiccant cooling cycles and deriving critical efficiencies of components. *Energy*, *35*(2), 1057–1067. doi:10.1016/j.energy.2009.06.021

Bowen, B., & Epstein, N. (1979). Fine particle deposition in smooth parallel plate channels. *Journal of Colloid and Interface Science*, *72*(1), 81–87. doi:10.1016/0021-9797(79)90184-X

Boynton, W. V., Feldman, W. C., Squyres, S. W., Prettyman, T. H., Brückner, J., Evans, L. G., Reedy, R. C., Starr, R., Arnold, J. R., Drake, D. M., Englert, P. A. J., Metzger, A. E., Mitrofanov, I., Trombka, J. I., d'Uston, C., Wänke, H., Gasnault, O., Hamara, D. K., Janes, D. M., ... Shinohara, C. (2002). Distribution of hydrogen in the near surface of Mars: Evidence for subsurface ice deposits. *Science*, *297*(5578), 81–85. doi:10.1126cience.1073722 PMID:12040090

Brasser, R., & Walsh, K. J. (2011). Stability analysis of the martian obliquity during the Noachian era. *Icarus*, *213*(1), 423–427. doi:10.1016/j.icarus.2011.02.024

Brunauer, S., Emmett, P. H., & Teller, E. (1940). Adsorption of gases in multi-molecular layers. *Journal of the American Chemical Society*, *60*(2), 309–319. doi:10.1021/ja01269a023

Burr, D. M., Grier, J. A., McEwen, A. S., & Keszthelyi, L. P. (2002a). Repeated Aqueous Flooding from the Cerberus Fossae: Evidence for Very Recently Extant, Deep Groundwater on Mars. *Icarus*, *159*(1), 53–73. doi:10.1006/icar.2002.6921

Butera, D., Tesoriere, L., Gaudio, D. F., Bongiorno, A., Allegra, M., Pintaudi, A. M., Kohen, R., & Livrea, M. A. (2002). Antioxidant activities of Sicilian prickly pear (Opuntia ficus-indica) fruit extracts and reducing properties of its betalains: Betanin and indicaxanthin. *Journal of Agricultural and Food Chemistry*, *50*(23), 6895–6901. doi:10.1021/jf025696p PMID:12405794

Bychkov, E. G. (2022). An integrated approach for designing Joule-Thomson refrigerators operating with mixtures. *Applied Thermal Engineering*, *202*, 117837. doi:10.1016/j.applthermaleng.2021.117837

Cabrol, N. A., & Grin, E. A. (2001). The evolution of lacustrine environments on Mars: Is Mars only hydrologically dormant? *Icarus*, *149*(2), 291–328. doi:10.1006/icar.2000.6530

Cabrol, N. A., Grin, E. A., Newsom, H. E., Landheim, R., & McKay, C. P. (1999). Hydrogeologic Evolution of Gale Crater and Its Relevance to the Exobiological Exploration of Mars. *Icarus*, *139*(2), 235–245. doi:10.1006/icar.1999.6099

Caley, E., & Axilrod, H. (1942). Separation of Lithium from potassium and sodium. *Industrial & Engineering Chemistry. Analytical Edition*, *14*(3), 242–244. doi:10.1021/i560103a018

Caliskan, H., Lee, D. Y., & Hong, H. (2019). Enhanced thermodynamic assessments of the novel desiccant air cooling system for sustainable energy future. *Journal of Cleaner Production*, *211*, 213–221. doi:10.1016/j.jclepro.2018.11.174

Cantwell, M. (1995). Post-harvest management of fruits and vegetable stems. In Agroecology, Cultivation and Uses of Cactus Pear. FAO Plant Production and Protection Paper.

Cao, X., Chen, D., Du, T., Liu, Z., & Ji, S. (2020). Numerical investigation and experimental validation of thermo-hydraulic and thermodynamic performances of helical baffle heat exchangers with different baffle configurations. *International Journal of Heat and Mass Transfer*, *160*, 120181. doi:10.1016/j.ijheatmasstransfer.2020.120181

Carpinlioglu, M. Ö. (2015). A comment on the concept of entransy (versus exergy) for The performance assessment of a desiccant cooling system. *Energy and Building*, *101*, 163–167. doi:10.1016/j.enbuild.2015.05.003

Carrier, B. L., Beaty, D. W., Meyer, M. A., Blank, J. G., Chou, L., DasSarma, S., ... Xu, J. (2020). *Mars extant life: What's next?* Conference Report.

Carr, M. (2001). Mars Global Surveyor observations of martian fretted terrain. *Journal of Geophysical Research*, *106*(E10), 23571–23594. doi:10.1029/2000JE001316

Carr, M. H., & Head, J. W. III. (2003). Basal melting of snow on early Mars: A possible origin of some valley networks. *Geophysical Research Letters*, *30*(24). Advance online publication. doi:10.1029/2003GL018575

Cassata, W. S., Shuster, D. L., Renne, P. R., & Weiss, B. P. (2012, September–October). Trapped Ar isotopes in meteorite ALH 84001 indicate Mars did not have a thick ancient atmosphere. *Icarus*, *221*(1), 461–465. doi:10.1016/j.icarus.2012.05.005

Chabrier, G., Ashcroft, N. W., & DeWitt, H. E. (1992). White dwarfs as quantum crystals. *Nature*, *360*(6399), 48–50. doi:10.1038/360048a0

Chamkha, A. J. (2002). Hydromagnetic combined convection flow in a vertical lid- driven cavity with internal heat generation or absorption. *Numerical Heat Transfer Part A-Application*, *41*(5), 529–546. doi:10.1080/104077802753570356

Chapman, M. G., & Tanaka, K. L. (1993). *Geologic map of theMTM-05152 and-10152 quadrangles, Mangala Valles region of Mars*. USGS Misc. Inv. Ser. Map I-2294 (1:500,000).

Charlton, S. R., & Parkhurst, D. L. (2011). Modules based on the geochemical model PHREEQC for use in scripting and programming languages. *Computers & Geosciences*, *37*(10), 1653–1663. doi:10.1016/j.cageo.2011.02.005

Chassefière, E., Leblanc, F., & Langlais, B. (2007). The combined effects of escape and magnetic field histories at Mars. *Planetary and Space Science*, *55*(3), 343–357. doi:10.1016/j.pss.2006.02.003

Chen, C. N., Han, J. T., Jen, T. C., Shao, L., & Chen, W. W. (2011). Experimental study on critical heat flux characteristics of R134a flow boiling in horizontal helically-coiled tubes. *International Journal of Thermal Sciences*, *50*(2), 169–177. doi:10.1016/j.ijthermalsci.2010.10.002

Chen, H., Yang, H., & Luo, Y. (2018). Investigation on solar assisted liquid desiccant dehumidifier and evaporative cooling system for fresh air treatment. *Energy*, *143*, 114–127. doi:10.1016/j.energy.2017.10.124

Chevrier, V., Poulet, F., & Bibring, J. P. (2007). Early geochemical environment of Mars as determined from thermodynamics of phyllosilicates. *Nature*, *448*(7149), 60–63. doi:10.1038/nature05961 PMID:17611538

Choi, S. U. S. (1995). Enhancing thermal conductivity of fluids with nano particles. *ASME FED*, *231*, 99–103.

Chougui, N., Tamendjari, A., Hamidj, W., Hallal, S., Barras, A., Richard, T., & Larbat, R. (1995). Oil composition and characterisation of phenolic compounds of Opuntiaficus-indica seeds. *Food Chemistry*, *139*(1-4), 796–803. doi:10.1016/j.foodchem.2013.01.054 PMID:23561175

Christensen, P. R., Bandfield, J. L., Hamilton, V. E., Ruff, S. W., Kieffer, H. H., Titus, T. N., Malin, M. C., Morris, R. V., Lane, M. D., Clark, R. L., Jakosky, B. M., Mellon, M. T., Pearl, J. C., Conrath, B. J., Smith, M. D., Clancy, R. T., Kuzmin, R. O., Roush, T., Mehall, G. L., ... Greenfield, M. (2001). The Mars Global Surveyor Thermal Emission Spectrometer experiment: Investigation description and surface science results. *Journal of Geophysical Research*, *106*(E10), 23823–23871. doi:10.1029/2000JE001370

Chung, K., Chang, K., & Chou, C. (2011). Wind loads on residential and large-scale solar collector models. *Journal of Wind Engineering and Industrial Aerodynamics*, *99*(1), 59–64. doi:10.1016/j.jweia.2010.10.008

Cirac, J. I., & Zoller, P. (2012). Goals and opportunities in quantum simulation. *Nature Physics*, *8*(4), 264–266. doi:10.1038/nphys2275

Clark, R. N., Carlson, R., Grundy, W., & Noll, K. (2013). Observed Ices in the Solar System. *The Science of Solar System Ices Astrophysics and Space Science Library*, *356*, 3–46. doi:10.1007/978-1-4614-3076-6_1

Clifford, S. M., & Parker, T. J. (2001). The evolution of the Martian hydrosphere: Implications for the fate of a primordial ocean and the current state of the northern plains. *Icarus, 154*(1), 40–79. doi:10.1006/icar.2001.6671

Connerney & Acuna. (1999). Magnetic lineations in the ancient crust of Mars. *Science, 284*, 794.

Connerney, J. E. P., Acuña, M. H., Wasilewski, P. J., Kletetschka, G., Ness, N. F., Rème, H., Lin, R. P., & Mitchell, D. L. (2001). The global magnetic field of Mars and implications for crustal evolution. *Geophysical Research Letters, 28*(21), 4015–4018. doi:10.1029/2001GL013619

Conté, I., & Peng, X. F. (2008). Numerical investigations of laminar flow in coiled pipes. *Applied Thermal Engineering, 28*(5-6), 423–432. doi:10.1016/j.applthermaleng.2007.05.009

Correa, P. C., Goneli, A. L. D., Junior, P. C. A., De Oliveira, G. H. H., & Valente, D. S. M. (2015). Moisture sorption isotherms and isosteric heat of sorption of coffee in different processing levels. *International Journal of Food Science & Technology, 45*(10), 2016–2022. doi:10.1111/j.1365-2621.2010.02373.x

Craddock, R. A., & Maxwell, T. A. (1993). Geomorphic evolution of the martian highlands through ancient fluvial processes. *Journal of Geophysical Research, 98*(E2), 3453–3468. doi:10.1029/92JE02508

Dai, N., & Li, S. (2018). Simulation and performance analysis on condenser coil in household heat pump water heater. *Sustainable Cities and Society, 36*, 176–184. doi:10.1016/j.scs.2017.10.020

Dai, N., Li, S., & Ye, Q. (2019). Performance analysis on the charging and discharging process of a household heat pump water heater. *International Journal of Refrigeration, 98*, 266–273. doi:10.1016/j.ijrefrig.2018.10.009

Daou, K., Wang, R. Z., & Xia, Z. Z. (2006). Desiccant cooling air conditioning: A review. *Renewable & Sustainable Energy Reviews, 10*(2), 55–77. doi:10.1016/j.rser.2004.09.010

Dawood, M. M. K., & Teamah, M. A. (2012). Hydro-Magnetic Mixed convection double diffusive in a lid driven square cavity. *European Journal of Scientific Research, 85*(3), 336–355.

De Hon, R. A. (1992). Martian lake basins and lacustrine plains. *Earth, Moon, and Planets, 56*(2), 95–122. doi:10.1007/BF00056352

de Pablo Hernández, M. Á., & Carrillo, J. D. C. (2012). *Geomorphological map of the lower NW flank of the Hecates Tholus volcano, Mars (scale 1:100,000).* doi:10.1080/17445647.2012.703902

De Vera, J. P., Boettger, U., Noetzel, R. T., Sánchez, F. J., Grunow, D., Schmitz, N., Lange, C., Hübers, H.-W., Billi, D., Baqué, M., Rettberg, P., Rabbow, E., Reitz, G., Berger, T., Möller, R., Bohmeier, M., Horneck, G., Westall, F., Jänchen, J., ... Spohn, T. (2012). Supporting Mars exploration: BIOMEX in Low Earth Orbit and further astrobiological studies on the Moon using Raman and PanCam technology. *Planetary and Space Science, 74*(1), 103–110. doi:10.1016/j.pss.2012.06.010

Declercq, J., Charles, J., Bowell, R., Warrender, R., & Barnes, A. (2017). Comparison of thermodynamic equilibrium and kinetic approach in the predictive evaluation of waste rock seepage quality in Northern Finland. Proceedings of International Mine Water Association, Mine Water and Circular Economy, 664-671.

Demina, N. Y., Andreev, A. O., & Nefedyev, Y. A. (2021, November). Meteorite hazard model for a space mission to Mars. *Journal of Physics: Conference Series*, *2103*(1), 012031. doi:10.1088/1742-6596/2103/1/012031

Deng, T. (2012). Stable and Metastable Phase Equilibria in the Salt-Water Systems. *Advances in Crystallization Processes*, 399-430.

Devanahalli, G. P., Timothy, J. R., & Raghavan, G. S. V. (2004). Natural convection heat transfer from helical coiled tubes. *International Journal of Thermal Sciences*, *43*(4), 359–365. doi:10.1016/j.ijthermalsci.2003.08.005

Dezfouli, M. M. S., Mat, S., Pirasteh, G., Sahari, K. S. M., Sopian, K., & Ruslan, M. H. (2014). Simulation analysis of the four confgurations of solar desiccant cooling system using evaporative cooling in tropical weather in Malaysia. *International Journal of Photoenergy*, *2014*, 1–14. doi:10.1155/2014/843617

Dickson, J. L., & Head, J. W. (2009). The formation and evolution of youthful gullies on Mars: Gullies as the late-stage phase of Mars' most recent ice age. *Icarus*, *204*(1), 63–86. doi:10.1016/j.icarus.2009.06.018

do Nascimento, C. A. R., Mariani, V. C., & dos Santos Coelho, L. (2020). Integrative numerical modeling and thermodynamic optimal design of counter-flow plate-fin heat exchanger applying neural networks. *International Journal of Heat and Mass Transfer*, *159*, 120097. doi:10.1016/j.ijheatmasstransfer.2020.120097

Dohm, J. M., Anderson, R. C., Baker, V. R., Ferris, J. C., Rudd, L. P., Hare, T. M., Rice, J. W. Jr, Casavant, R. R., Strom, R. G., Zimbelman, J. R., & Scott, D. H. (2001). Latent outflow activity for western Tharsis, Mars: Significant flood record exposed. *Journal of Geophysical Research*, *106*(E6), 12301–12314. doi:10.1029/2000JE001352

Dohm, J. M., Maruyama, S., Baker, V. R., Anderson, R. C., Ferris, J. C., & Hare, T. M. (2002). *Plate tectonism on early Mars: Diverse geological and geophysical evidence*. Academic Press.

Dolomatov, M. Y. (2021). Theory of systems with chaos of chemical composition in nature and technology. *Butlerov Communications A*, *1*(2), 2.

Dunn, K. (2019). *Process improvement using data*. Academic Press.

Eames, P. C., & Norton, B. (1998). The effect of tank geometry on thermally stratified sensible heat storage subject to low Reynolds number flows. Int. *J. Heat Mass Tran.*, *41*(14), 2131–2142. doi:10.1016/S0017-9310(97)00349-9

Eddington, A. (2022). Entropy in Science and Metaphor. *Entropic Philosophy: Chaos, Breakdown, and Creation, 17.*

Edgett, K.S. (2005). The sedimentary rocks of Sinus Meridiani: Five key observations from data acquired by the Mars Global Surveyor and Mars Odyssey orbiters. *Mars, 1*, 5–58.

Ehlmann, B. L., Berger, G., Mangold, N., Michalski, J. R., Catling, D. C., Ruff, S. W., Chassefière, E., Niles, P. B., Chevrier, V., & Poulet, F. (2013, January). Geochemical Consequences of Widespread Clay Mineral Formation in Mars' Ancient Crust. *Space Science Reviews, 174*(1-4), 329–364. doi:10.100711214-012-9930-0

El Maarry, M. R., Kodikara, J., Wijessoriya, S., Markiewicz, W. J., & Thomas, N. (2012, March). Desiccation mechanism for formation of giant polygons on Earth and intermediate-sized polygons on Mars: Results from a pre-fracture model. *Earth and Planetary Science Letters, 323–324*(15), 19–26. doi:10.1016/j.epsl.2012.01.016

Enteria, N., Yoshino, H., Mochida, A., Satake, A., & Takaki, R. (2015). Exergo-economic performances of the desiccant evaporative air-conditioning system at different regeneration and reference temperatures. *International Journal of Refrigeration, 56*, 81–98. doi:10.1016/j.ijrefrig.2014.11.007

Epstein, N. (1988). General Thermal Fouling Models. In L. F. Melo, T. R. Bott, & C. A. Bernardo (Eds.), *Fouling Science and Technology. NATO ASI Series* (Vol. 145). Springer. doi:10.1007/978-94-009-2813-8_2

Eriksson, L., Johansson, E., Kettaneh-Wold, N., Trygg, J., Wikström, C., & Wold, S. (2013). *Multi- and Megavariate Data Analysis: Basic Principles and Applications* (3rd ed.). MKS Umetrics.

Essefi, E. (2009). *Multidisciplinary study of Sidi El Hani Saline Environment: The Geological History and the Climatic Variability.* Academic Press.

Essefi, E. (2009). *Multidisciplinary study of Sidi El Hani Saline Environment: the History and the Climatic Variability* (Master thesis). Faculty of sciences of Sfax, University of Sfax.

Essefi, E. (2013). *Wet Aeolian Sedimentology and Sequence Stratigraphy within the Terrestrial Analogues in Eastern Tunisia: Implications for Wet Aeolian Sedimentology and Sequence Stratigraphy on Mars* (Doctoral dissertation). National Engineering School of Sfax, Sfax, Tunisia.

Essefi, E. (2013). *Wet Aeolian Sedimentology and Sequence Stratigraphy within the Terrestrial Analogues in Eastern Tunisia: Implications for Wet Aeolian Sedimentology and Sequence Stratigraphy on Mars.* National Engineering School of Sfax.

Essefi, E. (2021b). Geoeconomic interest versus environmental and health issues of the mineralogical assemblage of sebkha Oum El Khialate, southeastern Tunisia. *Arabian Journal for Science and Engineering, 46*(6), 5835–5845. doi:10.100713369-020-05244-5

Essefi, E., Komatsu, G., Fairén, A. G., Chan, M. A., & Yaich, C. (2013). Alignment of fault spring mounds at El-Guetiate, Southeast Tunisia: Terrestrial analogue implications for martian tectonics. *Lunar and Planetary Science*, 1229.

Essefi, E., Komatsu, G., Fairén, A. G., Chan, M. A., & Yaich, C. (2014). Groundwater influence on the aeolian sequence stratigraphy of the Mechertate–Chrita–Sidi El Hani system, Tunisian Sahel: Analogies to the wet–dry aeolian sequence stratigraphy at Meridiani Planum, Terby crater, and Gale crater, Mars. *Planetary and Space Science*, *95*, 56–78. doi:10.1016/j.pss.2013.05.010

Essefi, E., Mefteh, S., Medhioub, M., & Yaich, C. (2014). Magnetic Study of the Heated and Unheated Sedimentary Fillings of Sebkha Mhabeul, Southeast Tunisia: A Geophysical Method for Paleoclimatic Investigation and Tephrochronological Dating. *International Journal of Geophysics*, *2014*, 1–7. doi:10.1155/2014/908395

Essefi, E., Smida, N. B., Jandoubi, I., Othmani, M. A., & Tagorti, M. A. (2020). Progressive evaporation of brine of sebkhaMchiguig, central Tunisia: A geo-economical comparative study of salt and brine. *Carbonates and Evaporites*, *35*(2), 1–10. doi:10.100713146-020-00592-7

Essefi, E., & Tagorti, M. A. (2021a). Geoeconomic interest of minerals assemblage of sebkha El Melah, southeastern Tunisia. *Water Practice & Technology*, *16*(2), 633–647. doi:10.2166/wpt.2021.004

Essefi, E., Tagorti, M. A., Touir, J., & Yaich, C. (2012). Modeling of the chaotic behaviors at Sidi El Hani discharge playa, eastern Tunisia: Contribution of the philosophy of causality to solve complex chaotic systems in geology and biology. *Tunis. J. Med. Plants Nat. Prod*, *7*, 116–128.

Eugster, H. P., Harvie, C. E., & Weare, J. H. (1980). Mineral equilibria in a six-component seawater system, Na-K-Mg-Ca-SO4-Cl-H2O, at 25° C. *GeochimicaetCosmochimicaActa*, *44*(9), 1335–1347. doi:10.1016/0016-7037(80)90093-9

Eugster, H. P., & Maglione, G. (1979). Brines and evaporites of the Lake Chad basin, Africa. *GeochimicaetCosmochimicaActa*, *43*(7), 973–981. doi:10.1016/0016-7037(79)90087-5

Fang, X., Gao, Q., Zhang, J., Wang, Y., Guo, X., & Guo, Y. (2022). Entropy Enhancement of Chaotic Laser via Quantum Noise. *Chinese Journal of Lasers*, *48*(21), 2112001.

Farhan, M., Omar, Z., Mebarek-Oudina, F., Raza, J., Shah, Z., Choudhari, R. V., & Makinde, O. D. (2020). Implementation of one step one hybrid block method on nonlinear equation of the circular sector oscillator. *Computational Mathematics and Modeling*, *31*(1), 116–132. doi:10.100710598-020-09480-0

Fatah, A., Mahmud, H. B., Bennour, Z., Gholami, R., & Hossain, M. (2022). Geochemical modelling of CO_2 interactions with shale: Kinetics of mineral dissolution and precipitation on geological time scales. *Chemical Geology*, *592*, 120742. doi:10.1016/j.chemgeo.2022.120742

Feldman, W. C., Boynton, W. V., Tokar, R. L., Prettyman, T. H., Gasnault, O., Squyres, S. W., Elphic, R. C., Lawrence, D. J., Lawson, S. L., Maurice, S., McKinney, G. W., Moore, K. R., & Reedy, R. C. (2002). Global distribution of neutrons from Mars: Results from Mars Odyssey. *Science*, *297*(5578), 75–78. doi:10.1126cience.1073541 PMID:12040088

Fernández-Remolar, D. C. (2012). *Carbonate precipitation under bulk acidic conditions as a potential biosignature for searching life on Mars Earth and Planetary Science Letters*. Academic Press.

Finkbeiner, F., Gonard, T., & Filiol, B. (1993). Échangeurs Thermiques: Enjeux, Marchés, Technologie et Politique d'Innovation [Heat Exchangers: Challenges, Markets, Technology and Innovation Policy]. Éditions Européennes Thermique et Industrie (EETI).

Flint, G. (1971). Great Salt Lake Minerals and Chemicals Corporation (2nd ed.). Academic Press.

Foley, B. J., Bercovici, D., & Landuyt, W. (2012, May). The conditions for plate tectonics on super-Earths: Inferences from convection models with damage. *Earth and Planetary Science Letters*, *331–332*(15), 281–290. doi:10.1016/j.epsl.2012.03.028

Fraenkel, D. (2011). Monoprotic mineral acids analyzed by the smaller-ion shell model of strong electrolyte solutions. *The Journal of Physical Chemistry B*, *115*(3), 557–568. doi:10.1021/jp108997f PMID:21192660

Frank, M. P., & Shukla, K. (2021). Quantum Foundations of Classical Reversible Computing. *Entropy (Basel, Switzerland)*, *23*(6), 701. doi:10.3390/e23060701 PMID:34206044

Gac, J. Y., Al-Droubi, A., Paquet, H., Fritz, B., & Tardy, Y. (1979). Chemical model for origin and distribution of elements in salts and brines during evaporation of waters. Application to some saline lakes of Tibesti, Chad. *Physics and Chemistry of the Earth*, *11*, 149–158. doi:10.1016/0079-1946(79)90018-1

Gainey, S. R., Hausrath, E. M., & Hurowitz, J. A. (2022). Thermodynamic and kinetic analysis of transitions in clay mineral chemistry on Mars. *Icarus*, *372*, 114733. doi:10.1016/j.icarus.2021.114733

Gao, S. Y., Song, P. S., Xia, S. P., & Zheng, M. P. (2007). *Salt lake chemistry-new type borate salt lake*. Science Press.

Garbalinska, H., Bochnenek, M., Malorny, W., & Von Werder, J. (2017). Comparative analysis of the dynamic vapor sorption technique and the traditional method for sorption isotherms determination- Exemplified at autoclaved aerated concrete samples of four density classes. *Cement and Concrete Research*, *91*, 97–105. doi:10.1016/j.cemconres.2016.11.001

García-Pérez, J. V., Cárcel, J. A., Clemente, G., & Mulet, A. (2008). Water sorption isotherms for lemon peel at different temperatures and isosteric heats. *Lebensmittel-Wissenschaft + Technologie*, *41*(1), 18–25. doi:10.1016/j.lwt.2007.02.010

Gasteiger, J., & Engel, T. (2003). *Chemoinformatics: A Textbook*. Wiley-VCH. doi:10.1002/3527601643

Ghodake, H. M., Goswami, T. K., & Chakraverty, A. (2007). Moisture sorption isotherms, heat of sorption and vaporization of xithered leaves, black and green tea. *Journal of Food Engineering*, *78*(3), 827–835. doi:10.1016/j.jfoodeng.2005.11.023

Gibbs, J. W. (1902). *Elementary principles in statistical mechanics: Developed with especial reference to the rational foundations of thermodynamics*. C. Scribner's Sons.

Goh, L. H. K., Hung, Y. M., Chen, G. M., & Tso, C. P. (2021). Entropy generation analysis of turbulent convection in a heat exchanger with self-rotating turbulator inserts. *International Journal of Thermal Sciences*, *160*, 106652. doi:10.1016/j.ijthermalsci.2020.106652

Goncalves, P., Angrisani, G., Sasso, M., Rodrigues Gaspar, A., & Gamerio da Silva, M. (2014). Exergetic analysis of a desiccant cooling system: Searching for performance improvement opportunities. *International Journal of Energy Research*, *38*(6), 714–727. doi:10.1002/er.3076

Goneli, A.L.D., Correa, P.C., Oliveira, G.H.H., & Botelho, F.M. (2010). Water sorption isotherms and thermodynamic properties of okra seeds. Trans. ASABE. *American Society of Agricultural and Biological Engineering*, *53*, 191-197.

Goody, R. M. (2012). The Atmosphere of Mars. Weather, 12(1), 3–15. doi:10.1002/j.1477-8696.1957.tb00381.x

Greeley, R., & Guest, J.E. (1987). *Geologic map of the eastern equatorial region of Mars*. USGS Misc. Inv. Ser. Map I-1802B (1:15,000,000).

Greenspan, L. (1977). Humidity fixed points of binary saturated aqueous solutions. *Journal of Research of the National Bureau of Standards-A. Physics and Chemistry*, *81*, 92-93.

Gross, M. (2012). The search for life on Earth and other planets. *Current Biology, 22*(7). . doi:10.1016/j.cub.2012.03.040

Gudipati, M. S., & Cooper, P. D. (2013). *Chemistry in Water Ices: From Fundamentals to Planetary Applications Astrophysics and Space Science Library*. Academic Press.

Gueddari, M., Monnin, C., Perret, D., Fritz, B., & Tardy, Y. (1983). Geochemistry of brines of the chott El Jerid in southern Tunesia—Application of Pitzer's equations. *Chemical Geology*, *39*(1-2), 165–178. doi:10.1016/0009-2541(83)90078-5

Guo, J., Cheng, L., & Xu, M. (2010). Multi-objective optimization of heat exchanger design by entropy generation minimization. *Journal of Heat Transfer*, *132*(8), 081801. doi:10.1115/1.4001317

Guo, L. J., Feng, Z. P., & Chen, X. (2002). Transient convective heat transfer of steam–water two-phase flow in a helical tube under pressure drop type oscillations. *International Journal of Heat and Mass Transfer*, *45*(3), 533–542. doi:10.1016/S0017-9310(01)00178-8

Habeeb. (2012). Numerical Simulation of Convective Heat Transfer and Fluid Flow through Porous Media with Different Moving and Heated Walls. *International Journal of Mathematical, Computational, Physical, Electrical and Computer Engineering.*

Haberle, R. M., Forget, F., Head, J., Kahre, M. A., Kreslavsky, M., & Owen, S. J. (2013). Summary of the Mars recent climate change workshop NASA/Ames Research Center. *Icarus, 222*(1), 415–418.

Hacini, M., Kherici, N., & Oelkers, E. H. (2008). Mineral precipitation rates during the complete evaporation of the MerouaneChott ephemeral lake. *GeochimicaetcosmochimicaActa, 72*(6), 1583–1597.

Halsey, G. (1948). Physical adsorption on non-uniform surfaces. *The Journal of Chemical Physics, 16*(10), 931–937. doi:10.1063/1.1746689

Han, D. J., Pei, J., & Kamber, M. (2006). *Data Mining concepts and techniques. Southeast Asia Edition* (2nd ed.). Elsevier.

Hano, T. M. (1992). *Solvent extraction and ion exchange. Process Metallurgy*, 7(B).

Hargitai, H., & Gide, M. (2009). Three virtual globes of Mars: topographic, albedo and a historic globe. *EPSC Abstracts, 4.*

Harry, Y., & McSween, J. (2012). *What we have learned about Mars from SNC meteorites.* doi:10.1111/j.1945-5100.1994.tb01092.x

Hartmann, W. K. (2005, April). Martian cratering 8: Isochron refinement and the chronology of Mars. *Icarus, 174*(2), 294–320. doi:10.1016/j.icarus.2004.11.023

Hartmann, W. K., & Neukum, G. (2001). Cratering chronology and the evolution of Mars. In *Chronology and evolution of Mars* (pp. 165–194). Springer. doi:10.1007/978-94-017-1035-0_6

Harvey, J. P., Singh, S., Oishi, K., Acheson, B., Turcotte, R., Pilon, D., Lavoie, J., & Grange, B. (2021). Quantification of the chemical reactivity of molten nitrate salts with heat treatable aluminum alloys. *Materials & Design, 198*, 109293. doi:10.1016/j.matdes.2020.109293

Harvie, C. E., & Weare, J. H. (1980). The prediction of mineral solubilities in natural waters: The Na K Mg Ca Cl SO4 H2O system from zero to high concentration at 25° C. *Geochimica et Cosmochimica Acta, 44*(7), 981–997. doi:10.1016/0016-7037(80)90287-2

Hassini, L., Desmorieux, H., Torres, S. S., & Touil, A. (2015). Desorption isotherms and thermodynamic properties of prickly pear seeds. *Industrial Crops and Products, 97*, 457–465. doi:10.1016/j.indcrop.2015.01.078

Henderson, S. M. (1952). A basic concept of equilibrium moisture. *Transactions of the ASAE. American Society of Agricultural Engineers, 33*, 29–32.

Heredia, D. J. (2017). *Improvement of the numerical capacities of simulation tools for reactive transport modeling in porous media* (Doctoral dissertation). Université Rennes 1.

Hering, H. (1952). Séparation quantitative de trace de lithium contenues le calcium. *Analytica Chimica Acta*, *6*, 340–350. doi:10.1016/S0003-2670(00)86954-8

Hirt, C., Claessens, S.J., Kuhn, M., & Featherstone, W.E. (2012). Kilometer-resolution gravity field of Mars: MGM2011. *Planetary and Space Science*, *67*(1), 147–154. doi:10.1016/j.pss.2012.02.006

Hong, Y. (2007). Composite Fouling on Heat Exchanger Surfaces. Nova Science Books.

Hubinger, M., Menegalli, F. C., Aguerre, R. J., & Suarez, C. (1992). Water vapor adsorption isotherms of guava, mango and pineapple. *Journal of Food Science*, *57*(6), 1405–1407. doi:10.1111/j.1365-2621.1992.tb06869.x

Hull, R. W., Kharaka, Y. K., Maest, A. S., & Fries, T. L. (1985). Sampling and analysis of subsurface waters: A summary of current methodology. In *Proceedings First Canadian/American Conference on Hydrology*. National Water Well Association.

Hürdoğan, E., Büyükalaca, O., Yılmaz, T., & Hepbaşlı, A. (2010). Experimental investigation of a novel desiccant cooling system. *Energy and Building*, *42*(11), 2049–2060. doi:10.1016/j.enbuild.2010.06.014

Hu, Y. F., Chu, H. D., Li, J. G., Liu, Z. C., Peng, X. M., Ling, S., & Zhang, J. Z. (2011). Extension of the simple equations for prediction of the properties of mixed electrolyte solutions to the mixed ionic liquid solutions. *Industrial & Engineering Chemistry Research*, *50*(7), 4161–4165. doi:10.1021/ie1022496

Hwang, W. B., Choi, S., & Lee, D. Y. (2017). In-depth analysis of the performance of hybrid desiccant cooling system incorporated with an electric heat pump. *Energy*, *118*, 324–332. doi:10.1016/j.energy.2016.12.007

Hynek, B. M., & Phillips, R. J. (2001). Evidence for extensive denundation of the martian highlands. *Geology*, *29*(5), 407–410. doi:10.1130/0091-7613(2001)029<0407:EFEDOT>2.0.CO;2

Ibrahim, O., Fardoun, F., Younes, R., & Louahlia-Gualous, H. (2013). Air source heat pump water heater: Dynamic modeling, optimal energy management and mini-tubes condensers. *Energy*, •••, 1–15.

Iguedjtal, T., Louka, N., & Allaf, K. (2008). Sorption isotherms of potato slices dried and texturized by controlled sudden decompression. *Journal of Food Engineering*, *85*(2), 180–190. doi:10.1016/j.jfoodeng.2007.06.028

Ingham, D. B., & Pop, I. (Eds.). (2005). *Transport Phenomena in Porous Media*. Elsevier.

Ismael, M. A., Pop, I., & Chamkha, A. J. (2014). Mixed convection in a lid-driven square cavity with partial slip. *International Journal of Thermal Sciences*, *82*, 47–61. doi:10.1016/j.ijthermalsci.2014.03.007

Jakosky, B. M., Lin, R. P., Grebowsky, J. M., Luhmann, J. G., Mitchell, D. F., Beutelschies, G., Priser, T., Acuna, M., Andersson, L., Baird, D., Baker, D., Bartlett, R., Benna, M., Bougher, S., Brain, D., Carson, D., Cauffman, S., Chamberlin, P., Chaufray, J.-Y., ... Zurek, R. (2015). The Mars atmosphere and volatile evolution (MAVEN) mission. *Space Science Reviews*, *195*(1), 3–48. doi:10.100711214-015-0139-x

Jakosky, B., & Phillips, R. J. (2001). Mars' volatile and climate history. *Nature*, *412*(6843), 237–244. doi:10.1038/35084184 PMID:11449285

Jalalzadeh, S., da Silva, F. R., & Moniz, P. V. (2021). Prospecting black hole thermodynamics with fractional quantum mechanics. *The European Physical Journal C*, *81*(7), 1–13. doi:10.1140/epjc10052-021-09438-5

Jarzynski, C. (2011). Equalities and inequalities: Irreversibility and the second law of thermodynamics at the nanoscale. *Annual Review of Condensed Matter Physics*, *2*(1), 329–351. doi:10.1146/annurev-conmatphys-062910-140506

Javier, A. (2021). *Historical and Conceptual Foundations of Information Physics*. Academic Press.

Jiang, H., Jiang, Y., Wang, Y., Ma, Z., & Yao, Y. (2006). An experimental study on a modified air conditioner with a domestic hot water supply (ACDHWS). *Energy*, *31*(12), 1789–1803. doi:10.1016/j.energy.2005.07.004

Ji, Hu, Y., Li, L., Shi, D., Li, J., Nie, F., Song, F., Zeng, Z., Sun, W., & Liu, Z. (2016a). Lithium extraction with a synergistic system of dioctyl phthalate and tributyl phosphate in kerosene and FeCl3. *Hydrometallurgy*, *162*, 71–78. doi:10.1016/j.hydromet.2016.02.018

Ji, Li, L., Shi, D., Li, J., Liu, Z., Xu, D., & Song, X. (2016b). Extraction equilibria of lithium with N, N-bis (2-ethylhexyl)-3-oxobutanamide and tributyl phosphate in kerosene and FeCl 3. *Hydrometallurgy*, *164*, 304–312. doi:10.1016/j.hydromet.2016.06.022

Johnston, D. H., & Toksöz, M.N. (1977). *Internal structure and properties of Mars.* Academic Press.

Jones, B. F., & Deocampo, D. M. (2003). Geochemistry of saline lakes. *Treatise on Geochemistry, 5*, 605.

Jouannic, G., Gargani, J., Costard, F., Ori, G. G., Marmo, C., Schmidt, F., & Lucas, A. (2012). Morphological and mechanical characterization of gullies in a periglacial environment: The case of the Russell crater dune (Mars). *Planetary and Space Science*, *71*(1), 38–54. doi:10.1016/j.pss.2012.07.005

Joudi, K. A., & Dhaidan, N. (2001). Application of solar assisted heating and desiccant cooling systems for a domestic building. *Energy Conversion and Management*, *42*(8), 995–1022. doi:10.1016/S0196-8904(00)00111-4

Jubayer, C. M., & Hangan, H. (2012). Numerical Simulation of Wind Loading on Photovoltaic Panels. *Structures Congress*, 1180–1189. 10.1061/9780784412367.106

Júnior, P. F. (2021). On the Deduction of the Caratheodory's Axiom of the Second Law of Thermodynamics from the Clausius and Kelvin Principles. arXiv preprint arXiv:2110.02502.

Jurinak, J. J. (1982). *Open cycle solid desiccant cooling component models and system simulations* [PhD thesis]. University of Wisconsin-Madison, Madison, WI.

Kahlenberg, L., & Krauskopf, F. C. (1958). A new method of separating Lithium chloride from the chlorides of the other alkalis and the chloride of barium. *Journal of the American Chemical Society*, *30*(7), 1104–1120. doi:10.1021/ja01949a008

Kaltenegger, L. (2013). *Planetary Atmospheres and Chemical Markers for Extraterrestrial Life*. Astrochemistry and Astrobiology Physical Chemistry in Action. doi:10.1007/978-3-642-31730-9_5

Kanoglu, M., Çarpinlioglu, M. Ö., & Yildirim, M. (2004). Energy and exergy analyses of an experimental open-cycle desiccant cooling system. *Applied Thermal Engineering*, *24*(5-6), 919–932. doi:10.1016/j.applthermaleng.2003.10.003

Karen, C. R., Hellismar, W., Isneider, L., Samuel, G. F., Daniel, P., & Renato, S. (2020). Isotherms and thermodynamic properties of water adsorption in Cumari pepper seeds. *Revista Brasileira de Engenharia Agrícola e Ambiental*, *24*(4), 280–285. doi:10.1590/1807-1929/agriambi.v24n4p280-285

Kargel, J. S., Baker, V. R., Beget, J. E., Lockwood, J. F., Pewe, T. L., Shaw, J. S., & Strom, R. G. (1995). Evidence of ancient continental glaciation in the martian northern plains. *Journal of Geophysical Research*, *100*(E3), 5351–5368. doi:10.1029/94JE02447

Katterfel'd, G. N., Benesh, K., Khain, V. Ye., & Khodak, Yu. A. (1968). Problems of comparative planetologyInternational. *Dizhi Lunping*, *10*(9), 989–1017. doi:10.1080/00206816809474964

Kaymak-Ertekin, F., & Gedik, A. (2004). Sorption isotherms and isosteric heat of sorption for grapes, apricots, apples and potatoes. *Lebensmittel-Wissenschaft + Technologie*, *37*(4), 429–438. doi:10.1016/j.lwt.2003.10.012

Kern, D., & Seaton, R. (1959). A theoretical analysis of thermal surface fouling. British Chemical Engineering, 4(5), 258-262.

Khanafer, K., & Vafai, K. (2002). Double-diffusive mixed convection in a lid-driven enclosure filled with a fluid saturated porous medium. *Numerical Heat Transfer Part A*, *42*(5), 465–486. doi:10.1080/10407780290059657

Khanlari, A., Sözen, A., Variyenli, H. I., & Gürü, M. (2019). Comparison between heat transfer characteristics of TiO 2/deionized water and kaolin/deionized water nanofluids in the plate heat exchanger. *Heat Transfer Research*, *50*(5), 435–450. doi:10.1615/HeatTransRes.2018026288

Kharaka, Y. K., &Hanor, J. S. (2003). Deep fluids in the continents: I. Sedimentary basins. *Treatise on Geochemistry, 5*, 605.

Kharat, R., Bhardwaj, N., & Jha, R. S. (2009). Development of heat transfer coefficient correlation for concentric helical coil heat exchanger. *International Journal of Thermal Sciences*, *48*(12), 2300–2308. doi:10.1016/j.ijthermalsci.2009.04.008

Khatib, T., Abunajeeb, I., & Heneni, Z. (2020). Determination of Mars Solar-Belt by Modeling of Solar Radiation Using Artificial Neural Networks. *Journal of Solar Energy Engineering*, *142*(1), 011007. doi:10.1115/1.4044304

Klima, R. (2013). The Evolution of the Lunar Highlands Crust: A Complicated History. *Second Conference on the Lunar Highlands Crust*, *94*(2), 21–28. doi:10.1002/2013EO020009

Klose, M., & Shao, Y. (2013). Large-eddy simulation of turbulent dust emission. *Aeolian Research*, *8*, 49–58. doi:10.1016/j.aeolia.2012.10.010

Kohonen, T. (1995). *Self-Organizing Maps*. Springer. doi:10.1007/978-3-642-97610-0

Kolenda, Z., Donizak, J., & Hubert, J. (2004). On the minimum entropy production in steady state heat conduction processes. *Energy*, *29*(12-15), 2441–2460. doi:10.1016/j.energy.2004.03.049

Kreslavsky, M. A., & Head, J. W. III. (2002). Mars: Nature and evolution of young latitude-dependent water–ice-rich mantle. *Geophysical Research Letters*, *29*(15), 14–21. doi:10.1029/2002GL015392

Krug, R. R., Hunter, W. G., & Grieger, R. A. (1976a). Enthalpy entropy compensation. 1- some fundamental statistical problems associated with the analysis of Van't Hoff and Arrhenius data. *Journal of Physical Chemistry*, *80*(21), 2335–2341. doi:10.1021/j100562a006

Krug, R. R., Hunter, W. G., & Grieger, R. A. (1976b). Enthalpy entropy compensation. 2- Separation of the chemical from the statistical effect. *Journal of Physical Chemistry*, *80*(21), 2341–2351. doi:10.1021/j100562a007

Kumar, V., Saini, S., Sharma, M., & Nigam, K. D. P. (2006). Pressure drop and heat transfer study in tube-in-tube helical heat exchanger. *Chemical Engineering Science*, *61*(13), 4403–4416. doi:10.1016/j.ces.2006.01.039

Kurnia, J. C., Ghoreishi-Madiseh, S. A., & Sasmito, A. P. (2019). Heat transfer and entropy generation in concentric/eccentric double-pipe helical heat exchangers. *Heat Transfer Engineering*, *41*(18), 1552–1575. doi:10.1080/01457632.2019.1661666

Lach, A. (2015). *Modélisation thermodynamique des propriétés d'excès des saumures naturelles et industrielles* [Thermodynamic modelling of the excess properties of natural and industrial brines.] (Doctoral dissertation). Pau.

Lahsasni, S., Kouhila, M., Mahrouz, M., & Fliyou, M. (2002). Moisture adsorption-desorption isotherms of prickly pear cladode at different temperatures. *Energy Conversion and Management*, *44*, 923–936. doi:10.1016/S0196-8904(02)00094-8

Laskar, J., Robutel, P., Joutel, F., Gastineau, M., Correia, A., & Levrard, B. (2004). A long term numerical solution for the insolation quantities of the Earth. Astron. *Astrophys*. http://hal.ccsd.cnrs.fr/ccsd-00001603

Lebouc, A., Almanza, M., Yonnet, J. P., Legait, U., & Roudaut, J. (2014). Refrigeration magnetique Etat de l'art et developpements recents. *Symposium de Genie Electrique*, Cachan, France.

Lenardic, A., Nimmo, F., & Moresi, L. (2004). Growth of the hemispheric dichotomy and the cessation of plate tectonics on Mars. *Journal of Geophysical Research*, *109*(E2), E02003. doi:10.1029/2003JE002172

Li, M. Y., Wang, L. S., & Gmehling, J. (2011). Thermodynamics of phase equilibria in aqueous strong electrolyte systems. *Industrial & Engineering Chemistry Research*, *50*(6), 3621–3631. doi:10.1021/ie101428j

Li, M., & Lai, A. C. (2013). Thermodynamic optimization of ground heat exchangers with single U-tube by entropy generation minimization method. *Energy Conversion and Management*, *65*, 133–139. doi:10.1016/j.enconman.2012.07.013

Limaye, A. B. S., Aharonson, O., & Perron, J. T. (2012). Detailed stratigraphy and bed thickness of the Mars north and south polar layered deposits. *J. Geophys. Res.*, *117*. doi:10.1029/2011JE003961

Lin, S., & Chen, N. (2021). DEM Based Study on Shielded Astronomical Solar Radiation and Possible Sunshine Duration under Terrain Influences on Mars by Using Spectral Methods. *ISPRS International Journal of Geo-Information*, *10*(2), 56. doi:10.3390/ijgi10020056

Lucchitta, B. K. (1981). Mars and Earth: Comparison of cold-climate features. *Icarus*, *45*(2), 264–303. doi:10.1016/0019-1035(81)90035-X

Lu, X., Du, X., Zeng, M., Zhang, S., & Wang, Q. (2014). Shell-side thermal-hydraulic performances of multilayer spiral-wound heat exchangers under different wall thermal boundary conditions. *Applied Thermal Engineering*, *70*, 1216.

Lu, X., & Kieffer, S. W. (2009). Thermodynamics and mass transport in multicomponent, multiphase H2O systems of planetary interest. *Annual Review of Earth and Planetary Sciences*, *37*(1), 449–477. doi:10.1146/annurev.earth.031208.100109

Macedonio, F., Quist-Jensen, C. A., Al-Harbi, O., Alromaih, H., Al-Jlil, S. A., Al Shabouna, F., & Drioli, E. (2013). Thermodynamic modeling of brine and its use in membrane crystallizer. *Desalination*, *323*, 83–92. doi:10.1016/j.desal.2013.02.009

MacGregor, J. F. (2003). Multivariate statistical approaches to fault detection and isolation. *Fault Detection Supervision and Safety of Technical Processes*, *36*(5), 549–554. doi:10.1016/S1474-6670(17)36549-7

Maclaine-Cross, I. L., & Banks, P. J. (1972). Coupled heat and mass transfer in regenerators—Prediction using an analogy with heat transfer. *International Journal of Heat and Mass Transfer*, *15*(6), 1225–1242. doi:10.1016/0017-9310(72)90187-1

Mahmoudi, M., Tavakoli, M. R., Mirsoleimani, M. A., Gholami, A., & Salimpour, M. R. (2017). Experimental and numerical investigation on forced convection heat transfer and pressure drop in helically coiled pipes using TiO2/water nanofluid. *International Journal of Refrigeration*, *74*, 627–643. doi:10.1016/j.ijrefrig.2016.11.014

Mahmud, S., & Fraser, R. A. (2002). Second law analysis of heat transfer and fluid flow inside a cylindrical annular space. *Exergy*, *2*(4), 322–329. doi:10.1016/S1164-0235(02)00078-X

Mahmud, S., & Fraser, R. A. (2003). Mixed convection–radiation interaction in a vertical porous channel: Entropy generation. *Energy*, *28*(15), 1557–1577. doi:10.1016/S0360-5442(03)00154-3

Malin, M. C., & Edgett, K. S. (2000). Sedimentary rocks of Early Mars. *Science*, *290*(5498), 1927–1937. doi:10.1126cience.290.5498.1927 PMID:11110654

Mariah, L., Buckley, C. A., Brouckaert, C. J., Curcio, E., Drioli, E., Jaganyi, D., & Ramjugernath, D. (2006). Membrane distillation of concentrated brines—Role of water activities in the evaluation of driving force. *Journal of Membrane Science*, *280*(1-2), 937–947. doi:10.1016/j.memsci.2006.03.014

Mars, C. (2005). Sulfate formation and its relevence to environmental conditions on early. *Science, 307*, 1594-1597.

Matsumoto, K., Kondo, T., Ikeda, M., & Numazawa, T. (2011). Numerical analysis of active magnetic regenerators for hydrogen magnetic refrigeration between 20 and 77 K. *Cryogenics*, *51*(6), 353–357. doi:10.1016/j.cryogenics.2010.06.003

Mayhew, K. W. (2020). New Thermodynamics: Inelastic Collisions and Cosmology. *European Journal of Applied Physiology*, *2*(6).

McLennan, S. M. (2012). *Geochemistry of sedimentary processes on Mars*. Academic Press.

McLennan, S. M., Bell, J. F. III, Calvin, W. M., Christensen, P. R., Clark, B. C., de Souza, P. A., Farmer, J., Farrand, W. H., Fike, D. A., Gellert, R., Ghosh, A., Glotch, T. D., Grotzinger, J. P., Hahn, B., Herkenhoff, K. E., Hurowitz, J. A., Johnson, J. R., Johnson, S. S., Jolliff, B., ... Yen, A. (2005). Provenance and diagenesis of the evaporite-bearing Burns formation, Meridiani Planum, Mars. *Earth and Planetary Science Letters*, *240*(1), 95–121. doi:10.1016/j.epsl.2005.09.041

McMahon, S., & Cosmidis, J. (2022). False biosignatures on Mars: Anticipating ambiguity. *Journal of the Geological Society*, *179*(2), jgs2021-050. doi:10.1144/jgs2021-050

Meeussen, J. C. L., van der Sloot, H. A., Dijkstra, J. J., & Kosson, D. S. (2009). *Review of thermodynamic and adsorption databases*. Energy Research Centre of the Netherlands and Vanderbilt University/CRESP.

Mehrez, Z., Bouterra, M., El Gafsi, A., & Belghith, A. (2013). Heat transfer and entropy generation analysis of nanofluids flow in an open cavity. *Computers & Fluids*, *88*, 363–373. doi:10.1016/j.compfluid.2013.09.026

Mejia-Romero, S., Lugo, J. E., Bernardin, D., & Faubert, J. (2021). Evaluation of Bio-movements Using Nonlinear Dynamics. In *Proceedings of International Conference on Data Science and Applications* (pp. 197-208). Springer. 10.1007/978-981-15-7561-7_16

Melkikh, A. V. (2021). Can Quantum Correlations Lead to Violation of the Second Law of Thermodynamics? *Entropy (Basel, Switzerland)*, *23*(5), 573. doi:10.3390/e23050573 PMID:34067021

Melnikov, S., Sheldeshov, N., Zabolotsky, V., Loza, S., & Achoh, A. (2017). Pilot scale complex electrodialysis technology for processing a solution of lithium chloride containing organic solvents. *Separation and Purification Technology*, *189*, 74–81. doi:10.1016/j.seppur.2017.07.085

Mendoza, V. M., Mendoza, B., Garduño, R., Cordero, G., Pazos, M., Cervantes, S., & Cervantes, K. (2021). Thermodynamic simulation of the seasonal cycle of temperature, pressure and ice caps on Mars. *Atmosfera*, *34*(1), 1–23.

Meng, L. M. S., Gruszkiewicz, M. S., Deng, T., Guo, Y., & Li, D. (2015). Isothermal Evaporation Process Simulation Using the Pitzer Model for the Quinary System LiCl−NaCl−KCl−SrCl2−H2O at 298.15 K. *Industrial & Engineering Chemistry Research*, *54*(33), 8311–8318. doi:10.1021/acs.iecr.5b01897

Menter, F., Kuntz, M., & Langtry, R. (2003). Ten years of industrial experience with the SST turbulence model. *Turbulence, Heat and Mass Transfer.*

Mianping, Z., Yongsheng, Z., Xifang, L., Wen, Q. I., Fanjing, K., Zhen, N., & Yongjie, L. (2016). Progress and prospects of salt lake research in China. *ActaGeologicaSinica-English Edition*, *90*(4), 1195–1235. doi:10.1111/1755-6724.12767

Micadei, K., Peterson, J. P., Souza, A. M., Sarthour, R. S., Oliveira, I. S., Landi, G. T., Batalhão, T. B., Serra, R. M., & Lutz, E. (2019). Reversing the direction of heat flow using quantum correlations. *Nature Communications*, *10*(1), 1–6. doi:10.103841467-019-10333-7 PMID:31165732

Milliken, R. E., Grotzinger, J. P., & Thomson, B. J. (2010). Paleoclimate of Mars as captured by the stratigraphic record in Gale Crater. *Geophysical Research Letters*, *37*. doi:10.1029/2009GL041870

Mischna, M. A., Baker, V., Milliken, R., Richardson, M., & Lee, C. (2013). Effects of obliquity and water vapor/trace gas greenhouses in the early martian climate. *Journal of Geophysical Research. Planets*, *118*(3), 560–576. doi:10.1002/jgre.20054

Moawed, M. (2011). Experimental study of forced convection from helical coiled tubes with different parameters. *Energy Conversion and Management*, *52*(2), 1150–1156. doi:10.1016/j.enconman.2010.09.009

Moore, J. M., & Wilhelms, D. E. (2001). Hellas as a possible site of ancient ice-covered lakes on Mars. *Icarus*, *154*(2), 258–276. doi:10.1006/icar.2001.6736

Moores, J. E., Brown, R. H., Lauretta, D. S., & Smith P. H. (2012). Experimental and theoretical simulation of sublimating dusty water ice with implications for D/H ratios of water ice on Comets and Mars. *Planetary Science, 1*(2).

Moraga, G., Martinez, N., & Chiralt, A. (2004). Water sorption and phase transitions instrawberries: Influence of pre-treatment. *Journal of Food Engineering*, *62*(4), 315–321. doi:10.1016/S0260-8774(03)00245-0

Moraga, G., Martinez, N., & Chiralt, A. (2006). Water sorption and phase transitions inkiwi fruit: Influence of pre-treatment. *Journal of Food Engineering*, *72*, 156–174. doi:10.1016/j.jfoodeng.2004.11.031

Morales, P., Ramírez-Moreno, E., Sanchez-Mata, M. C., Carvalho, A. M., & Ferreira, I. C. F. R. (2012). Nutritional and antioxidant properties of pulp and seeds of twoxoconostle cultivars (Opuntia joconostle FAC Weber ex Diguet and Opuntia matudae Scheinvar) of high consumption in Mexico. *Food Research International*, *46*(1), 279–285. doi:10.1016/j.foodres.2011.12.031

Moreira, R., Chenlo, F., Torre, M. D., & Vallejo, N. (2008). Thermodynamic analysis of experimental sorption isotherms of loquat and quince fruits. *Journal of Food Engineering*, *88*(4), 514–521. doi:10.1016/j.jfoodeng.2008.03.011

Mussett, S. M. (2022). *Entropic Philosophy: Chaos, Breakdown, and Creation*. Rowman & Littlefield.

Muthtamilselvan, M., Kandaswamy, P., & Lee, J. (2009). Hydromagnetic mixed convection in a lid-driven cavity filled with a fluid-saturated porous medium. *International Journal of Applied Mathematics and Mechanics*, *5*(7), 28–44.

Naji, A., Idlimam, A., & Kouhila, M. (2010). Sorption isotherms and thermodynamic properties of powdered milk. *Chemical Engineering Communications*, *197*(8), 1109–1125. doi:10.1080/00986440903412936

Naphon, P. (2007). Thermal performance and pressure drop of thehelical-coil heat exchangers with and without helically crimped fins. *International Communications in Heat and Mass Transfer*, *34*(3), 321–330. doi:10.1016/j.icheatmasstransfer.2006.11.009

Nelson, P. R. C., MacGregor, J. F., & Taylor, P. A. (2006). The impact of missing measurements on PCA and PLS prediction and monitoring applications. *Chemometrics and Intelligent Laboratory Systems*, *80*(1), 1–12. doi:10.1016/j.chemolab.2005.04.006

Nemati, H., Moghimi, M. A., Sapin, P., & Markides, C. N. (2020). Shape optimisation of air-cooled finned-tube heat exchangers. *International Journal of Thermal Sciences*, *150*, 106233. doi:10.1016/j.ijthermalsci.2019.106233

Ness, N. F., Acuña, M. H., Connerney, J., Wasilewski, P., Mazelle, C., Sauvaud, J., Vignes, D., d'Uston, C., Reme, H., Lin, R., Mitchell, D. L., McFadden, J., Curtis, D., Cloutier, P., & Bauer, S. J. (1999). 14 colleagues MGS magnetic fields and electron reflectometer investigation: Discovery of paleomagnetic fields due to crustal remanence. *Advances in Space Research*, *23*(11), 1879–1886. doi:10.1016/S0273-1177(99)00271-9

Nicolaou, G., Wicks, R. T., Owen, C. J., Kataria, D. O., Chandrasekhar, A., Lewis, G. R., ... Bruno, R. (2021). Deriving the bulk properties of solar wind electrons observed by Solar Orbiter: A preliminary study of electron plasma thermodynamics. *Astronomy & Astrophysics, 656*, A10. doi:10.1073/iti0213110

Nield, D. A., & Bejan, A. (2006). *Convection in Porous Media*. Springer.

Niknia, I., Campbell, O., Christianse, T., Govindappa, P., Teyber, R., & Trevizoli, P. (2016). *A material screening technique for optimization performance of an AMR*. Seventh IIF-IIR International Conference on Magnetic Refrigeration at Room Temperature, Thermag VII, Torino, Italy.

Niles, P., Catling, D. C., Berger, G., Chassefière, E., Ehlmann, B. L., Michalski, J. R., Morris, R., Ruff, S. W., & Sutter, B. (2013, January). B. & 8 colleagues, 2013. Geochemistry of Carbonates on Mars: Implications for Climate History and Nature of Aqueous Environments. *Space Science Reviews*, *174*(1-4), 301–328. doi:10.100711214-012-9940-y

Nimmo, F., & Tanaka, K. L. (2005). Early crustal evolution of Mars. *Annual Review of Earth and Planetary Sciences*, *33*(1), 133–161. doi:10.1146/annurev.earth.33.092203.122637

Nisbet, R., Elder, J., & Miner, G. (2009). *Handbook of Statistical Analysis and Data Mining Applications*. Elsevier.

Noshad, M., Shahidi, F., Mohebbi, M., & Mortazav, S. (2012). Desorption isotherms and thermodynamic properties of fresh and osmotic ultrasonic dehydrated quince. *Journal of Food Processing and Preservation*, *37*, 1–12.

Numazawa, T., Kimura, H., Sato, M., & Maeda, H. (1993). Carnot magnetic refrigerator operating between 1.4 and 10 K. *Cryogenics*, *33*(5), 547–554. doi:10.1016/0011-2275(93)90252-J

Oehler, D. Z., & Allen, C. C. (2008). Ancient hydrothermal springs in Arabia Terra, Mars. *Lunar and Planetary Science*, 39.

Oehler, D. Z., & Allen, C. C. (2010). Evidence for pervasive mud volcanism in Acidalia Planitia, Mars. *Icarus*, *208*(2), 636–657. doi:10.1016/j.icarus.2010.03.031

Oehler, D. Z., & Allen, C. C. (2012, June). Giant Polygons and Mounds in the Lowlands of Mars: Signatures of an Ancient Ocean? *Astrobiology*, *12*(6), 601–615. doi:10.1089/ast.2011.0803 PMID:22731685

Oehler, J. F., Lénat, J. F., & Labazuy, P. (2008). Growth and collapse of the Reunion Island volcanoes. *Bull. Volcanol.*, *70*(6), 717–742. doi:10.100700445-007-0163-0

Ordóñez, J. C., & Bejan, A. (2000). Entropy generation minimization in parallel-plates counter-flow heat exchangers. *International Journal of Energy Research*, *24*(10), 843–864. doi:10.1002/1099-114X(200008)24:10<843::AID-ER620>3.0.CO;2-M

Oswin, C. R. (1946). The kinetics of package life. III. The isotherm. *Journal of the Society of Chemical Industry*, *6*(12), 419–421. doi:10.1002/jctb.5000651216

Padet, J. (2008). *Fluides en Écoulement Méthodes et modèles*. Seconde édition revue et augmenté, Reims.

Panaras, G., Mathioulakis, E., Belessiotis, V., & Kyriakis, N. (2010). Experimental validation of a simplifed approach for a desiccant wheel model. *Energy and Building*, *42*(10), 1719–2172. doi:10.1016/j.enbuild.2010.05.006

Pandelidis, D., Pacak, A., Cichon, A., Anisimov, S., Drąg, P., Vager, B., & Vasilijevc, V. (2018). Multi-Stage desiccant cooling system for moderate climate. *Energy Conversion and Management*, *177*, 77–90. doi:10.1016/j.enconman.2018.09.061

Park, C. Y., & Hrnjak, P. (2008). Experimental and numerical study on microchannel and round tube condensers in a R410A residential air-conditioning system. *International Journal of Refrigeration*, *31*(5), 822–831. doi:10.1016/j.ijrefrig.2007.10.007

Parker, T. J., Gorsline, D. S., Saunders, R. S., Pieri, D. C., & Schneeberger, D. M. (1993). Schneeberger D.M., Coastal geomorphology of the martian northern plains. *Journal of Geophysical Research*, *98*(E6), 11061–11078. doi:10.1029/93JE00618

Parker, T. J., Saunders, R. S., & Schneeberger, D. M. (1989). Transitional morphology in the west Deuteronilus Mensae region of Mars: Implications for modification of the lowland/upland boundary. *Icarus*, *82*(1), 111–145. doi:10.1016/0019-1035(89)90027-4

Park, H., Tamwattana, O., Kim, J., Buakeaw, S., Hongtong, R., Kim, B., Khomein, P., Liu, G., Meethong, N., & Kang, K. (2021). Probing lithium metals in batteries by advanced characterization and analysis tools. *Advanced Energy Materials*, *11*(15), 2003039. doi:10.1002/aenm.202003039

Parkhurst, D. L. (1995). User's guide to phreeqc - a computer program for speciation, reaction-path, advective-transport, and inverse geochemical calculations. U.S. Geological Survey Water-Resources Investigations Report.

Parkhurst, D. L., & Appelo, C. A. J. (2013). *Description of input and examples for PHREEQC version 3: a computer program for speciation, batch-reaction, one-dimensional transport, and inverse geochemical calculations (No. 6-A43)*. US Geological Survey.

Park, I., Kim, Y., Park, J., & Jeong, S. (2014). Investigation on the Two-stage Active Magnetic Regenerative Refrigerator for Liquefaction of Hydrogen. *Advances in Cryogenic Engineering*, *1573*, 74–81. doi:10.1063/1.4860685

Pasckert, J. H., Hiesinger, H., & Reiss, D. (2012). Rheologies and ages of lava flows on Elysium Mons, Mars. *Icarus*, *219*(1), 443–457. doi:10.1016/j.icarus.2012.03.014

Patankar, S. V., Pratap, V. S., & Spalding, D. B. (1974). Prediction of laminar flow and heat transfer in helically coiled pipes. *Journal of Fluid Mechanics*, *62*(03), 53–551. doi:10.1017/S0022112074000796

Pauli, W. (1928). *Über das H-Theorem vom Anwaschsen der Entropie vom Standpunkt der neuen* [On the H-theorem of washing up entropy from the point of view of the new]. Quantenmechanik in Probleme des Modernen Physik.

Pawar, S. S., & Sunnapwar, V. K. (2014). Experimental and CFD investigation of convective heat transfer in helically coiled tube heat exchanger. *Chemical Engineering Research & Design*, *92*(11), 2294–2312. Advance online publication. doi:10.1016/j.cherd.2014.01.016

Pengsheng, S., (2016). The phase diagram of salt-water systems and utilization of salt lake resources. *Journal of Salt Lake Research*, 3.

Perry, R. H., & Green, D. W. (2008). *Perry's Chemical Engineers' handbook*. McGraw-Hill.

Phillips, R. J., Zuber, M. T., Solomon, S. C., Golombek, M. P., Jakosky, B. M., Banerdt, W. B., Smith, D. E., Williams, R. M. E., Hynek, B. M., Aharonson, O., & Hauck, S. A. II. (2001). Ancient geodynamics and global-scale hydrology on Mars. *Science*, *291*(5513), 2587–2591. doi:10.1126cience.1058701 PMID:11283367

Pitzer, K. S. (1991) Ionic Interaction Approach: Theory and Data Correlation. In Activity Coefficients in Electrolyte Solutions. CRC Press.

Pitzer, K. S. (1973). Thermodynamics of electrolytes. I. Theoretical basis and general equations. *Journal of Physical Chemistry*, *77*(2), 268–277. doi:10.1021/j100621a026

Pitzer, K. S., & Mayorga, G. (1973). Thermodynamics of electrolytes. II. Activity and osmotic coefficients for strong electrolytes with one or both ions univalent. *Journal of Physical Chemistry*, *77*(19), 2300–2308. doi:10.1021/j100638a009

Plummer, L. N. (1988). A computer program incorporating Pitzer's equations for calculation of geochemical reactions in brines (Vol. 88, No. 4153). Department of the Interior, US Geological Survey.

Portyankina, G., Pommerol, A., Aye, K.-M., Hansen, C. J., & Thomas, N. (2012). Polygonal cracks in the seasonal semi-translucent CO2 ice layer in Martian polar areas. *Journal of Geophysical Research*, *117*(E2), E02006. Advance online publication. doi:10.1029/2011JE003917

Prabhanjan, D.G., Raghavan, G.S.V., & Rennie, T.J. (2002). Comparison of heat transfer rates between a straight tube heat exchanger and a helically coiled heat exchanger. *Int. Commun. Heat Mass Transf.*, *29*, 185–191. (02)00309-3 doi:10.1016/S0735-1933

Qi, C., Luo, T., Liu, M., Fan, F., & Yan, Y. (2019). Experimental study on the flow and heat transfer characteristics of nanofluids in double-tube heat exchangers based on thermal efficiency assessment. *Energy Conversion and Management*, *197*, 111877. doi:10.1016/j.enconman.2019.111877

Quintero, M. A., Hao, S., Patel, S. V., Bao, J. K., Zhou, X., Hu, Y. Y., Wolverton, C., & Kanatzidis, M. G. (2021). Lithium Thiostannate Spinels: Air-Stable Cubic Semiconductors. *Chemistry of Materials*, *33*(6), 2080–2089. doi:10.1021/acs.chemmater.0c04651

Qu, Y., Wang, F., Wang, Y., Wang, P., Li, T., & Feng Meng, Z. (2014). Simulation Optimization and Experiment of R1234-ze on the Heat Pump Water Heater Storage Tank. *Advanced Materials Research*, *1051*, 828–831. doi:10.4028/www.scientific.net/AMR.1051.828

Rafique, M., Gandhidasan, P., Luai, M., Hadhrami, A., & Rehman, S. (2016). Energy, exergy and anergy analysis of a solar desiccant cooling system. *Journal of Clean Energy Technologies*, *4*(1), 78–83. doi:10.7763/JOCET.2016.V4.257

Rahman, M. M., Billah, M. M., Mamun, M. A. H., Saidur, R., & Hasanuzzaman, M. (2010). Reynolds and Prandtl numbers effects on MHD mixed convection in a lid-driven cavity along with joule heating and a centred heat conducting circular block. *Int. J. of Mech. and Mat. Eng.*, (5), 163–170.

Rajan, A. G. (2021). *A Pedagogical Approach to Obtain the Combined First and Second Law of Thermodynamics from Classical Statistical Mechanics*. Academic Press.

Rapf, M., & Kranert, M. (2021). Irreversible entropy to account for environmental impacts and sustainability. *Procedia CIRP*, *98*, 601–606. doi:10.1016/j.procir.2021.01.161

Reddy, S. T., & Lewis, A. E. (2006). Water recovery and salt recovery from brine solutions. *Proc. 13th Int. Workshop Industrial Crystallization Univ. Technol.*, 328.

Rennie, T. J., & Raghavan, V. G. S. (2005). Experimental studies of a double-pipe helical heat exchanger. *Experimental Thermal and Fluid Science*, *29*(8), 919–924. doi:10.1016/j.expthermflusci.2005.02.001

Rennie, T. J., & Raghavan, V. G. S. (2006). Numerical studies of a double-pipe helical heat exchanger. *Applied Thermal Engineering*, *26*(11-12), 1266–1273. doi:10.1016/j.applthermaleng.2005.10.030

Rennie, T. J., & Raghavan, V. G. S. (2006b). Effect of fluid thermal properties on heat transfer characteristics in a double pipe helical heat exchanger. *International Journal of Thermal Sciences*, *45*(12), 1158–1165. doi:10.1016/j.ijthermalsci.2006.02.004

Rettig, S. L., Jones, B. F., & Risacher, F. (1980). Geochemical evolution of brines in the Salar of Uyuni, Bolivia. *Chemical Geology*, *30*(1-2), 57–79. doi:10.1016/0009-2541(80)90116-3

Revathy, S. M., Rangaraj, A. G., Srinath, Y., Boopathi, K., Shobana Devi, A., Balaraman, K., & Prasad, D. M. R. (2021). Impact on solar radiation parameters in India during COVID-19 lockdown: A case study. *International Journal of Sustainable Energy*, *40*(8), 806–820. doi:10.1080/14786451.2021.1893726

Ribeiro, M., Henriques, T., Castro, L., Souto, A., Antunes, L., Costa-Santos, C., & Teixeira, A. (2021). The entropy universe. *Entropy (Basel, Switzerland)*, *23*(2), 222. doi:10.3390/e23020222 PMID:33670121

Righter, K., Schönbächler, M., Pando, K., Rowland, I. I. R. II, Righter, M., & Lapen, T. (2020). Ag isotopic and chalcophile element evolution of the terrestrial and martian mantles during accretion: New constraints from Bi and Ag metal-silicate partitioning. *Earth and Planetary Science Letters*, *552*, 116590. doi:10.1016/j.epsl.2020.116590

Robertson, K., & Bish, D. (2012). Constraints on the distribution of $CaSO_4{\cdot}nH_2O$ phases on Mars and implications for their contribution to the hydrological cycle. *Icarus*.

Rodriguez-Manfredi, J. A., de la Torre Juárez, M., Alonso, A., Apéstigue, V., Arruego, I., Atienza, T., Banfield, D., Boland, J., Carrera, M. A., Castañer, L., Ceballos, J., Chen-Chen, H., Cobos, A., Conrad, P. G., Cordoba, E., del Río-Gaztelurrutia, T., de Vicente-Retortillo, A., Domínguez-Pumar, M., Espejo, S., ... Zurita, S. (2021). The Mars Environmental Dynamics Analyzer, MEDA. A suite of environmentalsensors for the Mars 2020 mission. *Space Science Reviews*, *217*(3), 1–86. doi:10.100711214-021-00816-9 PMID:34776548

Rosario, L., & Rahman, M. (2011). Analysis of magnetic refrigerator. *Applied Thermal Engineering*, *31*(6-7), 1082–1090. doi:10.1016/j.applthermaleng.2010.12.002

Rouquerol, J., Rouquerol, F., Llewellyn, P., Maurin, G., & Sing, K. S. (2014). *Adsorption by Powders and Porous Solids: Principles, Methodology and Applications* (2nd ed.). Academic press.

Row, A. (2012a). Therodynamics of active magnetic regenerators Part I. *Crygenics.*, *52*(2-3), 111–118. doi:10.1016/j.cryogenics.2011.09.005

Row, A. (2012b). Therodynamics of active magnetic regenerators: Part II. *Crygenics*, *52*(2-3), 119–128. doi:10.1016/j.cryogenics.2011.09.007

Roy, S., Poncet, R., & Sorin, M. V. (2016). *Multiobjective optimization of a reciprocating magnetic refrigerator using genetic algorithm.* Seventh IIF-IIR International Conference on Magnetic Refrigeration at Room Temperature, Thermag VII, Torino, Italy.

Rudraiah, N., Barron, R. M., Venkatachalappa, M., & Subbaraya, C. K. (1995). Effect of a magnetic field on free convection in a rectangular enclosure. *International Journal of Engineering Science*, *33*(8), 1075–1084. doi:10.1016/0020-7225(94)00120-9

Saha, L.K., Somadder, M.C., & Roy, N.C. (2015). Hydro-magnetic mixed convection flow in a lid-driven cavity with wavy bottom surface. *American Journal of Applied Mathematics*, *3*(1), 8-19.

Said, M., Nehari, D., Bouzit, M., & Chemlou, N.-E. S. (2013). Analysis of mixed convection in an inclined lid-driven cavity with a wavy wall. *Journal of Mechanical Science and Technology*, *27*(7), 2181–2190. doi:10.100712206-013-0533-9

Santanu-Basu, U. S., & Shivhare Mujumdar, A. S. (2006). Models for Sorption Isotherms for Foods: A Review. *Drying Technology*, *24*(8), 917–930. doi:10.1080/07373930600775979

Sarkar, S., Hubbard, J. B., Halter, M., & Plant, A. L. (2021). Information thermodynamics and reducibility of large gene networks. *Entropy (Basel, Switzerland)*, *23*(1), 63. doi:10.3390/e23010063 PMID:33401415

Sartre, J. P. (2022). Entropy in Ancient Greek Thought. *Entropic Philosophy: Chaos, Breakdown, and Creation*, 35.

Savicki, D. L., Vielmo, H., & Krenzinger, A. (2011). Three-dimensional analysis and investigation of the thermal and hydrodynamic behaviors of cylindrical storage tanks. *Renewable Energy*, *36*(5), 1364–1373. doi:10.1016/j.renene.2010.10.011

Sbárbaro, D., & Del Villar, R. (2010). *Advanced Control and Supervision of Mineral Processing Plants*. Springer. doi:10.1007/978-1-84996-106-6

Schellenberg, A., Maffei, J., Telleen, K., & Ward, R. (2013). Structural analysis and application of wind loads to solar arrays. *Journal of Wind Engineering and Industrial Aerodynamics Journal*, *123*, 261–272. doi:10.1016/j.jweia.2013.06.011

Scheridan, J. C., & Mitchell, J. W. (1985). A hybrid solar desiccant cooling system. *Solar Energy*, *34*(2), 187–193. doi:10.1016/0038-092X(85)90179-3

Scott, D. H. (1993). *Geologic map of the MTM 25057 and 25052 quadrangles, Kasei Valles region of Mars*. USGS Misc. Inv. Ser. I-Map 2208 (1:500,000).

Scott, D. H., & Carr, M. H. (1978). *Geologic map of Mars*. U.S. Geol. Survey, Misc. Inv. Map I-1083.

Scott, D. H., Dohm, J. M., & Rice, J. W., Jr. (1995). *Map of Mars showing channels and possible paleolake basins*. USGS Misc. Inv. Ser. Map I-2461 (1:30,000,000).

Scott, R. H., & Strasheim, A. (1970). Laser induced plasmas for analytical spectroscopy. *Spectrochimica Acta. Part B, Atomic Spectroscopy*, *25*(7), 311–332. doi:10.1016/0584-8547(70)80038-6

Sefton-Nash, E., Teanby, N. A., Montabone, L., Irwin, P. G. J., Hurley, J., & Calcutt, S. B. (2013, January). Climatology and first-order composition estimates of mesospheric clouds from Mars Climate Sounder limb spectra. *Icarus*, *222*(1), 342–356. doi:10.1016/j.icarus.2012.11.012

Sekulic, D. P. (1990). *The second law quality of energy transformation in a heat exchanger*. Academic Press.

Sekulic, D. P. (1986). Entropy generation in a heat exchanger. *Heat Transfer Engineering*, *7*(1-2), 83–88. doi:10.1080/01457638608939647

Selco, J. I. (1995). Chemical Thermodynamics on Mars. *Journal of Chemical Education*, *72*(7), 599. doi:10.1021/ed072p599

Seshadri, N., & Galperin, M. (2021). Entropy and information flow in quantum systems strongly coupled to baths. *Physical Review. B*, *103*(8), 085415. doi:10.1103/PhysRevB.103.085415

Shah, R. K., & Joshi, S. D. (1987). Convective heat transfer in curved ducts. In Handbook of single phase convective heat transfer. New York: Wiley Inter science.

Shah, T., & Hrnjak, P.S. (2014). *Linked Modelling of Heat Pump Water Heater Vapor Compression System and Water Tank*. Academic Press.

Shahsavar, A., Noori, S., Toghraie, D., & Barnoon, P. (2021). Free convection of non-Newtonian nanofluid flow inside an eccentric annulus from the point of view of first-law and second-law of thermodynamics. ZAMM-Journal of Applied Mathematics and Mechanics. *Zeitschrift für Angewandte Mathematik und Mechanik*, *101*(5), e202000266. doi:10.1002/zamm.202000266

Shahzad, M. K., Chaudhary, G. Q., Ali, M., Sheikh, N. A., Khalil, M. S., & Rashid, T. U. (2018). Experimental evaluation of a solid desiccant system integrated with cross flow Maisotsenko cycle evaporative cooler. *Applied Thermal Engineering*, *128*, 1476–1487. doi:10.1016/j.applthermaleng.2017.09.105

Sharif, M. A. R. (2007). Laminar mixed convection in shallow inclined driven cavities with hot moving lid on top and cooled from bottom. *Applied Thermal Engineering*, *27*(5-6), 1036–1042. doi:10.1016/j.applthermaleng.2006.07.035

Shi, D., Cui, B., Li, L., Peng, X., Zhang, L., & Zhang, Y. (2019). Lithium extraction from low-grade salt lake brine with ultrahigh Mg/Li ratio using TBP – kerosene – FeCl3 system. *Separation and Purification Technology*, *211*, 303–309. doi:10.1016/j.seppur.2018.09.087

Shi, D., Zhang, L., Peng, X., Li, L., Song, F., Nie, F., Ji, L., & Zhang, Y. (2018). Extraction of lithium from salt lake brine containing boron using multistage centrifuge extractors. *Desalination*, *441*, 44–51. doi:10.1016/j.desal.2018.04.029

Shokouhmand, H., & Salimpour, M. R. (2007). Entropy generation analysis of fully developed laminar forced convection in a helical tube with uniform wall temperature. *Heat and Mass Transfer*, *44*(2), 213–220. doi:10.100700231-007-0235-x

Sivanandam, S., Chamkha, A. J., Mallawi, F. O. M., Alghamdi, M. S., & Alqahtani, A. M. (2020). Effects of entropy generation, thermal Radiation and moving-wall direction on mixed convective flow of nanofluid in an Enclosure. *Mathematics*, *8*, 1471. doi:10.3390/math8091471

Smaili, A., Aït-Ali, S., & Chahine, R. (2011). Performance Predictions of First Stage Magnetic Hydrogen Liquefier. *International Journal of Hydrogen Energy*, *36*(6), 4169–4177. doi:10.1016/j.ijhydene.2010.09.006

Smith, L., Forster, C. B., & Evans, J. P. (1990). Interaction of fault zones, fluid flow, and heat transfer at the basin scale. Hydrogeology of Permeability Environments. *International Association of Hydrogeologists Symposium, 28th Int. Geol. Congress*, 2, 41-67.

Smith, D. E., Zuber, M. T., Frey, H. V., Garvin, J. B., Head, J. W., Muhleman, D. O., Pettengill, G. H., Phillips, R. J., Solomon, S. C., Zwally, H. J., Banerdt, W. B., Duxbury, T. C., Golombek, M. P., Lemoine, F. G., Neumann, G. A., Rowlands, D. D., Aharonson, O., Ford, P. G., Ivanov, A. B., ... Sun, X. (2001). Mars Orbiter Laser Altimeter: Experimental summary after the first year of global mapping of Mars. *Journal of Geophysical Research*, *106*(E10), 23689–23722. doi:10.1029/2000JE001364

Solomon, S. C., & Qian, L. (2012). The Interaction of Solar and Geomagnetic Activity with ClimateChange in the Thermosphere. EGU General Assembly 2012, Vienna, Austria.

Sphaier, L. A., & Nóbrega, C. E. L. (2012). Parametric analysis of components effectiveness on desiccant cooling system performance'. *Energy*, *38*(1), 157–166. doi:10.1016/j.energy.2011.12.019

Squyres, S. W. (1979). The distribution of lobate debris aprons and similar flows on Mars. *Journal of Geophysical Research*, *84*(B14), 8087–8096. doi:10.1029/JB084iB14p08087

Šrámek, O., & Zhong, S. (2012). Martian crustal dichotomy and Tharsis formation by partial melting coupled to early plume migration. *Journal of Geophysical Research, 117*(E1), E01005. Advance online publication. doi:10.1029/2011JE003867

Stanley, B. D., Schaub, D. R., & Hirschmann, M. M. (2012, November-December). CO_2 solubility in primitive martian basalts similar to Yamato 980459, the effect of composition on CO_2 solubility of basalts, and the evolution of the martian atmosphere. *The American Mineralogist, 97*(11-12), 1841–1848. doi:10.2138/am.2012.4141

Stivaletta, N., & Barbieri, R. (2009). Endolithic microorganisms from spring mound evaporite deposits (southern Tunisia). *Journal of Arid Environments, 73*(1), 33–39. doi:10.1016/j.jaridenv.2008.09.024

Strasberg, P., & Winter, A. (2021). First and second law of quantum thermodynamics: A consistent derivation based on a microscopic definition of entropy. *PRX Quantum, 2*(3), 030202. doi:10.1103/PRXQuantum.2.030202

Stumm, W., & Morgan, J. J. (1970). *Aquatic chemistry: An introduction emphasizing chemical equilibria in natural waters.* Academic Press.

Stumm, W., Morgan, J. J., & Drever, J. I. (1996). Aquatic chemistry. *Journal of Environmental Quality, 25*(5), 1162. doi:10.2134/jeq1996.2551162x

Sun, M. X. (1996). The method of solvent extraction kinetics—The interfacial area. *Chem. Bulletin., 7*, 50–52.

Sun, Y., Wang, Q., Wang, Y., Yun, R., & Xiang, X. (2021). Recent advances in magnesium/lithium separation and lithium extraction technologies from salt lake brine. *Separation and Purification Technology, 256*, 117807.

Taborek, J. (1979). Évolution of Heat Exchanger Design Techniques. *Heat Transfer Engineering, 1*(1), 15–29. doi:10.1080/01457637908939546

Talla, A., Jannot, Y., Nkeng, G. E., & Puiggali, J. R. (2005). Experimental Determination and Modelingof Sorption Isotherms of Tropical Fruits: Banana, Mango, and Pineapple. *Drying Technology, 23*(7), 1477–1498. doi:10.1081/DRT-200063530

Tanaka, K. L. (1986). The stratigraphy of Mars. *Journal of Geophysical Research, 91*(B13), E139–E158. doi:10.1029/JB091iB13p0E139

Tavakol, P., & Behbahaninia, A. (2018). Presentation of two new two-stage desiccant cooling cycles based on heat recovery and evaluation of performance based on energy and exergy analysis. *Journal of Building Engineering, 20*, 455–466. doi:10.1016/j.jobe.2018.08.002

te Vrugt, M. (2021). *The mereology of thermodynamic equilibrium.* arXiv preprint arXiv:2104.11140.

te Vrugt, M. (2021). The five problems of irreversibility. *Studies in History and Philosophy of Science, 87*, 136–146. doi:10.1016/j.shpsa.2021.04.006 PMID:34111817

Thadée, A. L. (2017). *Étude des saumures naturelles et industrielles: Approche expérimentale et par modélisation de l'extraction du lithium par évaporation* [Study of natural and industrial brines: Experimental and modelled approach to lithium extraction by evaporation] (Doctoral dissertation). Pau.

Thomas, D., & Grigull, U. (1974). Experimental Investigation of the Deposition of Magnetite from the Fluid Flow in Steam Generating Tubes. *Brennstoff, Wärme, Kraft*, *26*(3), 109–115.

Thomson, W. (1851). Article. *Transactions of the Royal Society of Edinburgh*, *20*, 261–268, 289–298.

Thonon, B. (2007). *Guide de l'encrassement des échangeurs de chaleur* [Heat Exchanger Fouling Guide]. Editions greth France.

Threlkeld, J. L., Ramsey, J. W., & Kuehn, T. H. (1998). Thermal environmental engineering (3rd ed.). Prentice-Hall.

Tian, P., Ning, P., Cao, H., & Li, Z. (2012). Determination and Modeling of Solubility for CaSO4·2H2O–NH4+–Cl—SO42—NO3—H2O System. *Journal of Chemical & Engineering Data*, *57*(12), 3664–3671. doi:10.1021/je300871p

Timmermann, E. O., Chirife, J., & Iglesias, H. A. (2001). Water sorption isotherms of foods and foodstuffs: BET or GAB parameters. *Journal of Food Engineering*, *48*(1), 19–31. doi:10.1016/S0260-8774(00)00139-4

Tlili, N., Bargougui, A., Elfalleh, W., Triki, S., & Nasri, N. (2011). Phenolic compounds, protein, lipid content and fatty acids compositions of cactus seeds. *Journal of Medicinal Plants Research*, *5*(18), 4519–4524.

Tovbin, Y. K. (2021). Second Law of Thermodynamics, Gibbs' Thermodynamics, and Relaxation Times of Thermodynamic Parameters. *Russian Journal of Physical Chemistry A*, *95*(4), 637–658. doi:10.1134/S0036024421020266

Tributsch, H. (2021). Time-Neutrality of Natural Laws Challenged: Time Is Not an Illusion but Ongoing Energy-Driven Information Loss. *Journal of Modern Physics*, *12*(03), 300–327. doi:10.4236/jmp.2021.123023

Tu, F. Q., Sun, B., Wan, M., & Huang, Q. H. (2021). Evolution of the universe from the perspective of entropy and information. *Modern Physics Letters A*, *36*(16), 2150111. doi:10.1142/S021773232150111X

Ulyanov, S. V. (2021). Quantum Algorithm of Imperfect KB Self-organization Pt I: Smart Control-Information-Thermodynamic Bounds. *Artificial Intelligence Advances, 3*(2).

Utaki, T., Kamiya, K., Nakagawa, T., Yamamoto, T. A., & Numazawa, T. (2007). Research on a Magnetic Refrigeration Cycle for Hydrogen Liquefaction. *International Cryocooler Conference, Inc*, Boulder, CO.

Vafai, K. (Ed.). (2005). *Handbook of Porous Media*. Taylor & Francis. doi:10.1201/9780415876384

Valente, G. (2021). Taking up statistical thermodynamics: Equilibrium fluctuations and irreversibility. *Studies in History and Philosophy of Science*, *85*, 176–184. doi:10.1016/j.shpsa.2020.10.006 PMID:33966773

Van Den Berg, C., & Bruim, S. (1981). Water activity and its estimation in food systems. In *Influence of food quality*. Academic Press.

Vanchurin, V., Wolf, Y. I., Koonin, E. V., & Katsnelson, M. I. (2021). *Thermodynamics of Evolution and the Origin of Life*. arXiv preprint arXiv:2110.15066.

Vaquiro, H. A., Simal, S., Reis de Carvalho, G., & Telis-Romero, J. (2011). Moisture desorption isotherms and thermodynamic properties of lime seeds. *European Drying Conference*, 26-28.

Varmuza, K., & Filzmoser, P. (2009). *Introduction to multivariate statistical analysis in chemometrics*. Academic Press.

Varmuza, K., & Filzmoser, P. (2009). *Introduction to Multivariate Statistical Analysis in Chemometrics* (1st ed.). CRC Press. doi:10.1201/9781420059496

Vaz, D. A. (2012). *Tectonic lineament mapping of the Thaumasia Plateau, Mars: Comparing results from photointerpretation and a semi automatic approach*. Academic Press.

VDI. (2010). *Heat Atlas. Gesellschaft Verfahrenstechnik und Chemieingenieurwesen (GVC). s.l* (2nd ed.). Springer Verlag.

Villanueva, G. L., Mumma, M. J., Novak, R. E., Radeva, Y. L., Käufl, H. U., Smette, A., Tokunaga, A., Khayat, A., Encrenaz, T., & Hartogh, P. (2013, March). A sensitive search for organics (CH4, CH3OH, H2CO, C2H6, C2H2, C2H4), hydroperoxyl (HO2), nitrogen compounds (N2O, NH3, HCN) and chlorine species (HCl, CH3Cl) on Mars using ground-based high-resolution infrared spectroscopy. *Icarus*, *223*(1), 11–27. doi:10.1016/j.icarus.2012.11.013

Vitányi, P. M. (2001). Quantum Kolmogorov complexity based on classical descriptions. *IEEE Transactions on Information Theory*, *47*(6), 2464–2479. doi:10.1109/18.945258

Von Neumann, J. (1955). Mathematical Foundations of Quantum Mechanics (R. T. Beyer, Trans.). Princeton University Press.

Von Neumann, J. (2018). *Mathematical foundations of quantum mechanics*. Princeton university press.

Wakabayashi, J., & Shervais, J. W. (2012). Initiation and Termination of Subduction: Rock Record, Geodynamic Models, and Modern Plate Boundaries. *Lithosphere*, *4*(6), 467–468. doi:10.1130/LINT1.1

Wang, X., Zhang, K., & Wang, J. (2021). *What can we learn about islands and state paradox from quantum information theory?* arXiv preprint arXiv:2107.09228.

Wang, D. D. (2006). *Research on Temperature Field and Flow Field in Tank of ASHPWH*. Beijing Institute of Civil Engineering and Architecture Doctor of Philosophy.

Wang, G., Wang, D., Deng, J., Lyu, Y., Pei, Y., & Xiang, S. (2019). Experimental and numerical study on the heat transfer and flow characteristics in shell side of helically coiled tube heat exchanger based on multi-objective optimization. *International Journal of Heat and Mass Transfer*, *137*, 349.

Wang, L. S. (2021). PROGRESS in entropy principle, as disclosed by nine Schools of thermodynamics, and its ecological implication. *Journal*, *16*(4), 359–372. doi:10.18280/ijdne.160403

Wang, Y., Yuan, Z., Liang, Y., Xie, Y., Chen, X., & Li, X. (2015). A review of experimental measurement and prediction models of crude oil fouling rate in crude refinery preheat trains. *Asia-Pacific Journal of Chemical Engineering*, *10*, 607–625. doi:10.1002/apj.1895

Warsido, W., Bitsuamlak, G., Barata, J., & Gan Chowdhury, A. (2014). Influence of spacing parameters on the wind loading of solar array. *Journal of Fluids and Structures*, *48*, 295–315. doi:10.1016/j.jfluidstructs.2014.03.005

Watkinson, A. P. (1968). *Particulate fouling of sensible heat exchangers (T)*. University of British Columbia. Retrieved from https://open.library.ubc.ca/collections/ubctheses/831/items/1.0059106

Weckesser, P., Thielemann, F., Hoenig, D., Lambrecht, A., Karpa, L., & Schaetz, T. (2021). Trapping, shaping, and isolating of an ion Coulomb crystal via state-selective optical potentials. *Physical Review. A*, *103*(1), 013112. doi:10.1103/PhysRevA.103.013112

Weinstein, G. (2021). *Demons in Black Hole Thermodynamics: Bekenstein and Hawking*. arXiv preprint arXiv:2102.11209.

Wen, P., Li, Y., Polkowski, L., Yao, Y. Y., Tsumoto, S., & Wang, G. (2009). Rough Sets and Knowledge Technology. In *4th International Conference, Gold coast, Australia, Proceedings*. Springer.

Westall, J.C. (1976). *MINEQL: A computer program for the calculation of chemical equilibrium composition of aqueous systems*. Mass. Inst. Technol.

Wheeler, J. A. (2018). *Information, physics, quantum: The search for links*. CRC Press.

Williamson, M. C. Johnson, C. L., & Sylvester, P. J. (2013). Introduction to Special Issue of. Canadian Journal of Earth Sciences: Canadian contributions to planetary geoscience. *Revue canadienne des sciences de la Terre*. . doi:10.1139/cjes-2012-0164

Willitsch, S., Bell, M. T., Gingell, A. D., & Softley, T. P. (2008). Chemical applications of laser- and sympathetically-cooled ions in ion traps. *Physical Chemistry Chemical Physics*, *10*(48), 7200–7210. doi:10.1039/b813408c PMID:19060963

Wineland, D. J., & Itano, W. M. (1979). Laser cooling of atoms. *Physical Review A*, *20*(4), 1521–1540. doi:10.1103/PhysRevA.20.1521

Wolery, T. J. (1979). *Calculation of chemical equilibrium between aqueous solution and minerals: the EQ3/6 software package (No. UCRL—52658)*. California University.

Wong, H., Liu, N., & Reichmanis, E. (2022). Single-Pot Fabrication of Cellulose-Reinforced Solid Polymer Lithium-Ion Conductors. *ACS Applied Polymer Materials*, *4*(3), 1948–1955. doi:10.1021/acsapm.1c01789

Xiaowen, Y., & Lee, W. L. (2009). The use of helical heat exchanger for heat recovery domestic Water cooled air-conditioners. *Energy Conversion and Management*, *50*(2), 240–246. doi:10.1016/j.enconman.2008.09.039

Xin, R. C., Awwad, A., Dong, Z. F., Ebadian, M. A., & Soliman, H. M. (1996). An investigation and comparative study of the pressure drop in air–water two-phase flow in vertical helicoidal pipes. *International Journal of Heat and Mass Transfer*, *39*(4), 735–743. doi:10.1016/0017-9310(95)00164-6

Xiong, P. Y., Jahanshahi, H., Alcaraz, R., Chu, Y. M., Gómez-Aguilar, J. F., & Alsaadi, F. E. (2021). Spectral entropy analysis and synchronization of a multi-stable fractional-order chaotic system using a novel neural network-based chattering-free sliding mode technique. *Chaos, Solitons, and Fractals*, *144*, 110576. doi:10.1016/j.chaos.2020.110576

Xiong, Z. Q., Dai, Y. J., & Wang, R. Z. (2010). Development of a novel two-stage liquid desiccant dehumidification system assisted by CaCl2 solution using exergy analysis method. *Applied Energy*, *87*(5), 1495–1504. doi:10.1016/j.apenergy.2009.08.048

Xu, X. (2021). *Gibbs Paradox in the View of Information Entropy*. arXiv preprint arXiv:2105.12566.

Xue, F., Wang, B., Chen, M., Yi, C., Ju, S., & Xing, W. (2019). Fe3O4-doped lithium ion-sieves for lithium adsorption and magnetic separation. *Separation and Purification Technology*, *228*, 115750. doi:10.1016/j.seppur.2019.115750

Yamamoto, H., Yamaji, H., Abe, Y., Harada, K., Waluyo, D., Fukusaki, E., Kondo, A., Ohno, H., & Fukuda, H. (2009). Dimensionality reduction for metabolome data using PCA, PLS, OPLS, and RFDA with differential penalties to latent variables. *Chemometrics and Intelligent Laboratory Systems*, *98*(2), 136–142. doi:10.1016/j.chemolab.2009.05.006

Yang, G., Dong, F., & Ebadian, M. A. (1995). Laminar forced convection in a helicoidal pipe with finite pitch. *International Journal of Heat and Mass Transfer*, *38*(5), 853–862. doi:10.1016/0017-9310(94)00199-6

Yang, L., Shao, L. L., & Zhang, Z. C. (2014). Modeling and optimization of air source heat pump water heaters using wrap-around micro-channel condenser. *J. Refrig.*, *35*, 66–70.

Yin, A. (2012). An episodic slab-rollback model for the origin of the Tharsis rise on Mars: Implications for initiation of local plate subduction and final unification of a kinematically linked global plate-tectonic network on Earth. *Lithosphere*, *4*(6), 553-593. doi:10.1130/L195.1

Yuan, D. N., Sjorgren, W. L., Konopliv, A. S., & Kucinskas, A. B. (2001). Gravity field of Mars: A 75th degree and order model. *Journal of Geophysical Research*, *106*(E10), 23377–23401. doi:10.1029/2000JE001302

Yu, X., Fan, X., Guo, Y., & Deng, T. (2019). Recovery of lithium from underground brine by multistage centrifugal extraction using tri-isobutyl phosphate. *Separation and Purification Technology*.

Zante, G., Boltoeva, M., Masmoudi, A., Barillon, R., & Trébouet, D. (2019). Lithium extraction from complex aqueous solutions using supported ionic liquid membranes. *Journal of Membrane Science*, *580*, 62–76. doi:10.1016/j.memsci.2019.03.013

Zhang, J., Wang, R. Z., & Wu, J. Y. (2007). System optimization and experimental research on air source heat pump water heater. *Applied Thermal Engineering*, *27*(5-6), 1029–1035. doi:10.1016/j.applthermaleng.2006.07.031

Zhu, H. F. (2011). *The new process study of extracting lithium from salt lake brine by solvent extraction*. Academic Press.

Zhu, J., & Chen, W. (2014). Energy and exergy performance analysis of a marine rotary desiccant air-conditioning system based on orthogonal experiment. *Energy*, *77*, 953–962. doi:10.1016/j.energy.2014.10.014

Zou, Y., Zheng, C., & Sheikhi, S. (2021). Role of ion exchange in the brine-rock interaction systems: A detailed geochemical modeling study. *Chemical Geology*, *559*, 119992. doi:10.1016/j.chemgeo.2020.119992

Zuber, M. T. (2000). Internal Structure and Early Thermal Evolution of Mars from Mars Global Surveyor Topography and Gravity. Science, 287(5459), 1788-1793. Doi:10.1126cience.287.5459.1788

About the Contributors

Elhoucine Essefi obtained his PhD form National engineering School of Sfax titled 'Wet Aeolian Sedimentology and Sequence Stratigraphy within the Terrestrial Analogues in Eastern Tunisia: Implications for Wet Aeolian Sedimentology and Sequence Stratigraphy on Mars', in 2013. In 2011, he obtained License in Linguistics and Anglophone Civilization from the Faculty of Arts and Humanities of Sfax. In 2009, he obtained the Master degree from the Faculty of Sciences of Sfax titled Multidisciplinary study of Sidi El Hani Saline Environment: the Geological History and the Climatic Variability. In 2003, he obtained the 'Maitrise' degree in Earth Sciences from the Faculty of Sciences of Sfax. He obtained his Habilitation from the university of Gabes in 2021.

Ibtissem Jendoubi is Assistant Professor at the Faculty of Sciences of Bizerte, University of Carthage; she is the General Secretary of the Regional Tunisian Physical Society of BIzerte. She obtained the PhD from the Faculty of Sciences of Monastir in 2015. She obtained the Master degree from the Faculty of Sciences of Monastir in 2006.

* * *

Sana Bedoui is a PhD student at ISSTEG in the Department of Earth Sciences.

Romdhane Ben Khalifa is an assistant professor at ISSAT GABES.

Romdhane Ben Slama is a professor of Electromechanics - Technology at the National Engineering School of Gabes.

Bechir Chaouachi is a Professor of Chemical Engineering at the University of Gabes.

Saoussen El Aguel is a Doctoral Engineer in chemical processes engineering at the National School of Engineers of Gabes.

Prasad G. is Assistant Professor (Sr.Gr.), Aerospace Engineering, Dayananda Sagar University, Bangalore, India. Prof. Prasad G. received his Masters degree and Bachelors degree in Aeronautical Engineering from Anna University, Chennai, India. He is pursuing his PhD at Anna University. He has 7 years of teaching experience. Prior to joining Dayananda Sagar University, Bangalore, India he worked at Anna University. He has published 11 articles in reputed Scopus Indexed International Journals with good Impact Factor. He has completed two funded project sponsored by The Institution of Engineers (India) and Tamilnadu State Council for Science and Technology. Awarded Indian National Science Academy (INSA) Visiting Scientist Programme 2019 and Awarded Science Academies' Summer Research Fellowship Programme (SRPF) 2019. Reviewer of Aircraft Engineering and Aerospace Technology, Reviewer International Journal of Engine Research, Reviewer Journal of The Institution of Engineers Series C and Chaired as Jury member-IEEE Project competition. He has attended several International Conferences, workshops and provided guest Lectures. He is a Professional Membership of American Institute of Aeronautics and Astronautics, Institution of Engineers, Life Member in Indian Cryogenic Council and Life Member in Shock Wave Society.

Reshma G. is a Project Scholar in the Department of Aerospace Engineering at Dayananda Sagar University in Bangalore, India.

Soumaya Hajji is an Engineer in Geology from National Engineering School in Sfax (ENIS), Tunisia. She obtained her PhD in Modeling in Hydrogeology in 2015. She obtained her HDR in Environnemntal Hydrogeology and geo-Informatic in 2021. She is a researcher at geochemistry and hydrogeology, modelling and geostatistics at 'Water-Energy-Environment' Research Laboratory (LR3E) in Engineering School Sfax, Tunisia. Her research activities deal with geochemistry, modeling, analysis, optimisation, simulation of groundwater resources, groundwater quality and. She is an Associate Professor in Geology Department in Faculty of Sciences of Sfax, Tunisia.

Younes Hamed is a professor at the Geological Sciences (Hydrogeology, radioactive isotope geochemistry and Climate change), Univ. of Gafsa Tunisia. He has an academic background in Groundwater Modelling, Environmental Geochemistry & Pollution combined with a practical work experience in geological sciences. His research interests include subsurface exploration, Environmental Contaminants, and Envi. Pollution & risks. He is interested in Reservoirs Production and Estimator quantity.

Mohamed-Razak Jeday is a university professor. Professor of Process Engineering at the National School of Engineers of Gabes (ENIG), Director and Founder of the "Processes, Energy, Environment & Electrical Systems" research laboratory at ENIG, he is also Chairman and Chief Executive Officer of the Management Company of the Industrial and Technological Pole of Gabès.

Nikitha K. N. is a Project Scholar in the Department of Aerospace Engineering at Dayananda Sagar University in Bangalore, India.

Hassan Khlissa is a PhD student in geosciences at the ISSTEG, University of Gabes.

Mourad Magherbi is a professor in the chemical and Process Engineering Department at the University of Gabes.

Zina Meddeb is an associate professor in the chemical and Process Engineering Department at the University of Gabes.

Pooja B. Reddy is a Project Scholar in Department of Aerospace Engineering, Dayananda Sagar University, Bangalore, India.

Gourav Nand Tiwary M. is a Project Scholar in the Department of Aerospace Engineering at Dayananda Sagar University in Bangalore, India.

Amira Touil is a Professor at the University of Carthage.

Soraya Trabelsi is a Professor at the University of Carthage.

Fethi Zagrouba is a Professor at the University of Carthage.

Index

G

H

I

L

M

N

P

Q

R

S

T

V

W

www.ingramcontent.com/pod-product-compliance
Lightning Source LLC
LaVergne TN
LVHW081336300926
849161LV00008B/1989